COSMOGONICAL PROCESSES

COSMOGONICAL PROCESSES

Proceedings of the Symposium held in
Boulder, Colorado, 25–27 March 1985

Edited by
W. D. ARNETT, C. HANSEN, J. TRURAN
and S. TSURUTA

VNU SCIENCE PRESS
Utrecht, The Netherlands
1986

VNU Science Press BV
P.O. Box 2073
3500 GB Utrecht
The Netherlands

First Published 1986

CIP-DATA KONINKLIJKE BIBLIOTHEEK, DEN HAAG
Cosmogonical processes: proceedings of the symposium held in Boulder, Colorado, 25–27 March 1985/ed.: W. D. Arnett . . . (*et al.*). – Utrecht: VNU Science Press
ISBN 90-6764-054-9 bound
SISO 552.2 UDC 524.8
Subject headings: cosmogony/cosmology.

Printed in Great Britain by A. Wheaton & Co. Ltd, Exeter

This book is dedicated to Al Cameron in honor of his 60th birthday and in thanks for his work and his perspectives, which have enriched us all.

Contents

Preface

Back in the early 1960s, the undersigned had the good fortune to have A. G. W. Cameron as their dissertation adviser. With patience, humor, instruction, and a bit of subtle prodding, he introduced us to the complexities of modern astrophysics. In the process he also passed on to us some of his sense of excitement in working in the subject. The 25 years which have passed since that time have not dulled that excitement. In small return for his efforts we decided to organize and host a symposium to celebrate Al's becoming a sexagenarian. The symposium, finally held in Boulder, Colorado, on 25–27 March 1985 (just three months short of Al's 60th), was to be entitled "Symposium on Cosmogonical Processes" at Al's suggestion. A frantic perusal of various dictionaries revealed to us that "cosmogonical" referred to no less than the study of the general origin and structure of the universe. After studying Al's vita, which lists some 300 publications, we decided that his suggested title was not only good but also entirely appropriate considering his breadth of interests. This left only the program and speakers to be decided upon. Since there were only four of us compared to one of Al, we again sought his advice, and the results are, in effect, summarized in this volume. We should like to thank those speakers who so kindly and effectively gave of their time and effort to make the symposium a success. Their personal friendships with Al clearly show in what they did. Due to the heavy demand of previous commitments Gerry Wasserburg was unable to contribute a written version of his excellent talk, which is a loss for this volume.

Before getting on with the invited papers, there is the matter of the introduction which follows this preface. We had fully intended to write this ourselves as a tribute to Al but happily decided against that course. Our version would most probably have been a dry recital of Al's accomplishments. What would have been missing, however, is the philosophy of how and why he went about doing what he did. Since only he could tell the story, we asked him to and he did so in record time. Let us add that his writing this for us is only right and proper; we still regard him as our adviser.

It is appropriate that this volume was put together using TeX, another of Al's interests. Susan Lancaster and Jennie Lightner learned the language, designed the format, and produced camera ready copy in good time. The volume itself attests to their skill and effort, for which we are deeply thankful.

W. David Arnett
Carl J. Hansen
James W. Truran
Sachiko Tsuruta

Some Autobiographical Notes

A. G. W. Cameron

Abstract

I give here an account of my scientific life, including the principal problems that have interested me and the principal activities other than research which have occupied significant amounts of my professional time.

Early Years

I have been very grateful to Dave Arnett, Carl Hansen, Jim Truran, and Sachiko Tsuruta for organizing this conference on cosmogonical processes in honor of my sixtieth birthday. The conference itself was very enjoyable for me, dealing, as it did, with the broad range of topics which have come to interest me during my scientific lifetime. The conference organizers were conspicuous in their modesty in choosing not to present papers themselves, and they have compounded this modesty in prevailing upon me to write the introduction to this volume of proceedings. Although such a contribution is quite unorthodox in a festschrift, orthodoxy has not been the guiding principle of my scientific life, and so I have accepted this task as an interesting challenge to present my own outlook on the range of problems that I collectively refer to as cosmogony.

I suggested to the organizers the title "Cosmogonical Problems" in order to pay a tribute on my own behalf to Dik ter Haar. Many years ago, I think in the late 1950s, he lent me a carbon copy of a book that he had prepared with this title, which ranged over many of the topics dealt with in this book, and gave an excellent historical account and critique of the ideas which had been put forward in an attempt to explain the origin of many of the diverse components of our universe. It is most unfortunate that this book was never published. It had been submitted to a publisher who had retained the manuscript for such a long time, before eventually rejecting it, that Dik felt it would need extensive revision to bring it up-to-date, and he had never found the time to do this. I found this book to be very inspirational, and it filled in my background on the history of many of the things that I was beginning to think about. Eventually I rescued a piece of the manuscript by incorporating it into a historical chapter in a proceedings volume on the origin of the solar system (Ter Haar and Cameron 1963). No volume put together by a variety of authors with a variety of objectives can match the unified treatment of the book that Dik put together, but I hope that we are using the title in the right spirit.

But I should start at the beginning. I was born on 21 June 1925 in Winnipeg, Canada. I was raised in that city and did my undergraduate work there at the University of Manitoba, "majoring" (a term not used by Canadians) in physics and mathematics. I obtained my Ph.D. at the University of Saskatchewan, specializing in nuclear physics.

I had an early interest in astronomy, but there seemed little opportunity to pursue this subject. When I entered the University of Manitoba, it was in the final years of the Second World War. Only one undergraduate course in astronomy was offered, which I took, but it turned out to be all navigation, which was then "practical astronomy." In retrospect, I am not sorry that I took this course, since it gave me a little practical experience in doing spherical trigonometry, which I would otherwise never have had. Thus it was that I started out to become a physicist.

I spent some time between the B.Sc. and Ph.D. degrees working at what was then the atomic energy project of the National Research Council of Canada, in Chalk River, Ontario. I started to work with nuclear emulsions to study nuclear tracks, working with C. H. ("Chas") Millar. Chas had started to devise techniques to develop these emulsions in such a way that nuclear tracks could be seen and measured in the presence of a substantial background fog produced in the emulsion by gamma-rays. My first publication dealt with a procedure in which deuterium was loaded into some home-made emulsions, and the relative yields of the gamma-rays from the first cluster of excited states (just above 6 MeV) in ^{16}O were determined by measuring the lengths of the photoprotons produced in the emulsion (Millar, Cameron, and Glicksman 1950). The experimental procedure was extremely simple. I laid the plates containing the emulsions next to the cooling water effluent pipe from the high neutron flux NRX reactor. High energy neutrons had converted some of the ^{16}O into ^{16}N in the cooling water, and this nitrogen isotope beta-decayed through some of the excited states in question; it was actually the relative yields of these beta-decay branches in which we were interested.

Since we had started along this path, it was natural to think about using other high energy photons to cause photodisintegrations in the natural constituents of an emulsion. We found out how to develop commercial nuclear emulsions that had been exposed to as much as 200 roentgens of radiation in such a way that we could still see and measure the tracks of alpha-particles. The University of Saskatchewan had just installed a 28 MeV commercial betatron for studies in nuclear physics and radiation medicine, and we arranged with Leon Katz, who had been a summer visitor at Chalk River, to expose emulsions for us. With these emulsions we were able to see ^{12}C and ^{16}O breaking up into alpha-particle "stars" as well as the emission of single alpha-particles from ^{14}N and ^{16}O, and we were able to study the mechanisms of these reactions to some extent (Millar and Cameron 1950, 1953).

Having thus started to work with the Saskatchewan betatron at long distance, it made a lot of sense to pursue my doctoral studies there using the betatron to produce photonuclear reactions in various materials, and so off I went to work with Leon Katz as my thesis adviser. I must pay a tribute to Leon, who was a dynamically aggressive experimenter but at the same time exhibited enormous patience with me when, often several times a week, I would burst in upon him with some new idea. We quickly became busy unraveling the mysteries of the newly discovered giant dipole resonance in nuclei, for which the energy range of the betatron was ideal (see, for example, Cameron 1951). Thus I was trained as an experimental nuclear physicist, but I did not go through the type of apprenticeship characteristic of most experimental workers, for I did not have to spend a lot of time constructing apparatus. The field was new, and little beyond standard detectors was needed. We did experiments intended to answer leading scientific questions in photonuclear physics, and I thrived in that atmosphere.

I did, however, spend a little time developing a procedure for converting a photonuclear activation curve, produced by the continuous spectrum of high energy photons, into the corresponding photonuclear cross-section curve (Katz and Cameron 1951*a*). Then we measured a variety of these cross sections (Katz and Cameron 1951*b*; Cameron, Harms, and Katz 1951; Cameron and Katz 1951; Katz, Kavanagh, Cameron, Bailey, and Spinks 1955). I also collaborated in some further nuclear emulsion work (Haslam, Cameron, Cook, and Crosby 1952). But our most intensive work involved the examination of small kinks in some photonuclear activation curves in light nuclei, which we interpreted as due to the excitation of discrete energy levels in the target nuclei (Katz, Haslam, Horsley, Cameron, and Montalbetti 1954).

Following the receipt of my Ph.D. in 1952 I accepted an assistant professorship at Iowa State College (now Iowa State University). This seemed to be a logical move, for Iowa State had recently acquired a 70 MeV synchrotron, and the energy range beyond 28 MeV was relatively little explored for photonuclear reactions. However, when I arrived I did not find the type of atmosphere in which I could thrive. The main interest of the staff was in improving the beam current in the machine, and little in the way of experimental resources existed.

Not long after I arrived, I was browsing in the reading room of the Ames Research Lab (at Iowa State), when I came upon an item in *Science News Letter* (now *Science News*) that immediately attracted my interest. Paul Merrill, a celebrated astronomical spectroscopist, had discovered lines of the unstable element technetium in red giant stars of spectral type S (Merrill 1952). My immediate thought was "where did the star get the neutrons?" At that point my interests completely switched to astrophysics and to the application of nuclear physics to astrophysical problems. First I had to learn something about astrophysics, and so I established a personal library in the subject by buying all the leading textbooks that I could lay my hands on, and I began some intensive reading.

By the following spring, in 1953, I was actively pursuing the source of the neutrons. The first and obvious thought was that deuterium burning might do it, producing neutrons by D-D reactions. But it turned out that the neutron yield from D-D reactions was small because of competition with D-P reactions, and it also turned out that the hydrogen had a large enough capture cross section so that, combined with its overwhelming abundance, it would grab almost all of the neutrons that were produced. So technetium could not have been a product of the hydrogen-burning stage of a star.

Thus I turned to the following stage in the evolution of a star, following hydrogen exhaustion, when helium thermonuclear reactions would become possible. It did not take long to discover that the $^{13}C(\alpha, n)^{16}O$ reaction was exothermic, and hence a likely source of the neutrons. That formed the topic of my first astrophysics paper, given at the Thanksgiving, 1953, meeting of the American Physical Society in Chicago (Cameron 1954; full publication, Cameron 1955).

Since the introduction of nuclear physics into astrophysics was almost a new field, the problem arose as to where I would best be able to pursue it. I felt that since I had learned a lot of astrophysics on my own, and since the state of nuclear physics was such that a great deal of development of particular parameters would be necessary for astrophysical needs, I would be better off in a nuclear physics environment. It also helped that I had become known in nuclear physics for my photonuclear work, and so it was clearly easier to get a job in that area. This led me back to the Canadian atomic energy project at Chalk River, Ontario (which had become Atomic Energy of Canada Limited), where I went in 1954. It is amusing that sometime later they let me know that I was considered a "cheap luxury."

During my stay in Iowa, I met my future wife Betsy in the neighboring state of Illinois. We were married not long after I returned to Canada.

Chalk River

At Chalk River I set out to do something that I had not wanted to do as an experimentalist: to construct for myself the proper tools. The overall goal was to understand what would happen to an initial solar composition of elements as the temperature and density of the medium were progressively increased. The ultimate result was quite clear, for Fred Hoyle had written a very definitive paper on nuclear statistical equilibrium which discussed in a clear way the conditions needed for it to hold (Hoyle 1946). In conditions of nuclear statistical equilibrium the most important nuclear quantity is the nuclear mass (better, the total nuclear binding energy). Of secondary importance are the nuclear level densities which enter through the nuclear partition function. However, for the reactions that lead up to nuclear statistical equilibrium, other quantities are necessary, including particle and radiation widths. For most of this required nuclear information there were no experimental data available and little prospect of any.

Therefore most of the nuclear physics research I did at Chalk River had the underlying goal of making accurate predictions of unmeasured nuclear quantities. For the most part this meant that I would take a reasonable theory for the basic properties of the nuclear quantity and adjust the parameters with semiempirical "fudge factors" to produce reasonably good fits to the available data.

I produced an improved nuclear mass formula by taking the liquid drop model of the nucleus, equipping it with a suitably fuzzy edge, and applying empirical corrections which were individual functions of the neutron and proton numbers (Cameron 1957*a*). It was a very happy discovery that the effects of the neutrons and protons could be reasonably well separated in this fashion, and furthermore that the individual shell corrections formed a pattern which brought out very well the shell structure of the nucleus.

This mass formula tool formed the basis for other tools. I produced a level density formula which was based on the shell and pairing corrections of the mass formula (Cameron 1958*a*, 1959*d*) (this was an interim tool because the later work with Arnold Gilbert [Gilbert, Chen, and Cameron 1965; Gilbert and Cameron 1965] produced a distinct improvement). These methods allowed the construction of relationships for nuclear radiation widths (Cameron 1957*b*). With all of these together, I was in the business of calculating nuclear reaction cross sections averaged over level density (Cameron 1958*b*, 1959*e*). I made some applications of these tools to nuclear problems, particularly the fission problem, which was undoubtedly an overoptimistic application of the level density tool (Cameron 1958*c*).

The main astrophysical applications of the tools were to determine the burning stages on the way to nuclear statistical equilibrium. I went through carbon burning (Cameron 1959*a*), neon and oxygen burning (Cameron 1959*b*), and a first crude cut at silicon burning (not published at the time but included in Cameron 1963*a*). In each of these stages the complexity lay not with the main burning reaction, but with the secondary reactions that followed it and soaked up the neutrons, protons, and alpha-particles that were liberated. In fact, in the case of silicon burning there is no primary reaction but only secondary reactions following some photodisintegrations. I had to worry about increasingly high temperatures, and hence about the effects of excited states on many of the processes, particularly on the acceleration of forbidden beta-decays by going through excited state allowed decays (Cameron 1959*c*). Since silicon burning leads to nuclear statistical equilibrium, this program had closed the gap in the nuclear-burning sequences and completed the survey I had undertaken.

The program I have just described started in 1954 and was completed in 1959. The path was not very clear in the beginning. And the rules of the game changed in midstream. At the beginning what was clear was that some stars exhibited signs of having undergone extensive nuclear processing in their interiors. The state of stellar evolution calculations was too primitive to cast much light on how this would come about. So the guiding principle had to be just to raise the temperature and density and see what would happen. Then Suess and Urey (1956) published their paper on abundances.

The Suess-Urey paper revolutionized nuclear astrophysics. They used the best data available to them on abundances from many natural sources, but particularly from the meteorites. They adjusted the data to produce a smooth variation in the abundances of the nuclei of odd mass number, a purely empirical principle, but one that can be seen in retrospect to be approximately correct. It was immediately possible to decompose their abundance distribution into subcomponents which could be related to nuclear reactions taking place in various stellar environments. I played that game, as did a number of others, most notably Burbidge, Burbidge, Fowler, and Hoyle (1957). Now it was clear not only that stars were capable of producing exotic nuclear products in their interiors, but also that these products were injected into space in great numbers and formed the elements that composed us all, and whose relative abundances we could study in the laboratory as well as with the astronomical spectroscope.

My views of the state of the art as of the spring of 1957 are recorded in an extensive set of lecture notes produced as a Chalk River report (Cameron 1957*c*). At that time I was invited to give a series of lectures at Purdue University during a two-week period, and the six lectures themselves simply provided the framework in which I tried to sketch out what was known and what could reasonably be conjectured. Other reviews were also prepared at roughly the same time (Cameron 1956, 1957*d*, 1958*d*).

It was not possible to think about these networks of nuclear reactions in stellar interiors in isolation from the accompanying evolution of the stars themselves. I did not engage in any stellar evolution studies (at that time), but I did think about some of the strange conditions that the stars were likely to encounter. What happens when nuclear fuels are present and the matter achieves very high density but stays cold? The motion associated with the zero-point energy may allow barrier penetration and hence nuclear reactions to take place. I coined the term "pycnonuclear reaction" for this process and worked out some estimates of reaction rates likely to be important (Cameron 1959*f*). What happens when a star has exhausted all of its nuclear fuels and the center collapses, but the mass is too great to be stabilized as a white dwarf? I revived the notion of a neutron star, dormant since the late 1930s except for the stubborn advocacy of Fritz Zwicky. I was interested in the large number of different types of baryons that could be formed in the interior of such a star at higher densities (Cameron 1959*g*).

Also at that time I began to think about the basic sources of our knowledge of elemental abundances that had been used by Suess and Urey in their classic paper. I attended the Gordon Conferences on Nuclear Chemistry in 1956 and 1958. The reason for my invitation to these meetings was that the attendees wanted to hear about nucleosynthesis, but they were very beneficial to me because the attendees also wanted to hear about meteoritic abundances and other relevant matters. This brought me into contact with people working on extraterrestrial materials and forced me to start wrestling with strange new types of jargon. It was a marvelous interdisciplinary education. It was clear that these abundances were of great intrinsic value in themselves, and I felt that it was now time to do Suess-Urey in reverse: to replace their empirical smoothing of odd mass number isotopic abundances with relative abundance estimates taken from the nuclear processes that had now been identified for nucleosynthesis. This led to the first of many tables of elemental abundances that I have produced (Cameron 1959*h*).

It is appropriate to mention computation at this point. In the first couple of years after I went to Chalk River I did a tremendous amount of slide rule manipulation as I tried to settle which reactions were important and which were not. Then I discovered the accounting machines. These were IBM 602A punched card relay calculators, which carried out calculations according to a program you were expected to wire into a plug board. One of my colleagues showed how to make these machines card-programmed calculators by putting together a general-purpose plug board with wires containing germanium diodes, which took their instructions from the cards passing through. You then created "DO loops" by putting together packets of identical instruction cards which were fed through the machine, which could process about six cards per minute, typically. Unfortunately you could not give any branching instructions, but the system was well suited for the straightforward integration of nonlinear differential equations such as one has in a simple nuclear reaction network. I became the biggest user of this facility, and I would leave multiple file drawers full of cards with instructions to the accounting staff to keep the hopper full of cards.

Then the head of the accounting branch phoned to tell me that IBM was bringing a demonstration 650 computer to Ottawa for two weeks, and was I interested in going along to see a demo? This was a vacuum tube machine that worked by reading and writing to a magnetic drum. I said I would go if I could try to run a program, so IBM sent along the instruction manual, and I wrote a program to compute a simplified version of the solution to the *s*-process equations; it had to be written in machine language and I even optimized the location of the instructions around the drum. We went to Ottawa, I fed in the cards, and the program ran without a hitch! This was very exciting, because this was how I found the large overabundances of the heavy elements which build up in the *s*-process. So I asked if I could come back the following week and run a longer problem. This time I programmed the full set of *s*-process equations. When we got to Ottawa and fed in the cards, the program did not run! So I had to learn about console debugging in real time. Then I ran the computer most of the night and went home with a car trunk full of output.

Displaying suitable gratitude, Chalk River then proceeded to buy a Datatron 205, which was similar to but easier to use than the IBM 650. For most of 1958 and into 1959 I was the primary user of that machine on nights and weekends. Being a vacuum tube machine, it not infrequently failed, and I learned how to administer the judicious kick and how to recognize a bad pattern of lights so that I could yank out one package of tubes and put in a replacement. Most of the time the machine ran unattended; I would leave instructions to the operator for taking my problem off in the morning. Characteristic instruction times were about a millisecond, so it was a very slow computer by modern standards but a vast improvement at the time. The tremendous spurt of publications which I had in 1958 and 1959 is directly attributable to the availability of that computer. In turn, the publications let to my election to the Royal Society of Canada.

Caltech

In 1959 I felt that the time had come when I would benefit from a closer association with astronomers. My interests had begun to spread beyond the strict confines of nuclear astrophysics. Sometime previously Jesse Greenstein had suggested that it might be profitable for me to spend some time at Caltech, so that when I approached him on this matter he invited me to spend a year there with support from his Air Force grant, and I accepted. I had some doubts about Chalk River as a long-term career choice, since it was narrowly specialized in nuclear physics and nuclear engineering. Therefore I came into the United States with a regular immigration visa rather than a visitor's visa.

Caltech is a Mecca for observational astronomers, and it was fairly common for theorists visiting there to be seduced into becoming observers by the availability of major instrumentation. Wal Sargent came there from England as a bright young theorist at the time I arrived, and he followed that path brilliantly. So some people were surprised when I showed not the slightest interest in observing, despite the fact that I had an experimental training. But the fact was that I was now pursuing some very fundamental problems and I did not see any prospect for solving these with any kind of short-term observational program. Astronomy then was and still is largely serendipitous in nature, and the most important evidence that bears on the more fundamental problems tends to come in unexpected ways or only after long observational campaigns.

Not long after I arrived at Caltech, Bruno Pontecorvo suggested that there was an analogy between processes which can emit photons in nature and processes in which neutrino-antineutrino pairs are emitted, so that one can guess the rates of the latter by multiplying the rates of the former by the ratio of the weak to the electromagnetic coupling constants. I used this technique to estimate quite a number of rates for neutrino-antineutrino pair emission. It was quite clear that many of these processes, if correct, would revolutionize our view of stellar evolution. I did not try to publish these estimates, but I circulated them to a few friends to try to generate some interest in working out the rates rigorously. This became an interesting field for the next two or three years.

The most interesting aspects of nuclear astrophysics seem to involve supernova explosions. Therefore it was very beneficial for me to be able to discuss these events with Walter Baade, Fritz Zwicky, and Rudolph Minkowski, all still at Caltech or Mount Wilson (i.e., Carnegie Institution) at that time. I spent quite a while pondering over the mysteries of the Type I supernova spectrum and its variation with time, but without making the slightest dent in the problem. That was as close as I came to becoming entangled directly with observations.

After I had been at Caltech for a few months, John Reynolds (at Berkeley) published his discovery of evidence for the extinct nuclide ^{129}I having been present in some meteorites at the time of their formation (Reynolds 1960). This radioactivity decays to the noble gas isotope ^{129}Xe with a half-life of 17 million years.

This discovery transformed the direction of much of my research. I was already interested in the meteorites as an interesting problem in connection with the abundances of the elements, and now it was clear that there were clues in the meteorites about the time scale on which they had been formed. Therefore I spent the rest of my time at Caltech thinking about problems of star formation and the origin of the solar system and their related time scales. The Caltech astronomy library was particularly useful for this purpose, and it was clear to me that I would never have been able to cover this ground had I been at Chalk River. Eventually all of this thinking was published in one lengthy paper (Cameron 1962), and this was the only significant publication to come out of my stay at Caltech. However, one of my principal research activities had now become the problem of the origin of the solar system.

As my time at Caltech came to a close, I began to feel increasingly that my interests had broadened too much for Chalk River, and although I returned to Chalk River for a few months, most of that time was spent negotiating with potential employers. A number of opportunities presented themselves, partly as a result of the rapid buildup of the U.S. space program that was now occurring. I was highly enthusiastic about the space program, and I was more than a little annoyed that the Canadian government was unwilling to invest in space research more than a small fraction of the per capita amount of the United States. Thus when the opportunity to become directly involved in the U.S. program was offered to me, I accepted it. In the spring of 1961 I moved to New York City and the Goddard Institute for Space Studies.

New York

When the U.S. space program was "ramped up" at a rapid rate following *Sputnik,* to use industrial production terminology, NASA inherited several centers from the aeronautics program and created others. The Goddard Space Flight Center (GSFC) in Greenbelt, Maryland, was one of the latter. At first the newly formed components of the GSFC existed in rented quarters scattered around the Washington area. Among these was the Theoretical Division temporarily located in Silver Spring. In April of 1961 this component spawned an offspring, the Goddard Institute for Space Studies (GISS), in New York City, which became independent of it but remained within the Goddard structure.

GISS was founded by its first director, Robert Jastrow, who persuaded the director of GSFC that a theoretical studies institute situated in an area convenient to several major universities, and with a responsibility to encourage interactions with the academic communities of these institutions, would be a major asset to NASA. This concept was a very good one, and for some years GISS was able to carry out this responsibility effectively. Eventually the entire essential original character of GISS was lost because intellectual ideals are terribly hard to sustain in a mission-driven agency that is in the process of becoming a rigid bureaucracy.

When I arrived in New York, GISS consisted of three offices hastily partitioned in a very large bare room in the Interchurch Center near Columbia University. Those offices were occupied by Jastrow, the new executive officer, and me, with two secretaries set up with desks immediately outside. The intent was that there should be only a small permanent staff and that most of the scientists should be visitors and postdoctoral research asssociates there for a relatively short time of a few months to a couple of years. It was not long before the staff filled not only the empty space, but also two other areas within the Interchurch Center and a renovated townhouse down the street. Eventually everyone moved into another building renovated by and rented from Columbia. The intent was also to have the finest available computer facilities, operated by and with programming support from a contractor. Apart from short-lived transitional machines, the computers available were an IBM 7094 followed by an IBM 360/95 (one of only two ever produced).

In the early years of the space program resources for worthwhile projects were fairly easy to obtain. The challenge was to use such resources wisely. One series of projects for which I was chiefly responsible was the holding of innovative and, for the most part, interdisciplinary conferences. After some consultation I would select and invite the principal speakers, and we would use GISS funds to cover travel expenses of the principal people who needed them and that we wanted particularly to come. During the conference we would tape record the presentations (toward the end they were videotaped) and we would copy the slides. Transcripts were typed of the talks. Then we would hire one or more graduate students (or postdoctoral research associates) to turn the transcripts into draft papers, working under my general supervision. These were then sent to the authors whom we badgered until revised versions of their papers were returned to us for book publication. The graduate students were encouraged to fill in the presentations with background material to the extent that they were able to do so. This arrangement materially helped with their living expenses and put their names on the covers of a book, which was good for their academic records.

These conferences were in themselves a further broadening experience for me. The topics included the origin of the solar system (Jastrow and Cameron 1963), the physics of nonthermal radio sources (Maran and Cameron 1964), the origin and evolution of atmospheres and oceans (Brancazio and Cameron 1964), stellar evolution (Stein and Cameron 1966), the dynamics of the Earth-Moon system (Marsden and Cameron 1966), nucleosynthesis (Arnett, Hansen, Truran, and Cameron 1968), infrared astronomy (Brancazio and Cameron 1968), and supernovae and their remnants (Brancazio and Cameron 1969). We experimented with a variety of different publishers for these books. I also organized two conferences which were highly topical and for which no proceedings were produced, one on neutron stars and celestial X-ray sources in 1964 and one on rapidly pulsating radio sources in 1968. The first of these was an exploratory attempt to see if a connection could be made between the concept of neutron stars and the newly discovered X-ray sources, a connection that could not be established for several subsequent years. The other was an attempt to make sense of all the rapidly increasing observational data on pulsars; in this respect the conference was probably more confusing than enlightening.

I was also involved in a different type of conference organizational activity. In 1961 I again attended the Gordon Conference on Nuclear Chemistry. There Ed Anders (University of Chicago) and I both felt that the interval between the sessions of the nuclear chemistry conferences devoted to cosmochemistry was too long and the degree of interdisciplinary attention that was being brought to bear on the subject was insufficient, so we laid plans to break this subject out of nuclear chemistry into a conference of its own. With support from like-minded people there, we proposed to the Gordon Conference committee within the American Association for the Advancement of Science the establishment of a Gordon Conference on the Chemistry and Physics of Space. This proposal was accepted, and the first of these was held in 1962 at Tilton, New Hampshire, with me as chairman and Ed as vice-chairman. These conferences, which were held yearly for just over a decade, turned out to be every bit as successful as I had hoped, and I attended all but the last. I also became vice-chairman in 1969 and chairman again in 1970. Eventually the competition from other conferences organized in connection with the lunar sample program caused the attendance at our Gordon Conferences to fall off too much, and they had to be discontinued.

NASA resources were also used to attract students into research of interest to the space program. Grants were arranged to some individuals in universities in the area to support space-related research and also to support graduate students from those institutions who wished to use the facilities of GISS, particularly the computer, and to work with GISS staff. This was the basic mechanism by which students from Yale, Columbia, and New York University came to work with me. The relationships with Columbia and NYU developed almost casually, on an individual student basis. That with Yale was a little more formal.

In the spring of 1962 Vernon Hughes, then chairman of the Yale Physics Department, visited New York and asked me if I would like to lecture at Yale on a regular basis and also to have some graduate students who wanted to work in theoretical astrophysics. Upon my acceptance, Yale made me a visiting lecturer, and for several years I came up approximately once every two weeks and lectured for about two hours. Before starting this I visited Yale to interview prospective graduate students in physics. There were four who expressed an interest: Dave Arnett, Carl Hansen, Jim Truran, and one other who soon decided that she wanted to work in a different field. Since the Yale NASA grant was going to support them during the first summer, I gave some thought to an assignment that would prove broadly educational and yet be a legitimate research project.

I asked them to calculate the chemical evolution of the Galaxy as stars are formed from interstellar gases, evolve, and die with ejection of nucleosynthesized material back into the interstellar medium. The work they did that summer formed the basis for work that was later developed into a serious study (Truran, Hansen, and Cameron 1965, 1966). Because of the enormous uncertainties that surround the assumptions that need to be made in such a study, I have always described this work as something of a game. Jim Truran and I played the game a few more times (Truran and Cameron 1970; Cameron and Truran 1971; Truran and Cameron 1971). Then several other people started to play the game much more seriously, most notably Beatrice Tinsley. She and I began a vigorous discussion of the subject one evening at a party in Dallas, concerning the possibility that the formation of comets accompanying star formation might be an important sink for heavy elements formed in stellar nucleosynthesis, and this led to a joint publication with her (Tinsley and Cameron 1974). She never took this idea as seriously as I did and do.

The lectures that I gave at Yale formed an important part of my education in astrophysics. The first year I talked mainly about my research experiences in a course on nuclear astrophysics. After that I embarked on an ambitious scheme to describe the universe, from the center of the Earth to the cosmological horizon, through the eyes of a physicist and for other physicists. In the first year of a two-year sequence I dealt with the solar system, and in the second year with galactic and stellar physics. All of these lectures were tape recorded. I would take the tape with me back to New York, where it would be transcribed. The tape and the transcription would then go back to New Haven, where Arnett, Hansen, and Truran would take turns making each topic into a chapter of lecture notes. These were then reproduced by GISS and distributed fairly widely (Cameron 1963*a*, 1964, 1965*b*). The preparation of these lectures took a large amount of time during these years, but it was well worth it, since it induced me to learn the fundamentals of several branches of physics to which I had not paid a great deal of attention until then, but which are important in the study of the cosmos. I still find frequent use for those lecture notes.

I undertook to improve the calculations I had done on the approach to nuclear statistical equilibrium. This involved some more experimentation with the mass formula (Cameron and Elkin 1965). I have already mentioned the work of Arnold Gilbert on level densities (Gilbert, Chen, and Cameron 1965; Gilbert and Cameron 1965). Arnold was a student from the Columbia Physics Department. I developed a fission systematics (Cameron 1965*a*). When the tools were all prepared, we used improved methods for solving the nuclear reaction network problem for the approach to nuclear statistical equilibrium (Truran, Hansen, Cameron, and Gilbert 1966; Truran, Cameron, and Gilbert 1966).

Many other studies were going on at the same time. I first met Dilhan Ezer when she was a visitor to Chalk River, where she started to do some stellar evolution calculations. When I went to Caltech she went to the University of Indiana for a while, and then she came to New York to GISS not long after it opened its doors. She and I collaborated on studies of the pre-main-sequence evolution of the Sun and other stars, gradually extending the mass range and then looking at stars without heavy elements and making some guesses about the solar neutrino problem (Ezer and Cameron 1963, 1965, 1966, 1967*a*, *b*, 1968, 1971*a*, *b*, 1972).

On a personal level, I made a major transition in 1963 by becoming an American citizen. The waiting time for this had been shortened because I had married an American citizen.

Shortly after my arrival in New York, I became involved in a number of editorial activities. Zdenek Kopal started the journal *Icarus* and invited me to serve as an associate editor. Since then, both for this and other journals, I have spent a certain amount of time receiving manuscripts and seeing that they were properly refereed. In 1962 W. A. Benjamin involved me with his new book company, where my principal responsibility was to suggest prospective authors for books. Other book companies claimed my services later, including W. B. Saunders, Gordon and Breach, and Elsevier. The number of journals with which I became associated also proliferated. These included the *Journal of Geophysical Research*, *Earth and Planetary Science Letters*, *Physics Today*, and *Annual Review of Astronomy and Astrophysics.* For Gordon and Breach I created and edited several journals, including *Comments on Astrophysics and Space Physics*, *Earth and Extraterrestrial Sciences*, and *Fundamentals of Cosmic Physics.* Zdenek Kopal started a new journal, *Astrophysics and Space Science*, and asked me to serve as coeditor. This was a Reidel journal, and in due course it acquired a companion, *The Moon*, which I also agreed to coedit. This later became *The Moon and the Planets*, and still later *Earth, Moon, and Planets.* These editorial activities reached a peak in the late 1960s and early 1970s and have been tapering off since then.

I became involved with the search for extraterrestrial life as a result of the W. A. Benjamin connection. As an early project for them I thought of collecting together the various scientifically based speculations that had occurred on the subject and publishing them as a book. This was the genesis of *Interstellar Communication* (Cameron 1963*b*).

These early years in New York also saw my initiation into government advisory activities, since in 1962 I was appointed to the Astronomy Subcommittee of NASA's Space Science Steering Committee.

All of these various activities were in sharp contrast to my prior activities, which were strictly involved with research, nearly entirely research done by me.

As I am sure is obvious, they cut into the time I had for personally doing research in a serious way. There were many compensations, for all of these activities served to broaden my horizons. But it meant that most of the research that I did by myself consisted of "back-of-the-envelope" calculations resulting in brief publications. My more serious research was carried out in collaboration with a variety of talented students.

One of these was Sachiko Tsuruta, a graduate student in the Columbia Physics Department. She chose to do her thesis on the subject of neutron stars. She wrote a massive thesis, over 400 pages, which was reproduced and distributed by GISS. The underlying theme was the composition of matter at high density (Tsuruta and Cameron 1965*a*), the effect of nuclear forces on the models (Tsuruta and Cameron 1966*a*), and their rate of cooling and detectability (Tsuruta and Cameron 1965*b*, 1966*b*). She continued to publish some occasional work on neutron stars with me in the years that followed (Tsuruta, Wright, and Cameron 1965; Tsuruta and Cameron 1966*c*, 1970, 1976).

My long involvement with the calculation of neutron capture cross sections led to another interesting scientific diversion starting in 1964 and lasting for several years. There was a continuing effort at the Los Alamos and Livermore Laboratories to create heavy elements by multiple neutron capture using nuclear weapons technology, following the accidental discovery that just this had happened in the "Mike" thermonuclear explosion (see Cameron 1959*e*). I became a consultant to Livermore on this project. The experiments that were carried out produced some interesting results, but the goal of very heavy element production was ultimately foiled by neutron-induced fission in the neutron-enriched product nuclei.

By 1966 I was beginning to become restless. The civil service structure within NASA was becoming more rigid, and demands for paper work to justify this or that were continually increasing. I became envious of the academic atmosphere in which I was immersed and in which the paperwork jungle was not nearly so intrusive. So Jastrow and I agreed that I would move to an academic position, but he encouraged me to maintain a close relationship with GISS. For the 1965–66 academic year I had accepted an appointment as adjunct professor of space physics at the Belfer Graduate School of Science (BGSS) of Yeshiva University in northern Manhattan, and in 1966 I moved there as my principal place of employment. At the time BGSS was growing fairly rapidly and had attracted a high-level group of physicists, chemists, and mathematicians. In fact, I moved at the height of the expansion in government funding for the sciences, and what appeared to be a very promising prospect in 1966 became progressively less so in subsequent years. But this decline set in slowly. And I normally spent at least two days per week at GISS.

The reliance on graduate students (and some postdoctoral research associates) to do serious calculations meant that I had no occasion to use the GISS computer. Meanwhile, computational techniques were changing. Everyone was now using Fortran rather than assembly or machine language. It became embarrassing to have students show me their program listings and only partly to understand them. So I undertook a project to teach myself Fortran and to do a project on the computer. This took the form of the construction of a numerical model of a disk of gas such as might represent the primitive solar nebula. I began to wrestle with the instabilities that can arise in a disk where the gravitational potential is nonspherical and its derivatives depend on the local distribution of matter in a very sensitive way. I found that a uniformly rotating disk of gas can collapse (in principle) to form either of two different disk density distributions, as Leon Mestel had been telling me, and I was able to model these distributions. I did not publish anything on this at the time, but the exercise prepared the way for the work that was later done with Milton Pine.

My original group of students at Yale had finished by now. Jim Truran and Dave Arnett came to spend some time at GISS as postdoctoral research associates. Dave Arnett had done his thesis on supernova hydrodynamics. Jim Truran solved the thermonuclear reaction network to make the statistical equilibrium peak. Carl Hansen showed that neutron star oscillations would quickly be damped. I acquired a second group of students at Yale, including George Greenstein, Bill Langer, and Dave Watkins. George worked mostly on cosmological topics (e.g., Greenstein, Truran, and Cameron 1967), Bill on the effects of hyperons on the damping of neutron star oscillations (Langer and Cameron 1969), and Dave strictly on nuclear physics. Bill Langer came to GISS after getting his degree, and George Greenstein joined me at BGSS.

A Columbia student, Bob Stein, worked at GISS with me as a nominal supervisor, but he did most of his work with Ed Spiegel, then at NYU. His thesis was on the generation of waves in the solar envelope and atmosphere, and I learned more from him than vice versa. Mike Mock was another Columbia student who worked on stellar accretion and X-ray emission (Cameron and Mock 1967). At NYU I had as students Pete Brancazio, Dick Weiss, Mike Delano, and Len Rosen. Brancazio investigated the possible role of nuclear reactions in producing the surface abundance anomalies in peculiar A stars (Brancazio, Gilbert, and Cameron 1967; Brancazio and Cameron 1967). Weiss investigated nuclear symmetry energy in infinite nuclear matter and in finite nuclei (Weiss and Cameron 1969*a*, *b*). Delano worked on the nuclear reactions that would take place among the lightest nuclei in the supernova ejection of neutron-rich material (Delano and Cameron 1971). Rosen worked on diffusion processes in neutron star surfaces (Rosen and Cameron 1972).

About a year after I went to BGSS, pulsars were discovered, and late in 1968 these were shown unambiguously to be neutron stars. This gave legitimacy to the work that we had been doing on the physics of neutron stars for many years. Until then, the reaction of typical astronomers had been that I was wasting my time, since "neutron stars do not exist because no one has ever observed one." Physicists were more open-minded and were more willing to believe that neutron stars were worthy objects of theoretical investigation because they were a theoretically possible form of matter. But it was not very long before it became evident that we had not thought of all the relevant phenomena, since the pulsars were slowing down and occasionally showing large discontinuities in the rate of slowing down ("glitches"). George Greenstein and I wrestled with the problem (Greenstein and Cameron 1969). The ideas of superfluidity came along, followed by the ideas about starquakes. I never liked the starquake theory, and I was much happier when it was later replaced by the theory of pinning and unpinning of superfluid vortices. Meanwhile, with a group at GISS I considered the transition from white dwarfs to neutron stars and some improved neutron star models (Cameron and Cohen 1969; Cohen, Lapidus, and Cameron 1969; Langer, Rosen, Cohen, and Cameron 1969; Cohen and Cameron 1969; Cohen, Langer, Rosen, and Cameron 1970; Cohen and Cameron 1971).

Throughout most of the time I was in New York my response to the quasar problem was to postulate a glowing cloud of gas at the center of a compact galactic core. With Stirling Colgate I wondered if the high luminosity and the light fluctuations might be due to the conversion of supernova ejecta kinetic energy to light by the high density gas (Colgate and Cameron 1963; Cameron 1965*c*). In time this led to a detailed investigation of the hydrodynamics of head-on stellar collisions (Seidl and Cameron 1972). This problem took a large amount of time on the GISS computer, but it led to my first attempt to make a movie of the output, which was shown at a meeting of the American Astronomical Society.

The Texas Symposia on Relativistic Astrophysics were organized in an attempt to discuss and understand the quasar problem. I attended the first two, in December 1963 and December 1964, in Dallas and Austin, respectively. At the second meeting it was decided not to meet the following year. Thus it happened that I agreed to serve on the organizing committee of a "replacement" conference organized by Fred Singer in Miami Beach. This was a Conference on Observational Aspects of Cosmology, held in 1965. After going to BGSS in the summer of 1966, I proposed to the organizers of the Texas Symposia that the next one be held in New York, with BGSS as the host, and they agreed. Thus it was that I took the principal responsibility for putting together the Third Texas Symposium on Relativistic Astrophysics, which was actually held in January 1967. I have served on the general organizing committee for the Texas Symposia since then. In the earlier years I contributed very actively to putting the program together, but after that my participation in the planning has become minimal.

In 1967 I agreed to participate in the organization of a Symposium on the Origin and Distribution of the Elements, to be held by the newly formed International Association of Geochemistry and Cosmochemistry (IAGC) in Paris. So I helped to put the meeting together and to invite people to present papers. What I did not realize until I got to Paris was that I was expected to serve on the Council of the IAGC itself. So it was that I was asked to attend the Council meeting of the IAGC in Prague at the meeting of the International Geological Congress in August 1968. My wife and I were therefore in Prague during the Soviet invasion on 22 August. My personal response to that event has been to refuse all invitations to attend meetings behind the Iron Curtain since that time.

My interest in nucleosynthesis continued, with Truran and Arnett at GISS and Truran later coming to BGSS as an assistant professor. We worked on the relationship of supernova explosions to nucleosynthesis, particularly the formation of heavy nuclei in the *r*-process (Cameron, Truran, and Arnett 1967; Truran and Cameron 1967; Truran, Arnett, and Cameron 1967; Arnett and Cameron 1967; Truran, Arnett, Tsuruta, and Cameron 1968). Later I returned to the problem of the mass formula, and all of the nuclear parameters that depend on it, as part of an attempt to understand the *r*-process (Truran, Cameron, and Hilf 1970; Cameron, Delano, and Truran 1970). Jim Truran and I thought about the *p*-process (Truran and Cameron 1972). With Thaddeus ("Ted") Mazurek, a Yeshiva student, we investigated supernova explosions in degenerate stars (Mazurek, Truran, and Cameron 1974). Willy Fowler and I collaborated on a paper which examined the possible relationship between substantial stellar lithium abundances and the proton-proton chains (Cameron and Fowler 1971).

In the later part of my stay at BGSS an increasing amount of my time became devoted to calculations relevant to the origin of the solar system. Milton Pine, a Yeshiva student, computed solar nebula models by adapting techniques I had developed to compute disk models (Cameron and Pine 1973). These were static models, since we did not know how to formulate the physics of disk dissipation. Against the background of these models I made some lengthy estimates of the rates of various accumulation processes within the solar nebula (Cameron 1973*a*). With Fausto Perri, another Yeshiva student, I investigated the conditions under which gas concentrates toward a gravitating center in the solar nebula and eventually at a critical core mass collapses onto the core (Perri and Cameron 1974). At that time numerical models of the giant outer planets Jupiter and Saturn were produced using the assumption that the planetary parameters should be fit by adjusting the helium/hydrogen ratio in the planets, a procedure that produces a helium-rich mixture. With Morris Podolak, another Yeshiva student, I introduced a different procedure: the helium/hydrogen ratio was taken to be the solar mixture and the parameters were fit by introducing a planetary core of heavier materials (Podolak and Cameron 1974). At the time very accurate measurements of isotopic ratios in meteorites were leading to predictions that differences in the ratios represented differences in formation times of millions of years among the meteorite classes. This was in conflict with my views on the time scales relevant to the formation of the solar system, so I raised the question whether the isotopic ratio differences really represented time differences, or whether they might be real initial differences due to incomplete mixing between the products of nucleosynthesis sources (Cameron 1973*b*). This doubt was probably vindicated when real isotopic differences not due to time differences were discovered in meteorites not long afterward.

By 1973 New York was becoming unattractive to me as a place to work and BGSS at Yeshiva University was rapidly deteriorating. More and more I came to see the city as dirty and crime-ridden. Government support for science was now diminishing quite rapidly. BGSS suffered from this and also from serious financial problems that Yeshiva University was encountering as a whole. Therefore I welcomed an invitation to come to Harvard College Observatory in Cambridge, Massachusetts, which I accepted, moving in the summer of 1973.

Harvard

For some years Harvard College Observatory and the Smithsonian Astrophysical Observatory had been living together in the same set of buildings under separate administrations. Relations between them had become strained, and this led to some concern on the part of their parent organizations. Now, in 1973, major changes were to take place. Both of the observatories were to be placed under a single director, and George Field had been chosen for this role. George decided to set up a joint administrative structure, called the Harvard-Smithsonian Center for Astrophysics, to unite the various research programs carried out within the observatories. Eight divisions were established within the center to coordinate programs in different areas, and each was headed by an associate director. I became associate director for planetary sciences. Thus, although I was a Harvard employee, I found myself administering the work of a group of Smithsonian scientists.

This administrative burden was quite light, since the associate administrators had direct control over relatively little funding, as the two observatories maintained their own administrative infrastructures. However, the amount of civil service red tape gradually increased, and the associate administratorship gradually became less fun. I was glad to give it up in 1981. Much more demanding tasks were soon to come.

My involvement with national scientific committees was gradually increasing. From 1969 to 1971 I had served on the Astronomy Survey Committee of the National Academy of Sciences (the Greenstein Committee). Here I was chairman of the panel on space astronomy; this was the last time that all the techniques for doing astronomy in space would be lumped together in a single treatment. There was a serious overlap with most of the other panels.

Then, in the years 1971 to 1975, I served on a series of four committees reporting to NASA Headquarters and dealing with plans for the exploration of the outer planets of the solar system. During the last of these years I was actually the chairman of a steering group that coordinated the work of four other advisory groups trying to hammer out a unified strategy for outer planets exploration. This kept me quite busy, since I felt the need to monitor the meetings of the other four groups, and so I did a lot of traveling in that final year.

In 1973 I became a member of the Committee on Planetary and Lunar Exploration (COMPLEX) of the Space Science Board of the National Academy of Sciences. Here I became involved in an exercise to prioritize planetary missions for the next few years within a given cost envelope. This was a very interesting exercise, but in general it is not the best way for an academy committee to give advice to the government. Then in 1974 I became a member of the Space Science Board (SSB) itself, when Richard Goody was chairman. Throughout this essay I have described various ways in which my outlook on science was broadened; I think that my many years of SSB experience were more broadening than any of the others. Planetary exploration, astronomy and astrophysics, space plasma physics and solar-terrestrial relationships, earth sciences, medical sciences in space, and planetary biology: we had to encourage all of them and yet reach decisions on priorities (in time sequencing although generally not in absolute merit). A very profound influence on my thinking about these matters occurred when Jerry Wasserburg became chairman of COMPLEX. Jerry's approach to scientific planning emphasized strategic considerations: state what scientific measurements you want to get for a given object and with what priorities, but do not try to spell out in detail the mission that is to get them. This emphasizes the hard core of the scientific advice that is to be given to NASA and gives NASA good guidelines around which missions can be planned. There is a great difference between this approach and the previous way of saying "fly a mission to planet X." We put together a strategy for the exploration of the outer planets.

In the spring of 1976 I was elected to the National Academy of Sciences. In the summer of that year Goody stepped down as chairman of the SSB. I was asked to take his place. With considerable reluctance I agreed to do so, for I knew that the job would be very time-consuming. The SSB gives advice to the administrator of NASA, and thus the chairman must see that gentleman occasionally and his associate administrator for space science quite frequently. The chairman must stay on top of the work of the SSB's own committees, generally about eight in number. The chairman must coordinate with related boards and committees in the National Academy of Sciences and with other parts of the academy's infrastructure. He must maintain relations with the SSB's counterpart in Europe, the Space Science Committee of the European Science Foundation, and with senior officials in ESA, the European Space Agency. He must give testimony to congressional committees. He has a small staff in Washington to help him. I was also elected to the Cosmos Club in Washington in 1976, and I came to know it well.

I spent about half of my time on SSB affairs during the first year, when I was learning the job, and about one-third of my time in subsequent years. The major problem facing space science was the increasing fiscal austerity in the nation which was cutting back on the pace of space exploration. It was necessary to get the scientific house in order, so that each discipline had a well-prioritized strategy (i.e., Jerry Wasserburg's approach within COMPLEX had to be applied by the other committees, and COMPLEX had to do more of the same for the rest of the solar system). It took a fair amount of time to persuade the other committees of the merits of this strategic approach, and in a few cases this was not successfully done until after the rotation of a committee chairman. Thus putting together the strategies was a major preoccupation of the SSB for the next six years. My term as chairman was for three years, but when this was up I was persuaded to serve another term in order to finish getting the strategies prepared. Thus the SSB remained a major activity of my life until 1982. After I stepped down NASA gave me its Distinguished Public Service Medal.

Just to add a little spice to life, in 1976 I was also asked to become chairman of the Harvard Astronomy Department. Here also I was asked to serve a second term, and thus my departmental duties extended to 1982 in parallel with my SSB activities. This chairmanship was fortunately not too demanding, since the duties dealt entirely with the academic program and not with research (which was the responsibility of the director of Harvard College Observatory). But of necessity I had to learn something of the workings of the Harvard administration.

One of the main problems in dealing with a fairly high-pressure environment, as I had to do in the SSB job, is to cope with a very high rate of information flow. One quickly learns to look for the "bottom line" and to ignore most of the details. Unfortunately, this mental attitude spills over into one's scientific work, and I would find myself impatient with the papers at the few scientific meetings that I was able to attend, disregarding details and looking for the "bottom line." I did not realize how bad this disease had become until after I had left the SSB, when I found that I was once again able to appreciate the means by which an author had arrived at his final result.

With all of the above competition for my time, my scientific research program slowed down very appreciably. However, I regarded remaining viable in research as the highest priority which must be maintained at any cost, so that a number of major problems were addressed during my SSB years. I have already mentioned in the bibliographical references several extensions of previous work which were carried out during my earliest years at Harvard.

One new piece of work which attracted quite a lot of attention was the suggestion by Jim Truran (now at the University of Illinois) and me that the explosion of a supernova had caused the collapse of a nearby interstellar cloud and had thus triggered the formation of the solar system (Cameron and Truran 1977). This was proposed as a way to get short-lived ^{26}Al from its presumed source in the supernova into the primitive solar system in at most a few half-lives. Jim and I also estimated the supernova production rate of the ^{26}Al (Truran and Cameron 1978).

The above theory was not destined to last, however. Observations, particularly by radio astronomers, and later from the *IRAS* infrared satellite, showed that the common way to form stars is from the collapse of local density enhancements, called "cores," in dense molecular clouds. So I had to rethink the possible sources of the ^{26}Al and the time scales for introducing it, and also the ^{107}Pd and ^{129}I extinct radioactivities into the primitive solar nebula in a molecular cloud scenario. This led to a proposal that the source was a star of roughly solar mass entering its asymptotic giant branch, which also seemed to give a nice solution for the time scales (Cameron 1984, 1985*a*).

The discovery that there were anomalies in the abundances of stable isotopes in some meteoritic materials, as well as traces of extinct radioactivities, raised the question as to whether strange nucleosynthetic processes were producing the observed anomalies. If this were to be true, then it would be precisely the kind of phenomenon that I should be investigating. Indeed, within the anomalies in calcium isotopes I found evidence for just such a new process of nucleosynthesis, the neutron-rich silicon-burning process (Cameron 1979*a*). With a postdoctoral fellow, Guy Consolmagno, I looked at the other "FUN" anomalies, and we decided that all of the anomalies could be produced within the errors by slightly changing the mix of the products of normal nucleosynthetic processes in the observed materials (Consolmagno and Cameron 1980). Therefore I did not find a revitalized career in nuclear astrophysics as a result of these new meteorite measurements.

However, there was much unfinished business in the traditional nucleosynthesis to keep me busy. With a postdoctoral fellow, John Cowan (who later went to the University of Oklahoma), Jim Truran and I renewed our attack on the *r*-process. We set up a numerical simulation of the *r*-process that can take place during rapid helium burning in a helium zone if there is ^{13}C present when the temperature is rapidly raised in the zone. We had in mind the helium shell in a supernova, but later we also looked at the possibility that this might happen in a helium core flash if a stellar core could undergo two such flashes with the introduction of ^{13}C in between. At first we were optimistic about the results (Truran, Cowan, and Cameron 1978*a*, *b*; Cowan, Cameron, and Truran 1980, 1982, 1983, 1984). After improvements in the nuclear parameters of the problem, we concluded that it would not be feasible to get an adequate neutron source in the environments that we had been considering (Cameron, Cowan, and Truran 1985; Cowan, Cameron, and Truran 1985). However, using a steady-flow approximation, we did find new conditions in which the *r*-process would have to be operating when it would reach a freezing point (Cameron, Cowan, Klapdor, Metzinger, Oda, and Truran 1983; Cameron, Cowan, and Truran 1983).

A substantial amount of the work I did after coming to Harvard concerned problems of the origin of the solar system. I have already mentioned modeling the introduction of now-extinct radioactivities into the early solar system as part of the star formation process (Cameron 1984). In 1974 a proper formulation of the dissipation of viscous disks was published (Lynden-Bell and Pringle 1974). I started to apply this theory to models of the primitive solar nebula (Cameron 1978*a*, *b*, *c*, 1985*a*). With Bruce Fegley, a postdoctoral fellow, I looked at the conditions required for condensation of solids in models of the inner part of the nebula (Cameron and Fegley 1982). These inner parts turned out in the calculations to be very hot, and I saw in this a possible way for Mercury to have attained its high density by evaporating silicates and having them swept away by the surrounding solar nebula (Cameron 1985*b*).

One of the conclusions I soon reached was that, in the early stages of the development of the primitive solar nebula, it was likely that there would be local gravitational instabilities in the solar nebula gas. I called the resulting objects "giant gaseous protoplanets," and the first task, done with a student, Bill DeCampli, was to make models of these objects, treating them as if they were very small stars (DeCampli and Cameron 1979). The precipitation of melted solids toward the center of such objects was treated by Wayne Slattery, a postdoctoral fellow (Slattery, DeCampli, and Cameron 1980). The next step was to look at the evolution of giant gaseous protoplanets subject to boundary conditions typical of the primitive solar nebula, from which it was found that protoplanets in the inner solar system would be evaporated (Cameron, DeCampli, and Bodenheimer 1982). I also looked at the gravitational interaction between the giant gaseous protoplanets and the primitive solar nebula (Cameron 1979*b*).

The Moon is a unique body within the solar system, and I had long been puzzled about its mode of origin. With Bill Ward, a postdoctoral fellow, I noted that imparting the angular momentum of the Earth-Moon system to the proto-Earth by means of a collision requires, at the minimum, that the colliding body have as much mass as Mars (Cameron and Ward 1976). Later we concluded that the hot mineral accretion disk that would be created by this collision would be very unstable, and hence the dissipation would be rapid (Ward and Cameron 1978). I further discussed this point as part of a paper on terrestrial planet atmospheres (Cameron 1983). Later I ran a crude simulation of the collision in the form of particle-in-cell hydrodynamics (Cameron 1985*c*, *d*). Late in 1984 I discussed this at Los Alamos, arousing an interest in the problem in Willy Benz and Wayne Slattery, who are presently achieving some very interesting results after turning loose the full power of a Cray XMP on the problem.

I was also curious as to how the early evolution of the Sun, if formed from the dissipation of an accretion disk, would differ from the case where it was formed without angular momentum. This problem was started by Richard Epstein, a postdoctoral fellow. It was finished by "Jas" Mercer-Smith, another postdoctoral fellow. This exercise gave us important information about the time scale for dissipation of the solar nebula, since in order to fit H-R diagrams of young stars we had to assume that the nebula dissipation time was less than 10^5 yr (Mercer-Smith, Cameron, and Epstein 1984).

My Harvard years also saw the beginning of a rapid rate of change in the way in which I did my computing. This was a consequence of the rapid evolution of computing technology. Shortly after arriving in Cambridge, I acquired a

Hewlett-Packard programmable desk calculator with attached hard disk. This operated in interpreted Basic, and it was about the same speed of computation that I had had available at Chalk River. After doing a Monte Carlo calculation on the accumulation of grains into dust balls (Cameron 1975), I did the primitive solar accretion disk calculations with it (Cameron 1978*b*). However, this required several months of solid computation, and I concluded that something a lot faster was required. I started to read a lot of technical material regarding minicomputers, and in 1976 we purchased a Data General 3/D Nova minicomputer. This was intended to support students and postdoctoral fellows as well as me.

At the time the Center for Astrophysics had a CDC 6400 computer; my minicomputer ran about 25% as fast as the CDC, but computing with it was very much cheaper than using the CDC. Thus I could get a lot more done for the same computation dollar, at the expense of some aggravation in maintaining and operating the minicomputer. Not long afterward the center formed a committee to consider how it should replace its aging 6400, and I served on this. It was an excellent educational experience; I represented one extreme of the spectrum of opinion, since I felt that it was cheaper to go with a distribution of smaller machines. It was just as well that these arguments stretched out the decision, because it gave time for the Digital Equipment Corporation to develop the VAX 11/780 computer, and we were among the first to order one of these. It had the same speed as the 6400, so I remained with a large cost advantage with my Nova, although not quite as large as previously.

In due course our activities saturated the Nova 3, and it became necessary to supplement it with a Nova 4/X, which ran about 1.8 times as fast as the Nova 3. At about this time the disgraceful state of data management for NASA space science experiments had become very evident, and to study the matter and make recommendations I started a new committee within the Space Science Board, the Committee on Data Management and Computation (CODMAC), chaired by Ralph Bernstein from IBM. I participated in the meetings of this committee, which was also highly educational, and the ultimate report was widely distributed.

Eventually both Novas became saturated. The question now arose as to whether I should buy more memory for each machine in order to ease the pressure. However, it turned out that I could buy two microcomputers for the same amount of money, which would allow the offloading from the Novas of some activities, such as word processing and some program development. This started my transition to microcomputers. Eventually I acquired several IBM PC compatible computers, and the Novas have now been shut down. In one IBM microcomputer I have an "accelerator" board which runs at 25% of the speed of the VAX 11/780, and thus performs at the same speed as the Nova 3. Much higher speeds will be available on cheap microcomputers shortly. These microcomputers are technically much superior to the Novas in the way in which they handle memory, and they have very much more memory, and so they are superior for the purpose of running large Fortran programs.

Concluding Remarks

This brings me to 1985 and the age of 60. I have regained the momentum in my research that was significantly diminished during my Space Science Board days. In passing this milestone I realize all too clearly that I will be subject to compulsory retirement in just a decade, and with so many exciting new and old research projects to do, I expect that I will be able to resist most requests for large commitments of time on committee-related work in order to emphasize the research program. I am thankful that in another decade the capabilities of cheap desk-top computers will be well in advance of the present state of the art, and so I will be able to continue research projects even though in a state of unemployment.

It has been remarked that I have a broad range of scientific interests. This did not come about by deliberate design. Throughout this essay I have emphasized those events that led to a broadening of these interests. More than anything else it is because nature is complicated, and to understand all of the ramifications of a subject one must enquire into these interrelationships. To the extent that I may be different than most scientists, it is because I not only want to know about the current state of an object, but I also want to know how it got that way.

The pursuit of origins is certainly broadening. It is also something that exerts a discipline on what one does. Many times, in thinking about the origin of something that I have not previously considered, consistency with my view on related subjects usually shapes the hypothesis that develops. But hypotheses are fragile and are continually in a state of remolding. Along may come an observation or a calculation that demolishes one hypothesis. Not only must that subject be rethought, but the need for interrelated consistency will frequently demand that other hypotheses be revised as well. That rethinking seems to be a normal state of affairs to me. That explains why many of my friends are never sure what my current thoughts on a subject are and why I am ready at a moment's notice to abandon a favorite hypothesis when presented with a good reason. But I am also stubborn at times, if I think that some piece of experimental or observational evidence may properly be interpreted in an unconventional way which would be consistent with my views. Many of my short papers have been conceived in this way.

I have not mentioned all of the papers I have written, nor all the students and postdoctoral fellows I have had. Papers have been omitted for being trivial, for duplicating mentioned papers, or for being off the subject themes I have developed here. Students and postdocs have been omitted if they have been sufficiently independent in their work that they did not publish with me. Other omissions, if any, have been accidental and I apologize for them.

I have enjoyed the papers presented at the symposium. In some subjects they show a state of knowledge much beyond the point at which I stopped working on that subject. In other subjects I still have personal views and expect to make further contributions. In all cases I wish to thank the authors who made this occasion very memorable for me. And I was delighted to see again so many old friends. The preparation of this paper has been supported in part by NASA grants NGR 22-007-269, NGR 22-007-272, and NAG 9-89.

References

Arnett, W. D., and Cameron, A. G. W. 1967, Supernova hydrodynamics and nucleosynthesis, *Canadian J. Phys.*, **45**, 2953.

Arnett, W. D.; Hansen, C. J.; Truran, J. W.; and Cameron, A. G. W.; eds. 1968, *Nucleosynthesis* (New York: Gordon and Breach).

Brancazio, P. J., and Cameron, A. G. W., eds. 1964, *Origin and Evolution of Atmospheres and Oceans* (New York: Wiley).

——. 1967, Abundance anomalies produced by nuclear reactions in stellar surface layers. II. Results and comparison with anomalies in peculiar A stars, *Canadan J. Phys.*, **45**, 3297.

——, eds. 1968, *Infrared Astronomy* (New York: Gordon and Breach).

——, eds. 1969, *Supernovae and Their Remnants* (New York: Gordon and Breach).

Brancazio, P. J., Gilbert, A., and Cameron, A. G. W. 1967, Abundance anomalies produced by nuclear reactions in stellar surface layers. I. Nuclear reaction rates, *Canadian J. Phys.*, **45**, 3275.

Burbidge, E. M.; Burbidge, G. R.; Fowler, W. A.; and Hoyle, F. 1957, Synthesis of the elements in stars, *Revs. Modern Phys.*, **29**, 547.

Cameron, A. G. W. 1951, Interpretation of photoneutron yields, *Phys. Rev.*, **82**, 272.

——. 1954, Origin of anomalous abundances of the elements in giant stars, *Phys. Rev.*, **93**, 932.

——. 1955, Origin of anomalous abundances of the elements in giant stars, *Ap. J.*, **121**, 144.

——. 1956, Nuclear reactions in stars and nucleogenesis, Chalk River report CRP-652.

——. 1957*a*, A revised semiempirical atomic mass formula, *Canadian J. Phys.*, **35**, 1021.

——. 1957*b*, Nuclear radiation widths, *Canadian J. Phys.*, **35**, 666.

——. 1957*c*, Stellar evolution, nuclear astrophysics, and nucleogenesis, Chalk River report CRL-41.

——. 1957*d*, Nuclear reactions in stars and nucleogenesis, *Pub. A.S.P.*, **69**, 201.

——. 1958*a*, Nuclear level spacings, *Canadian J. Phys.*, **36**, 1040.

——. 1958*b*, Computation of the average neutron capture cross sections, *Proceedings International Conference on the Neutron Interactions with the Nucleus*, U. S. A. E. C. report TID-7547.

——. 1958*c*, Semiempirical correlation of fission yields and kinetic-energy distributions, *Revs. Modern Phys.,* **30,** 553.

——. 1958*d*, Nuclear astrophysics, *Ann. Rev. Nucl. Sci.,* **8,** 299.

——. 1959*a*, Carbon thermonuclear reactions and the formation of heavy elements, *Ap. J.,* **130,** 429.

——. 1959*b*, Neon and oxygen thermonuclear reactions, *Ap. J.,* **130,** 895.

——. 1959*c*, Photobeta reactions in stellar interiors, *Ap. J.,* **130,** 452.

——. 1959*d*, Spin dependence of nuclear level spacings, *Canadian J. Phys.,* **37,** 244.

——. 1959*e*, Multiple neutron capture in the Mike fusion explosion, *Canadian J. Phys.,* **37,** 322.

——. 1959*f*, Pycnonuclear reactions and nova explosions, *Ap. J.,* **130,** 916.

——. 1959*g*, Neutron star models, *Ap. J.,* **130,** 884.

——. 1959*h*, A revised table of abundances of the elements, *Ap. J.,* **129,** 676.

——. 1962, The formation of the Sun and planets, *Icarus,* **1,** 13.

——. 1963*a*, Nuclear astrophysics, notes from a course at Yale University, compiled by W. D. Arnett, C. J. Hansen, and J. W. Truran.

——, ed. 1963*b*, *Interstellar Communication* (New York: W. A. Benjamin).

——. 1964, Physics of the solar system, notes from a course at Yale University, compiled by W. D. Arnett, C. J. Hansen, and J. W. Truran.

——. 1965*a*, Systematics of fission based on the exponential mass formula, *Canadian J. Phys.,* **43,** 1723.

——. 1965*b*, Galactic and stellar physics, notes from a course at Yale University, compiled by W. D. Arnett, C. J. Hansen, and J. W. Truran.

——. 1965*c*, Gas clouds in dense star clusters, a possible quasar model, *Nature,* **207,** 1140.

——. 1973*a*, Accumulation processes in the primitive solar nebula, *Icarus,* **18,** 407.

——. 1973*b*, Are large time differences in meteorite formation real? *Nature,* **246,** 30.

——. 1975, Clumping of interstellar grains during formation of the primitive solar nebula, *Icarus,* **24,** 128.

——. 1978*a*, The primitive solar accretion disk and the formation of the planets, in *The Origin of the Solar System,* ed. S. F. Dermott (New York: Wiley).

——. 1978*b*, Physics of the primitive solar accretion disk, *Moon and Planets,* **18,** 5.

——. 1978*c*, Physics of the primitive solar nebula and of giant gaseous protoplanets, in *Protostars and Planets,* ed. T. Gehrels (University of Arizona Press).

—— 1979*a*, The neutron-rich silicon-burning and equilibrium processes of nucleosynthesis, *Ap. J. (Letters),* **23,** L53.

——. 1979*b*, The interaction between giant gaseous protoplanets and the primitive solar nebula, *Moon and Planets,* **21,** 173.

——. 1983, Origin of the atmospheres of the terrestrial planets, *Icarus,* **56,** 195.

——. 1984, Star formation and extinct radioactivities, *Icarus,* **60,** 416.

——. 1985*a*, Formation and evolution of the primitive solar nebula, in *Protostars and Planets II,* ed. D. C. Black and M. S. Matthews (University of Arizona Press).

——. 1985*b*, The partial volatilization of Mercury, *Icarus,* in press.

——. 1985*c*, Formation of the prelunar accretion disk, *Icarus,* **62,** 319.

——. 1985*d*, The impact theory for origin of the Moon, in *Origin of the Moon,* ed. W. K. Hartmann, R. J. Phillips, and G. J. Taylor (Houston: Lunar and Planetary Institute), in press.

Cameron, A. G. W., and Cohen, J. M. 1969, The instability of stellar structures intermediate between white dwarfs and neutron stars, *Ap. Letters,* **3,** 3.

Cameron, A. G. W.; Cowan, J. J.; Klapdor, H. V.; Metzinger, J.; Oda, T.; and Truran, J. W. 1983, Steady flow approximations to the helium *r*-process, *Ap. Space Sci.,* **91,** 221.

Cameron, A. G. W., Cowan, J. J., and Truran, J. W. 1983, The waiting-point approximation in *r*-process calculations, *Ap. Space Sci.,* **91,** 235.

——. 1985, Neutron source requirements for the helium *r*-process, in *Nucleosynthesis: Challenges and New Developments,* ed. W. D. Arnett and J. W. Truran (Chicago: University of Chicago Press).

Cameron, A. G. W., DeCampli, W. M., and Bodenheimer, P. H. 1982, Evolution of giant gaseous protoplanets embedded in the primitive solar nebula, *Icarus,* **49,** 298.

Cameron, A. G. W., Delano, M. D., and Truran, J. W. 1970, Dynamics of the rapid neutron capture process, in *Proceedings of the International Conference on the Properties of Nuclei Far from the Valley of Beta-Stability,* **2,** CERN.

Cameron, A. G. W., and Elkin, R. M. 1965, Role of the symmetry energy in atomic mass formulas, *Canadian J. Phys.,* **43,** 1288.

Cameron, A. G. W., and Fegley, M. B. 1982, Nucleation and condensation in the primitive solar nebula, *Icarus,* **52,** 1.

Cameron, A. G. W., and Fowler, W. A. 1971, Lithium and the *s*-process in red giant stars, *Ap. J.,* **164,** 111.

Cameron, A. G. W., Harms, W., and Katz, L. 1951, Photoproton reactions in lead, *Phys. Rev.,* **83,** 1264.

Cameron, A. G. W., and Katz, L. 1951, Minimum gamma-gamma cross section in gold, *Phys. Rev.,* **84,** 608.

Cameron, A. G. W., and Mock, M. 1967, Stellar accretion and X-ray emission, *Nature,* **215,** 464.

Cameron, A. G. W., and Pine, M. R. 1973, Numerical models of the primitive solar nebula, *Icarus,* **18,** 377.

Cameron, A. G. W., and Truran, J. W. 1971, Chemical evolution of the Galaxy, *J. Roy. Astr. Soc. Canada,* **65,** 1.

——. 1977, The supernova trigger for formation of the solar system, *Icarus,* **30,**

447.

Cameron, A. G. W., Truran, J. W., and Arnett, W. D. 1967, Formation of heavy nuclei in nucleosynthesis, *Arkiv for Fysik,* **36,** 521.

Cameron, A. G. W., and Ward, W. R. 1976, The origin of the Moon, in *Lunar Science VII* (Houston: Lunar Science Institute).

Cohen, J. M., and Cameron, A. G. W. 1969, Rotational properties of neutron star models, *Nature,* **222,** 862.

——. 1971, Neutron star models including the effects of hyperon formation, *Ap. Space Sci.,* **10,** 227.

Cohen, J. M.; Langer, W. D.; Rosen, L. C.; and Cameron, A. G. W. 1970, Neutron star models based on an improved equation of state, *Ap. Space Sci.,* **6,** 217.

Cohen, J. M., Lapidus, A., and Cameron, A. G. W. 1969, Treatment of pulsating white dwarfs including general relativistic effects, *Ap. Space Sci.,* **5,** 113.

Colgate, S. A., and Cameron, A. G. W. 1963, Enhancement of light output from a supernova, *Nature,* **200,** 870.

Consolmagno, G. J., and Cameron, A. G. W. 1980, The origin of the FUN anomalies and the high temperature inclusions in the Allende meteorite, *Moon and Planets,* **23,** 3.

Cowan, J. J., Cameron, A. G. W., and Truran, J. W. 1980, Seed abundances for *r*-processing in the helium shells of supernovae, *Ap. J.,* **241,** 1090.

——. 1982, The thermal runaway *r*-process, *Ap. J.,* **252,** 348.

——. 1983, Explosive helium-burning in supernovae: A source of *r*-process elements, *Ap. J.,* **265,** 429.

——. 1984, *R*-process production in low-mass stars, in *Stellar Nucleosynthesis,* ed. C. Chiosi and A. Renzini (Dordrecht: Reidel).

——. 1985, *R*-process nucleosynthesis in dynamic helium-burning environments, *Ap. J.,* **294,** 656.

DeCampli, W. M., and Cameron, A. G. W. 1979, Structure and evolution of isolated giant gaseous protoplanets, *Icarus,* **38,** 367.

Delano, M. D., and Cameron, A. G. W. 1971, Nucleosynthesis in neutron rich supernova ejecta, *Ap. Space Sci.,* **10,** 203.

Ezer, D., and Cameron, A. G. W. 1963, The early evolution of the Sun, *Icarus,* **1,** 442.

——. 1965, A study of solar evolution, *Canadian J. Phys.,* **43,** 1497.

——. 1966, Solar evolution with varying *G*, *Canadian J. Phys.,* **44,** 493.

——. 1967*a*, Early and main sequence evolution of stars in the range 0.5 to 100 solar masses, *Canadian J. Phys.,* **45,** 3429.

——. 1967*b*, Evolution of stars of low mass, *Canadian J. Phys.,* **45,** 3461.

——. 1968, Solar spin-down and neutrino fluxes, *Ap. Letters,* **1,** 177.

——. 1971*a*, Pre–main sequence stellar evolution with mass loss, *Ap. Space Sci.,* **10,** 52.

——. 1971*b*, The evolution of hydrogen helium stars, *Ap. Space Sci.,* **14,** 399.

——. 1972, Effects of sudden mixing in the solar core on solar neutrinos and ice ages, *Nature Phys. Sci.,* **240,** 180.

Gilbert, A., and Cameron, A. G. W. 1965, A composite nuclear level density formula with shell corrections, *Canadian J. Phys.,* **43,** 1446.

Gilbert, A., Chen, F. S., and Cameron, A. G. W. 1965, Level densities in lighter nuclei, *Canadian J. Phys.,* **43,** 1248.

Greenstein, G. S., and Cameron, A. G. W. 1969, Spin-down effects in neutron stars, *Nature,* **222,** 862.

Greenstein, G. S., Truran, J. W., and Cameron, A. G. W. 1967, The helium deficiency in old halo B stars, *Nature,* **213,** 871.

Haslam, R. N. H.; Cameron, A. G. W.; Cooke, J. A.; and Crosby, E. H. 1952, Alpha-particles from the photodisintegration of silver and bromine, *Canadian J. Phys.,* **30,** 349.

Hoyle, F. 1946, The synthesis of the elements from hydrogen, *M.N.R.A.S.,* **106,** 343.

Jastrow, R., and Cameron, A. G. W., eds. 1963, *Origin of the Solar System* (New York: Academic Press).

Katz, L., and Cameron, A. G. W. 1951*a*, The solution of X-ray activation curves for photonuclear cross sections, *Canadian J. Phys.,* **29,** 518.

——. 1951*b*, Photonuclear cross sections in aluminum and magnesium, *Phys. Rev.,* **84,** 1115.

Katz, L.; Haslam, R. N. H.; Horsley, R. J.; Cameron, A. G. W.; and Montalbetti, R. 1954, Fine structure in the $C^{12}(\gamma,\alpha)C^{11}$ and $O^{16}(\gamma,\alpha)O^{15}$ activation curves, *Phys. Rev.,* **95,** 464.

Katz, L.; Kavanagh, T. M.; Cameron, A. G. W.; Bailey, E. C.; and Spinks, J. W. T. 1955, Photofission of U^{238}, *Phys. Rev.,* **99,** 987.

Langer, W. D., and Cameron, A. G. W. 1969, Effects of hyperons on the vibrations of neutron stars, *Ap. Space Sci.,* **5,** 213.

Langer, W. D.; Rosen, L. C.; Cohen, J. M.; and Cameron, A. G. W. 1969, An equation of state at subnuclear densities, *Ap. Space Sci.,* **5,** 259.

Lynden-Bell, D., and Pringle, J. E. 1974, The evolution of viscous discs and the origin of the nebular variables, *M.N.R.A.S.,* **168,** 603.

Maran, S. P., and Cameron, A. G. W., eds. 1964, *Physics of Non-thermal Radio Sources* (Washington: Government Printing Office).

Marsden, B. G., and Cameron, A. G. W., eds. 1966, *The Earth-Moon System* (New York: Plenum Press).

Mazurek, T. J., Truran, J. W., and Cameron, A. G. W. 1974, Electron capture in carbon dwarf supernovae, *Ap. Space Sci.,* **27,** 261.

Mercer-Smith, J. A., Cameron, A. G. W., and Epstein, R. I. 1984, On the formation of stars from disk accretion, *Ap. J.,* **279,** 363.

Merrill, P. W. 1952, Spectroscopic observations of stars of class S, *Ap. J.,* **116,** 21.

Millar, C. H., and Cameron, A. G. W. 1950, Photo-alpha-reactions in oxygen and nitrogen, *Phys. Rev.,* **78,** 78.

——. 1953, Photo-alpha reactions in nuclear emulsions, *Canadian J. Phys.,* **31,** 723.

Millar, C. H., Cameron, A. G. W., and Glicksman, M. 1950, Gamma-ray studies using deuterium-loaded photographic plates, *Canadian J. Research,* **A28,** 475.

Perri, F., and Cameron, A. G. W. 1974, Hydrodynamic instability of the solar nebula in the presence of a planetary core, *Icarus,* **22,** 416.

Podolak, M., and Cameron, A. G. W. 1974, Models of the giant planets, *Icarus,* **22,** 123.

Reynolds, J. H. 1960, Determination of the age of the elements, *Phys. Rev. Letters,* **4,** 8.

Rosen, L. C., and Cameron, A. G. W. 1972, Surface composition of magnetic neutron stars, *Ap. Space Sci.,* **15,** 137.

Seidl, F. G. P., and Cameron, A. G. W. 1972, A numerical hydrodynamic study of coalescence in head-on collisions of identical stars, *Ap. Space Sci.,* **15,** 44.

Slattery, W. L., DeCampli, W. M., and Cameron, A. G. W. 1980, Protoplanetary core formation by rain-out of minerals, *Moon and Planets,* **23,** 381.

Stein, R. F., and Cameron, A. G. W., eds. 1966, *Stellar Evolution* (New York : Plenum Press).

Suess, H. E., and Urey, H. C. 1956, Abundances of the elements, *Revs. Modern Phys.,* **28,** 53.

Ter Haar, D., and Cameron, A. G. W. 1963, Historical review of theories of the origin of the solar system, in *Origin of the Solar System,* ed. R. Jastrow and A. G. W. Cameron (New York: Academic Press).

Tinsley, B., and Cameron, A. G. W. 1974, Possible influence of comets on the chemical evolution of the Galaxy, *Ap. Space Sci.,* **31,** 31.

Truran, J. W., Arnett, W. D., and Cameron, A. G. W. 1967, Nucleosynthesis in supernova shock waves, *Canadian J. Phys.,* **45,** 2315.

Truran, J. W.; Arnett, W. D.; Tsuruta, S.; and Cameron, A. G. W. 1968, Rapid neutron capture in supernova explosions, *Ap. Space Sci.,* **1,** 129.

Truran, J. W., and Cameron, A. G. W. 1967, Properties of nuclei in the superheavy region, *Arkiv for Fysik,* **36,** 509.

——. 1970, The galactic halo, *Nature,* **225,** 710.

——. 1971, Evolutionary models of nucleosynthesis in the Galaxy, *Ap. Space Sci.,* **14,** 179.

——. 1972, The *p*-process in explosive nucleosynthesis, *Ap. J.,* **171,** 89.

——. 1978, ^{26}Al production in explosive carbon burning, *Ap. J.,* **219,** 226.

Truran, J. W., Cameron, A. G. W., and Gilbert, A. 1966, The approach to nuclear statistical equilibrium, *Canadian J. Phys.,* **44,** 563.

Truran, J. W., Cameron, A. G. W., and Hilf, E. 1970, Construction of mass formulas designed to be valid for neutron-rich nuclei, in *Proceedings of the International Conference on the Properties of Nuclei Far from the Valley of*

Beta-Stability, **1**, CERN.

Truran, J. W., Cowan, J. J., and Cameron, A. G. W. 1978*a*, The helium-driven *r*-process in supernovae, *Ap. J. (Letters),* **222**, L63.

——. 1978*b*, The helium-driven *r*-process and possible superheavy element production, in *Superheavy Elements,* ed. M. A. K. Lodhi (New York: Pergamon Press).

Truran, J. W., Hansen, C. J., and Cameron, A. G. W. 1965, The helium content of the Galaxy, *Canadian J. Phys.,* **43**, 1616.

——. 1966, Formation of helium in the Galaxy, in *Stellar Evolution,* ed. R. F. Stein and A. G. W. Cameron (New York: Plenum Press).

Truran, J. W.; Hansen, C. J.; Cameron, A. G. W.; and Gilbert, A. 1966, Thermonuclear reactions in medium and heavy nuclei, *Canadian J. Phys.,* **44**, 151.

Tsuruta, S., and Cameron, A. G. W. 1965*a*, Composition of matter in nuclear statistical equilibrium at high densities, *Canadian J. Phys.,* **43**, 2056.

——. 1965*b*, Cooling of neutron stars, *Nature,* **207**, 364.

——. 1966*a*, Some effects of nuclear forces on neutron star models, *Canadian J. Phys.,* **44**, 1895.

——. 1966*b*, Cooling and detectability of neutron stars, *Canadian J. Phys.,* **44**, 1863.

——. 1966*c*, The rotation of neutron stars, *Nature,* **211**, 356.

——. 1970, URCA shells in dense stellar interiors, *Ap. Space Sci.,* **7**, 374.

——. 1976, The URCA process in convective cores, *Ap. Space Sci.,* **39**, 397.

Tsuruta, S., Wright, J. P., and Cameron, A. G. W. 1965, Oscillation periods of neutron stars, *Nature,* **206**, 1137.

Ward, W. R., and Cameron, A. G. W. 1978, Disc evolution within the Roche limit, in *Lunar Planetary Sci. IX* (Houston: Lunar Planetary Institute).

Weiss, R. A., and Cameron, A. G. W. 1969*a*, Equilibrium theory of the nuclear symmetry energy of infinite nuclear matter, *Canadian J. Phys.,* **47**, 2171.

——. 1969*b*, Equilibrium theory of the symmetry energy of finite nuclei, *Canadian J. Phys.,* **47**, 2211.

Part I

Galaxies

1

Cosmic Dark Matter and Solar Neutrinos: The Invisible Explains the Missing (or Vice Versa)

William H. Press

Abstract

Arguments are given, in the form of otherwise unexplained numerical coincidences involving physical and astronomical parameters, that (i) the dark galactic halo is composed of weakly interacting, massive particles in the GeV range, (ii) a number concentration of order 10^{-12} of these particles has been captured by the Sun, and (iii) the thermal transport due to these particles is sufficient to suppress the Sun's SNU neutrino signature. The particles cannot be the massive neutrinos or photinos of presently fashionable particle models. They could be undiscovered particles produced cosmologically with a new excess of a conserved quantum number (as were baryons), or Majorana particles with nearly perfect s-wave suppression of annihilation. The particles, if they exist, may in the future be directly detectable by laboratory experiment.

1.1. Introduction

It is both a pleasure and an honor to be given the opportunity to contribute to this volume honoring A. G. W. Cameron. It was my misfortune not to be one of Al's graduate students; nor to work with him as a postdoc; nor even as an assistant professor. Perhaps feeling that my misfortune ought not to continue, Al did help bring me to Harvard in 1976, shortly after which he became my department chairman. Since then I have had all manner of pleasant and useful interaction with him, not only as regards science but also as regards the related areas of microcomputers and department chairmanship (I having succeeded Al as chairman in 1982 and acquired a personal computer at about the same time).

I say "related areas" because, as Al taught me so well, the two arts of microcomputer and departmental management require mostly the same skills: patience in the face of repetitive tasks, and wily craftiness.

However, I want to relate my contribution here to a different talent of Al's, one that bears more directly on his many scientific contributions. Al is a superb storyteller. I refer not to fictive stories, but to the scientific variety. Almost always, Al's work goes after the "big picture," and he is by no means shy about weaving the individual threads of his science into broad, composite tapestries, breathtaking in their extent. All of us participating in this symposium have had the experience of hearing Al's grand discourses on "how it all fits together," whether "it" happens to be nucleosynthesis, the origin of the solar system, or another of his major areas of research.

In this contribution, I want to tell a scientific story of my own, based on recent and continuing collaborative work with David Spergel, Katherine Freese, Lawrence Krauss, and Ronald Gilliland (Spergel and Press 1985; Press and Spergel 1985; Krauss et al. 1985; Gilliland et al. 1986) and drawing on present and previous work by other investigators, notably Steigman et al. (1978) and Faulkner and Gilliland (1985). My story does not have quite the grand scope of some of Al Cameron's epics, but I hope that Al will nevertheless take its telling here in its intended spirit of salutatory good wishes to him on the occasion of his 60th birthday.

1.2. Some Cosmic Coincidences

Let me begin, as the universe did, with particle physics. Much interest recently has been focused on the electro-weak interactions of hypothetical massive particles that may be predicted by one or another model of particle physics. The weak interaction is not so weak, it turns out, between massive particles. As emphasized by Goodman and Witten (1985), the electro-weak scattering cross section of particles of mass m which exchange a particle of mass M is typically on the order of

$$\sigma_s \sim \frac{\hbar^2}{c^2}\frac{\alpha^2}{M^4} \approx 2.5 \times 10^{-36}\,\mathrm{cm}^2 \left(\frac{m}{10\,\mathrm{GeV}}\right)^2 \left(\frac{M}{30\,\mathrm{GeV}}\right)^{-4}. \tag{1}$$

Some cases that exemplify equation (1) are: scattering of massive neutrinos from quarks in baryons (exchanging a Z); likewise, scattering of supersymmetric scalar neutrinos off of baryons; and scattering of photinos—supersymmetric counterparts of the photon—from quarks (with an intermediate squark).

Corresponding annihilation cross sections are also about the same order of magnitude. With that in mind, we turn to the early universe and ask, as a function of the cosmic temperature T, what density of weakly interacting particles do we expect to have been present?

When the kT is much larger than mc^2, particles can be created or destroyed in thermal equilibrium, so the density of each species is simply an appropriate statistical weight times the familiar blackbody density at^4. As the universe expands, the temperature falls below mc^2. At this point the density of particles implied by thermal equilibrium starts to fall off exponentially in mc^2/kt. However, stable particles can disappear only when they are able to find their appropriate antiparticles on which to annihilate. As the expanding universe dilutes their density, it becomes harder and harder for them to do so. At a certain expansion, annihilations thus effectively turn off, and particle density thereafter decreases only as the geometrical dilution factor $R^{-3} \propto T^3$.

We know the temperature of the universe today, namely about 3 K. The density of an annihilating particle species in the universe today is thus completely determined by the temperature at which the particle "gets off" the thermal equilibrium curve and "gets onto" the geometrical dilution curve. Evidently this depends only on the particle's annihilation cross section: low cross-section particles freeze out earlier onto a higher density track, while high cross-section particles end up (after additional annihilations) on a parallel lower density track. The details of this have been calculated by Lee and Weinberg (1977), Wolfram (1979), and Silk and Srednicki (1985), among others. The results can be roughly summarized as an equation that relates the current fraction of critical cosmological density Ω_x contributed by a particle x, to its annihilation cross section σ_a,

$$\Omega_x h_{50}^2 \approx \frac{5 \times 10^{-37}\,\mathrm{cm}^2}{\sigma_a}. \tag{2}$$

Here h_{50} is the Hubble constant in units of $50\,\mathrm{km\,s^{-1}Mpc^{-1}}$, a number between about 1 and 2.

It is remarkable that the mass m of the particle does not appear in equation (2). That equation says that any particle (capable of annihilation) which contributes, say, Ω_x of 0.1 or 0.2 *must* have an annihilation cross section on the order of $2 \times 10^{-36}\,\mathrm{cm}^2$. *But this cross section is the same as that of equation (1).* Let me call this "Cosmic Coincidence #1," that the natural electro-weak annihilation cross sections of particles in the multi-GeV range can give them plausible cosmological abundances today.

Let us now turn from the long ago and far away to the here and now of our own Sun. We ask a seemingly unrelated question: What cross section is implied by the condition that a cosmic particle, in traversing the solar interior, scatters off of a solar proton about one time, i.e., on average neither gets through unscathed nor scatters many times? Call this cross section σ_{crit}. Simple geometry gives

$$\sigma_{\mathrm{crit}} \approx R_\odot \frac{m_p}{M_\odot}, \tag{3}$$

where m_p is the proton mass, and $M_\odot$ and $R_\odot$ are the solar mass and radius. The numerical value of the right-hand side of equation (3) is about $4 \times 10^{-36}\,\mathrm{cm}^2$. And the fact that this value is very nearly the same as the two cross sections previously discussed is Cosmic Coincidence #2!

In astrophysics, coincidences by themselves never *prove* anything. Nevertheless they do delineate fertile areas for thought: one way of minimizing the implausibility of extant coincidences in a field is to show that they allow explanations of other puzzling phenomena. Coincidences thus "explained" do not have to be multiplied into your personal lifetime quota; instead, you get to lay them off onto the poor observer who discovered the puzzle in the first place. We proceed with this goal in mind. There are more Cosmic Coincidences in store, but we defer them to subsequent sections.

1.3. Capture of Cosmions by the Sun

Cosmic particles, which we will here call "cosmions," with cross sections around $\sigma_{\rm crit}$ will usually be captured by the Sun when they try to pass through it. One way of seeing this is to note some relative escape velocities from the Sun $617\,{\rm km\,s^{-1}}$ at the solar surface, $\sim 2100\,{\rm km\,s^{-1}}$ at the Sun's half-mass radius, and $\sim 3000\,{\rm km\,s^{-1}}$ at the solar center. In a scattering, a cosmion loses on average a certain *fraction* of its kinetic energy, roughly the ratio of its mass to what it scatters off of, or the reciprocal of that ratio, whichever is smaller. A cosmion's velocity far from the Sun is that appropriate to a bound galactic population, say $300\,{\rm km\,s^{-1}}$ in round numbers. Therefore, for cosmions more massive than a proton and up to a mass of about $m_p \times \left(2100\,{\rm km\,s^{-1}}/300\,{\rm km\,s^{-1}}\right)^2$, everything that "hits," "sticks." A more accurate calculation is found in Press and Spergel (1985, hereafter cited as Paper II).

We can also easily estimate the flux of cosmions through the Sun: Since the Sun is spherical and the flux is presumably isotropic, there is no harm in imagining that all cosmions are moving in a single direction, say with mean velocity $\bar{v}$ (the $300\,{\rm km\,s^{-1}}$ above), and with mean number density n_x. Then the Sun sweeps out, per unit time, a cylindrical tube containing $n_x\bar{v}(\pi R_\odot^2)$. but we should not neglect gravitational focusing of orbits, which is well known to give an additional factor $(GM_\odot)/R_\odot\bar{v}^2)$. A detailed and accurate calculation in Paper II shows that the above estimate is wrong only in omitting the additional factor $\sqrt{6/\pi}$.

The age of the Sun is rather accurately known. So in terms of the above parametrization, we can calculate the number of previously captured cosmions in the Sun as a fraction of the number of baryons. The result is approximately

$$\begin{pmatrix}\text{solar abundance}\\ \text{by number}\end{pmatrix} = 3\times 10^{-9}\left(\frac{\rho_x}{1M_\odot/{\rm pc}^3}\right)\frac{v_{\rm esc}}{\bar{v}}\frac{\sigma_{\rm min}}{\sigma_{\rm crit}}\frac{m_p}{m_x}. \qquad (4)$$

Here ρ_x is the mass density of cosmions in the Galaxy, $v_{\rm esc}$ is the surface solar escape velocity, and $\sigma_{\rm min} \equiv \min(\sigma_s, \sigma_{\rm crit})$. If cosmions make up the missing matter in the galactic halo, then a plausible value for the density is $\rho_x = 0.01$ $M_\odot$ ${\rm pc}^{-3}$. Taking also $m_p/m_x = 0.2$, we get a number density in the Sun on the order of 10^{-11}.

1.4. Effect of Cosmions on Solar Energy Transport

So what? Isn't 10^{-11} so small as to be uninteresting? The perhaps surprising answer, originally noted by Faulkner and Gilliland in unpublished work that is summarized in Steigman et al. (1978) (and unintentionally rediscovered by Spergel and Press [1985, hereafter cited as Paper I]), is that cosmions with even this low abundance can have a profound effect on the Sun's thermal energy transport.

One way to see this is to calculate some relevant time scales: Consider, as

*We eschew the alternative term "WIMPs" (weakly interacting massive particles) proposed by eponymous colleagues.

two different cases, the volume of the whole Sun, and the volume of that much smaller central core in which energetic neutrinos (which can be detected as SNUs by the ^{37}Cl solar neutrino experiment) are produced. It is worth noting that this latter volume (about 3×10^9 cm in radius) lies well within the region of the Sun that produces most of the solar luminosity (a radius of about 8×10^9 cm). Now for each volume, two-dimensional time scales can be defined. First, there is a dynamic time scale, the orbital time for a captured, non-interacting particle with aphelion at the edge of the volume. Second, there is the thermal or Kelvin time scale, the time for thermal energy to diffuse from the center of the Sun to the edge of the volume under consideration. The approximate values of these time scales are shown in the following table:

	Orbital Time Scale	Kelvin Time Scale	Ratio
Whole Sun	10^4 s	10^8 yr $\sim 3 \times 10^{15}$ s	3×10^{-12}
Neutrino Region	10^3 s	10^6 yr $\sim 3 \times 10^{13}$ s	3×10^{-11}

We can interpret these numbers as follows: If a cosmion orbiting within the Sun interacts with its surroundings about once per orbital time scale (which it will do if its scattering cross section is about σ_{crit}), then it can transport about kT of energy per dynamical time. By contrast, the time that it would have taken this same kT to leak out by normal diffusion is some 10 or 11 orders of magnitude longer. Therefore a concentration of cosmions as small as the last column in the table can carry order unity of the thermal flux, or, equivalently, can reduce the effective opacity and thermal temperature gradient of the Sun.

Note in particular that a number concentration of 3×10^{-11} will have order-unity effect on the Sun's neutrino producing region. We have stumbled on Cosmic Coincidence #3: The number of cosmions needed to lower substantially the temperature gradient in the SNU (solar neutrino unit) producing region of the Sun is about the same number as will naturally be captured by the Sun from a plausible galactic population. In other words, the solutions to the neutrino problem in the Sun and the missing mass problem in the Galaxy may be one and the same!

This is a rather striking conclusion, which thus needs to be backed up rather carefully by detailed calculations. Some of these calculations have already been done. Paper I, for example, develops a fairly accurate nonlocal formalism for calculating the energy flux carried by a population of cosmions which have come to thermal equilibrium within the Sun. The formalism uses the collisional Boltzmann equation and derives a self-consistent effective temperature T_x for the cosmion population, and, also, as a function of the cosmion's mass m_x, a dependence of the flux on radius. Overly massive cosmions will sink to equilibrium in too small a central core to affect the neutrino flux, while ones that are too light will evaporate from the Sun. The mass range $5\,\text{GeV} < m_x < 60\,\text{GeV}$ is free of either difficulty.

A local approximation to the more exact treatment, but one which furnishes some insight, gives the following formula for the luminosity L_x carried by cosmions:

$$L_x = \frac{n_x}{n_p}\min\left(\frac{\sigma_{\rm crit}}{\sigma_x}, \frac{\sigma_x}{\sigma_{\rm crit}}\right)\sqrt{\frac{m_p}{m_x}}\left[\frac{4\pi M_\odot kT_c}{m_p R_\odot}\sqrt{\frac{kT_c}{m_p}}\right]. \tag{5}$$

Notice that the factor in square brackets makes no reference to the properties of the cosmions. It has a numerical value $\approx 1.6\times 10^{46}$ ergs $\mathrm{s}^{-1} \approx 4.1\times 10^{12}\ L_\odot$, where $L_\odot$ is the solar luminosity. We see that some 12 orders of magnitude are available to be absorbed in the cosmion-dependent factors in equation (5), still leaving the cosmions transporting of order an entire solar luminosity! The cosmion-dependent factors are (i) their fraction abundance by number relative to protons, (ii) the discrepancy between their cross section and $\sigma_{\rm crit}$, and (iii) a weak dependence on their mass relative to the proton mass.

Another aspect which requires detailed calculation is to find how big an effect on the measured SNU flux will be produced by a given level of cosmion-induced thermal transport. Detailed solar models are required to answer this. The work of Faulkner and Gilliland (1985) gives some preliminary indications on this point, and work by Gilliland et al. (1986) gives a more definitive result. The results are that an abundance by number of 10^{-12} of cosmions of mass 5 GeV will reduce the SNU flux by about a factor of 2.

1.5. Are Cosmions an Already Hypothesized Particle?

It is interesting to ask whether any massive particles with a basis in current particle theory can fulfill the astrophysical requirements that we have deduced. The answer is "probably not," and it is an interesting story why this is the case.

Any given particle model implies a relationship between scattering cross sections (which affect both cosmion capture and cosmion energy transport) and cosmion-cosmion annihilation cross sections. If the density of cosmions in the Sun is too high, then the time scale for them to annihilate mutually can be shorter than the age of the Sun. That is, so high a density cannot be reached by slow accumulative capture. A lower, equilibrium density will be reached at which captures and annihilations are in balance.

Krauss et al. (1985, hereafter cited as Paper III) develop detailed calculations along these lines. The annihilation rate turns out to be given by

$$A = \frac{N^2 \langle\sigma_a v\rangle_\odot}{V}, \tag{6}$$

where N is the total number of cosmions in the Sun, $\langle\sigma_a v\rangle_\odot$ is the annihilation rate factor averaged over the cosmions' approximately Maxwellian velocity distribution, and V is an effective volume given by

$$V = \frac{\left(\int n(r)d^3r\right)^2}{\int n^2(r)d^3r} \tag{7}$$

with $n(r)$ the local cosmion number density. For an index 3 polytropic model of the Sun, with the isothermal distribution of cosmions detailed in Paper I, one gets $V \approx 3\times 10^{31}\ m_x^{-1.5}\mathrm{GeV\ cm}^3$.

The equilibrium abundance is obtained by equating capture and annihilation rates. The result can be written relative to the baryon number in the Sun $N_\odot$,

$$\frac{N}{N_\odot} = 1\times 10^{-14}\left(\frac{\rho}{0.01M_\odot/\mathrm{pc}^3}\right)^{0.5}\left(\frac{v}{300\,\mathrm{km\,s^{-1}}}\right)^{-0.5}\left(\frac{\sigma_{\min}c}{\langle\sigma_a v\rangle_\odot}\right)^{0.5} m_{x\,\mathrm{GeV}}^{-1.25}. \tag{8}$$

Since we need a fractional abundance not much less than 10^{-12}, for reasonable values of m_x (say 5 GeV) this leaves about three orders of magnitude to be made up by the three scaled factors in equation (8). Not more than one can be gained from the astrophysical factors ρ_{halo} and v, even if the cosmions are assumed to be concentrated in the galactic disk, which is not likely since they are dissipationless. Therefore (taking into account the square root), we need the annihilation cross section to be suppressed relative to the scattering cross section by at least

$$\langle\sigma_a v\rangle_\odot \lesssim 10^{-4}\sigma_{\min}c. \tag{9}$$

Another way of writing equation (8) is to use equation (2) relating the cosmion annihilation cross section in the early universe to Ω_x, the cosmion density, today. Then equation (8) becomes

$$\begin{aligned}\frac{N}{N_\odot} &\approx 3\times 10^{-15}\left(\frac{\rho}{0.01M_\odot/\mathrm{pc}^3}\right)^{0.5}\left(\frac{v}{300\,\mathrm{km\,s^{-1}}}\right)^{-0.5}\left(\frac{\langle\sigma_a v\rangle_{\mathrm{hot}}}{\langle\sigma_a v\rangle_\odot}\right)^{0.5}\\ &\quad\times\left(\frac{\sigma_{\min}}{\sigma_{\mathrm{crit}}}\right)^{0.5} m_{x\,\mathrm{GeV}}^{-1.25}\,h_{50}^{-1}\,\Omega_x^{-0.5}.\end{aligned} \tag{10}$$

To affect the solar neutrino flux, we now have the requirement

$$\langle\sigma_a v\rangle_\odot \lesssim 10^{-5}\langle\sigma_a v\rangle_{\mathrm{hot}}. \tag{11}$$

Notice that equations (9) and (11) are physically different constraints on a particle model, the first bounding cold annihilation relative to scattering, the second bounding cold annihilation relative to hot annihilation.

These conditions on annihilation cross sections are quite restrictive on particle models. In Paper III, a number of different possible candidates for the cosmion are ruled out: massive Majorana neutrinos, massive Dirac neutrinos, supersymmetric scalar neutrinos (sneutrinos), and supersymmetric photons (photinos).

While photinos fail, there is an interesting bit of numerology that might possibly presage the viability of some as-yet-unproposed, photino-like candidate:

Photinos are Majorana particles. This has the consequence that their *s*-wave annihilation, which gives a velocity-independent annihilation rate factor, is suppressed with respect to *p*-wave annihilation, which has a v^2 dependence in the annihilation rate factor. The photino annihilation rate has been calculated in the nonrelativistic limit (Goldberg 1983; Ellis et al. 1984; Krauss 1983) to be

$$(\sigma_a|v|)_{\tilde\gamma} = \frac{1}{2\pi}\sum_f q_f^4\left(1-\frac{m_f^2}{m_{\tilde\gamma}^2}\right)^{1/2}\left[\frac{1}{3}\frac{m_{\tilde\gamma}^2-m_f^2}{m_{\tilde f}^4}|v|^2+\frac{m_f^2}{m_{\tilde\gamma}^4}\right] \tag{12}$$

(units with $\hbar = c = 1$). The sum f ranges over the quarks (u, d, s, c, etc.) with $\tilde{f}$ the corresponding squarks, and over the leptons (and corresponding sleptons).

The first term inside the square brackets in equation (12) is the p-wave term. Because of its v^2 dependence, this term implies a much smaller annihilation rate factor at low kinetic energies than at high energies. By itself, this term could in fact satisfy constraints (9) and (11), leaving the photino at least a marginally viable candidate for solving the solar neutrino problem, if its scattering cross section is maintained sufficiently large. In fact, we can call this "Cosmic Coincidence #4," that the ratio of v^2 implied by the two energies 100 MeV (temperature of cosmological annihilation) and 1 keV (temperature in the Sun) is just the same factor of 10^5 needed according to the astrophysical requirements that led to equation (11). In other words, a perfectly s-wave-suppressed Majorana particle could be the perfect cosmion.

The second term in the sum in equation (12), with m_f^2 is the suppressed s-wave term. for f ranging only over the two light quarks (u, d), the s-wave term is small enough to keep the photino marginally viable. Unfortunately, we live in a universe with heavier quarks also. One might hope to suppress the terms corresponding to annihilation into heavy quarks and leptons by postulating a hierarchy of squark masses similar to that of quark masses, so that $m_f^2/m_{\tilde{f}}^4$ is small for all f. However, in the range we are interested in ($M_{\tilde{f}} \approx 50\,\mathrm{GeV}$), the squark masses must generally be degenerate (Dugan, Grinstein, and Hall 1985; Kane and Haber 1985). Otherwise their contribution to the $K^0 - \overline{K^0}$ mixing is not canceled by the GIM mechanism, leading to a large $K_L - K_S$ mass splitting, in conflict with observation.

1.6. Prospects for Cosmions

On the theoretical side, what is required is a more natural way of avoiding annihilations within the Sun. A Majorana particle with nearly perfect s-wave suppression would probably do the trick. An alternative would be a particle which, like baryons, is produced cosmologically with an excess of some conserved quantum number. In this case the Galaxy and the Sun would contain cosmions with no anticosmions, there would be no annihilations, and equation (4) would hold. Of course one then loses Cosmic Coincidence #1 (which related abundance today to annihilation cross sections) and would have to have some other explanation of why the surviving excess corresponds to a reasonable value of Ω today. (Or, if one takes the view that $\Omega = 1$ on theoretical grounds, and that cosmions dominate the density, then one must explain the coincidence that Ω_{baryon} is within shouting distance of unity.)

Equally interesting if not more so is the fact that it may be possible to detect galactic cosmions by direct laboratory experiment, independent of any theoretical prejudices. For example, Drukier, Freese, and Spergel (1985) have described "superconducting colloid detectors" (SUPERCDs) which might serve. The SUPERCD consists of superconducting grains of a few microns' diameter, embedded in a dielectric material and placed in a magnetic field. The grains are maintained just below their superconducting transition temperature. When an incident cosmion scatters off a baryon in a grain, it imparts a recoil energy of a few hundred eV, sufficient to make the grain change from the superconducting to the normal state. The instant penetration of the grain by the ambient magnetic field produces an electromagnetic signal which can be detected and recorded.

In conclusion, it is fun to tell tall tales. Even more rewarding, however, is the telling of tales which, like so many of those told over the years by A. G. W. Cameron, might possibly turn out to be verifiably true. It seems to me that the story of cosmions in the Sun is in this latter category.

Acknowledgments

I thank my collaborators David Spergel, Katherine Freese, Ronald Gilliland, and Lawrence Krauss for permission to summarize their work. For helpful discussion and correspondence, I thank John Faulkner, Gary Steigman, Alan Guth, John Bahcall, and Willy Fowler. This work was supported in part by NSF grant PHY-83-06693.

References

Drukier, A. K., Freese, K., and Spergel, D. N. 1985, preprint.

Dugan, M., Grinstein, B., and Hall, L. 1985, *Nucl. Phys. B,* in press.

Ellis, J.; Hagelin, J. S.; Nanopoulos, D. V.; Olive, K.; and Srednicki, M. 1984, *Nucl. Phys. B,* **238,** 453.

Faulkner, J., and Gilliland, R. L. 1985, *Ap. J.,* **299,** 994.

Gilliland, R. L.; Faulkner, J.; Press, W. H.; and Spergel, D. N. 1986, *Ap. J.,* in press.

Goldberg, H. 1983, *Phys. Rev. Lett.,* **50,** 1419.

Goodman, M. W., and Witten, E. 1985, *Phys. Rev. D,* **30,** 3059.

Kane, G. L., and Haber, H. E. 1985, *Phys. Reports,* **117,** 77.

Krauss, L. M. 1983, *Nucl. Phys. B,* **227,** 556.

Krauss, L. M.; Freese, K.; Spergel, D. N.; and Press, W. H. 1985, *Ap. J.,* **299,** 1001 (Paper III).

Lee, B. W., and Weinberg, S. 1977, *Phys. Rev. Lett.,* **39,** 165.

Press, W. H., and Spergel, D. N. 1985, *Ap. J.,* **296,** 679 (Paper II).

Silk, J., and Srednicki, M. 1985, ITP preprint.

Spergel, D. N., and Press, W. H. 1985, *Ap. J.,* **294,** 663 (Paper I).

Steigman, G.; Sarazin, C. L.; Quintana, H.; and Faulkner, J. 1978, *Astr. J.,* **83,** 1050.

Wolfram, S. 1979, *Phys. Lett.,* **82B,** 65.

2

Aspects of Galaxy Formation

Martin J. Rees

Abstract

Our understanding of galaxies and how they evolve is still tentative and fragmentary—perhaps at the same level as the study of stellar structure several decades ago. The "seeds" from which galaxies formed were small amplitude fluctuations imprinted during the initial high energy phases of the big bang; but it is conventional physics—gas dynamics, atomic physics, and Newtonian gravity—that determines their characteristic sizes and shapes. Quantitative modeling of galactic evolution is impeded by our poor understanding of how stars form from protogalactic gas. Another stumbling block has been the realization that 90% of the gravitating material in galaxies is in some "hidden" form: the conspicuous stars and gas are merely "sediment" in a potential well 10 times larger contributed by material of unknown nature. The hidden mass could be small faint stars or the collapsed remnants of massive stars; alternatively, it could be some species of particle surviving from the big bang. Until we understand galaxy formation better, the large scale distribution of matter in the universe will remain uncertain; in particular, we cannot yet decide whether the mean cosmic density has the "critical" value ($\Omega = 1$) favored by inflationary cosmology.

2.1. The Dimensions of Galaxies

Our present knowledge of how galaxies formed is on the same rudimentary level as was our understanding of stellar evolution 30 years ago, before Al Cameron and other luminaries gave impetus to that subject. The most basic questions are still unresolved. In particular, we do not know why such things as galaxies should exist at all—why these assemblages of stars and gas with standardized properties are the most conspicuous large scale features of the cosmos. Furthermore, about 90% of the mass associated with galaxies is hidden. The luminous stars and gas contribute only about a tenth of the gravitating stuff inferred from dynamical arguments. What the rest consists of is still a mystery. We are perplexed about these issues, just as our predecessors were perplexed about how stars evolved.

Astronomers have developed an elaborate taxonomy for classifying galaxies. Most galaxies can, however, be crudely categorized as either "disk" or "elliptical" systems; a simple classification can be based simply on the relative prominence of disk and elliptical components (Ostriker and Meisels 1985). A "cartoon"

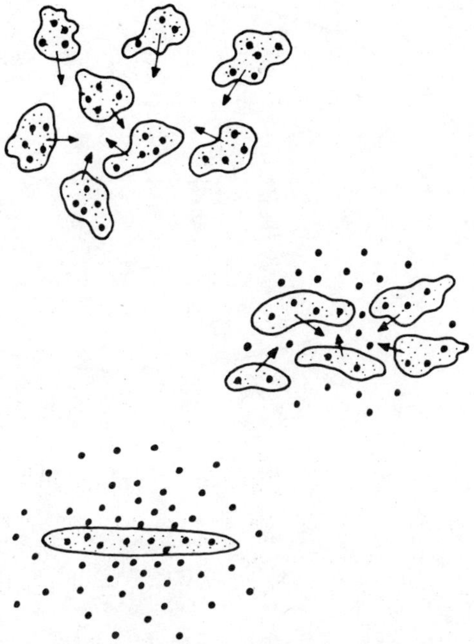

Figure 2.1. Schematic illustration of a collapsing protogalaxy. A gravitationally bound cloud for which $t_{\text{cool}} < t_{\text{dyn}}$ would contract at almost the free-fall rate. Inhomogeneities would develop into dissipative internal shocks, and the gas would settle into a thin rotating disk. (The cloud's original angular momentum would have been derived from tidal interactions with its neighbors.) Stars which condense from gas after formation of the disk will themselves retain a disklike distribution; but stars forming during the preceding free fall would retain orbits filling the entire original volume. Elliptical galaxies are those in which star formation proceeds so efficiently during the initial infall that rather little gas survives to form a disk.

of how galaxies might have formed is depicted in Figure 2.1. Suppose that a galaxy started life as an irregularly shaped gas cloud contracting under gravity and that the collapse was highly dissipative. The end result would be a rotating disk: this is the lowest energy state that the cloud can reach if it does not lose or redistribute its angular momentum. On the other hand, stars do not collide with each other, and are unable to dissipate energy in the same fashion as gas clouds. So the *rate of conversion of gas into stars* seems the crucial feature determining the type of galaxy that results, elliptical galaxies being those in which the conversion is fast, so that most of the stars have already formed before the gas has had time to settle down in a disk; contrariwise, the disk galaxies are those in which the star formation is delayed until the gas has already settled into a disk, or where gaseous infall is prolonged. According to this picture, ellipticals and spirals may have the same age. Disk galaxies are those with slower "metabolism," which have not yet got so close to the final state in which essentially all the gas is tied up in low mass stars or dead remnants. The "irregular galaxies" are even more extreme cases of arrested development: these

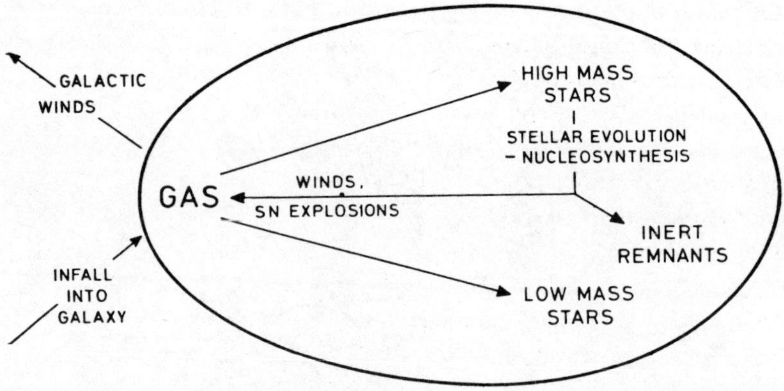

Figure 2.2. Star-gas interactions during early stages of galactic evolution.

are systems in which maybe less than half of the gas has so far been incorporated into stars.

Progress in understanding individual galaxies depends on quantifying the simple flow chart in Figure 2.2. Over a galaxy's lifetime star formation depletes the original gas: stars of low mass ($< 0.5\ M_\odot$) have hydrogen-burning lifetimes exceeding 2×10^{10} yr; higher mass stars may by now have died and ejected processed material back into interstellar space. Our poor understanding of star formation is the main stumbling block preventing us from quantifying how galaxies evolve. We do not know what determines the initial mass function (IMF) of stars forming now in our own Galaxy; still less do we know how the IMF depends on physical conditions. Nor do we know much about the *efficiency* of star formation: this determines how much gas turns into stars on the free-fall time scale and how much enriched gas can be recycled into new stars rather than being expelled from the system.

Galaxy formation straddles the interface between cosmology and astrophysics. The origin of the clouds envisaged in Figure 2.1 falls in the cosmologist's province. On the other hand, once galaxies have formed, the phenomena within them that interest astrophysicists proceed more or less regardless of the broader cosmic context.

We still do not know *why* galaxies should have the characteristic masses (10^{10}–$10^{12}\ M_\odot$) and radii (10^4–10^5 light-years) that we observe. Galaxies are no less standardized than stars. We understand why stars—"gravitationally bound fusion reactors"—exist only for a certain range of masses. Ordinary stellar masses lie within an order of magnitude of the Chandrasekhar mass $[\hbar c/Gm_p^2]^{3/2} m_p$, where m_p is the proton mass. (The dimensionless number in square brackets is the reciprocal of the gravitational fine structure constant α_G. The fact that α_G^{-1} is so huge ($\sim 10^{38}$), and stars consequently so large, reflects the weakness of gravity on the microscopic scale.)

A suggestive clue that galactic dimensions may be determined by astrophysical processes is the following (cf. Rees and Ostriker 1977; Silk 1977; Binney

1977). Two important time scales are important in determining how a self-gravitating gas cloud evolves. The first of these is the dynamical, or free-fall, time $t_{\rm dyn}$: this is of order $(G\rho)^{-1/2}$, its precise value depending on the geometry of the collapse. The second is the radiative cooling time scale: this depends on the gas temperature T_g and can be written $t_{\rm cool} = T_g/\rho\Lambda(T_g)$, where $\Lambda(T_g)$ depends on the composition and ionization of the gas and can be calculated from atomic physics. (The cooling rate per unit volume is proportional to $\rho^2\Lambda(T_g)$. At temperatures such that $\alpha_f^2 m_e c^2 \lesssim kT \lesssim m_e c^2$, nonrelativistic bremsstrahlung cooling dominates, and $\Lambda(T_g) \propto \sim T_g^{1/2}$; when $kT \lesssim \alpha_f^2 m_e c^2 = 2$ rydbergs, recombination and line emission from H and He are more important.)

If $t_{\rm cool} > t_{\rm dyn}$, a cloud of mass M and radius R can be in quasi-static equilibrium with T_g equaling the virial temperature $kT_v = GMm_p/R$. But if $t_{\rm cool} < t_{\rm dyn}$, such equilibrium is impossible: the cloud cools below the virial temperature and undergoes free-fall collapse or fragmentation. The criterion $t_{\rm cool} \gtrless t_{\rm dyn}$ also determines whether a shock developing during infall is isothermal (permitting a density enhancement of order the Mach number squared) or adiabatic (with a density enhancement less than ~4). We would expect clouds to collapse and fragment in the fashion depicted in Figure 2.1 only if they enter the part of the M-R plane where $t_{\rm cool} < t_{\rm dyn}$. This happens below a mass-independent radius.

$$R_{\rm crit} = \left(\frac{e^2}{m_e c^2}\right)\left[\alpha_G^{-1}\alpha_f^2\left(\frac{m_p}{m_e}\right)^{1/2}\right] \cong 75 \text{ kpc} \qquad (1)$$

for all masses such that the virial temperature (when $R = R_{\rm crit}$) lies in the range $m_e c^2 \gtrsim kT_v \gtrsim \alpha_f^2 m_e c^2$. The mass for which the virial temperature at $R_{\rm crit}$ is $kT_v = \alpha_f^2 m_e c^2$ is

$$M_{\rm crit} = \left[\alpha_G^{-2}\alpha_f^5\left(\frac{m_p}{m_e}\right)^{1/2}\right]m_p \cong 10^{12}\ M_\odot. \qquad (2)$$

Masses below $M_{\rm crit}$ cool efficiently by H and He recombination and line emission, even at radii somewhat larger than $R_{\rm crit}$. Clouds with mass below $M_{\rm crit}$ will readily fragment; but above $M_{\rm crit}$ fragmentation is impossible unless the cloud contracts to $R \leq R_{\rm crit}$. (See Figure 2.3 and legend.)

These radii and masses (equations [1] and [2]) have values relevant to large galaxies and play a role in many schemes for galaxy formation. However, some prior mechanism in the early universe must have given rise to fluctuations which evolved into gas clouds spanning the range around mass $M_{\rm crit}$. Only then can mass-dependent cooling processes single out a preferred scale for galaxies. The foregoing argument nevertheless assures us that, irrespective of cosmological details, we would never expect galaxies as massive as $\sim 10^{14}\ M_\odot$ to have formed as single units.

2.2. Dark Matter in the Universe

2.2.1 Distribution of Dark Matter

An impediment to understanding galaxies is that the stars and gas we observe may be little more than a tracer for the material that is dynamically

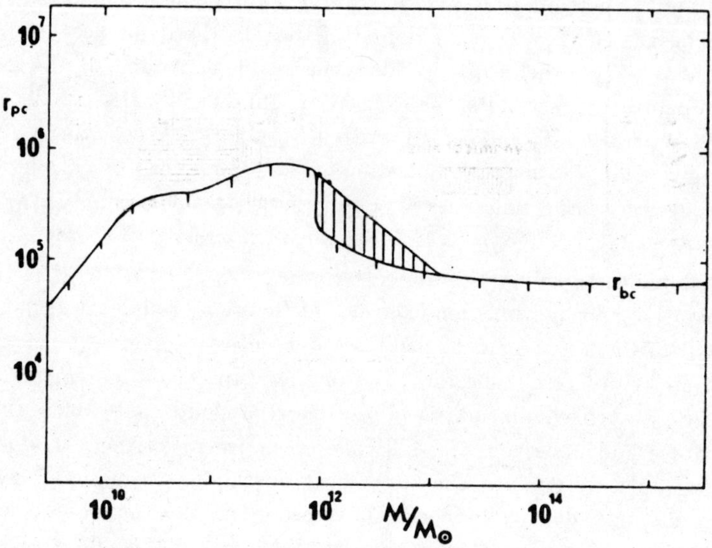

Figure 2.3. This diagram, adopted from Rees and Ostriker (1977), delineates the mass-radius relation for which $t_{\rm cool} = t_{\rm dyn}$ (assuming cooling due only to bremsstrahlung, H and He recombination, and line emission). A cloud of given mass whose radius was initially very large would deflate quasi-statically (because $t_{\rm cool} > t_{\rm dyn}$) until it crossed the critical line; it would then collapse in free fall and evolve in the fashion depicted in Figure 2.1. This simple argument (which can readily be modified to allow for nonspherical geometry, a nonbaryonic component of mass, etc.) suggests why, irrespective of the cosmological details, no galaxies form with baryonic masses greater than 10^{12} $M_\odot$ and radii greater than 10^5 pc.

dominant. The evidence for "dark matter" is now compelling (for a review, see Kormendy and Knapp 1986); what it consists of is, however, still a mystery.

The masses inferred from *relative motions* of galaxies exceed by a factor ~10 those inferred from the *internal dynamics* of the luminous parts of galaxies. This apparent discrepancy could be resolved if galaxies were embedded in extensive dark "halos." This hypothesis can be checked in some edge-on disk galaxies, where emission from gas can be observed at radii far exceeding the extent of the conspicuous stellar disk. The mass of this gas is itself negligible, but rotation velocities derived from its spectral lines do not fall of as $r^{-1/2}$, as would be expected if the gas were orbiting a mass distribution concentrated at much smaller radii. Instead, the velocity remains almost constant, implying $M(< r) \propto \sim r$, out to ~80 kpc in some cases. This implies local values of M/L exceeding 300 $(M_\odot/L_\odot)$ at the periphery and an extensive dark halo more massive than the disk component. Direct lower limits on M/L can be obtained for the matter in the outlying parts of galaxies with measured rotation curves and for the halos of edge-on galaxies. Estimates of the masses of galaxy clusters come from the virial theorem—a technique first applied to the Coma cluster by Zwicky (1933). This method is now complemented by X-ray studies of thermal emission from

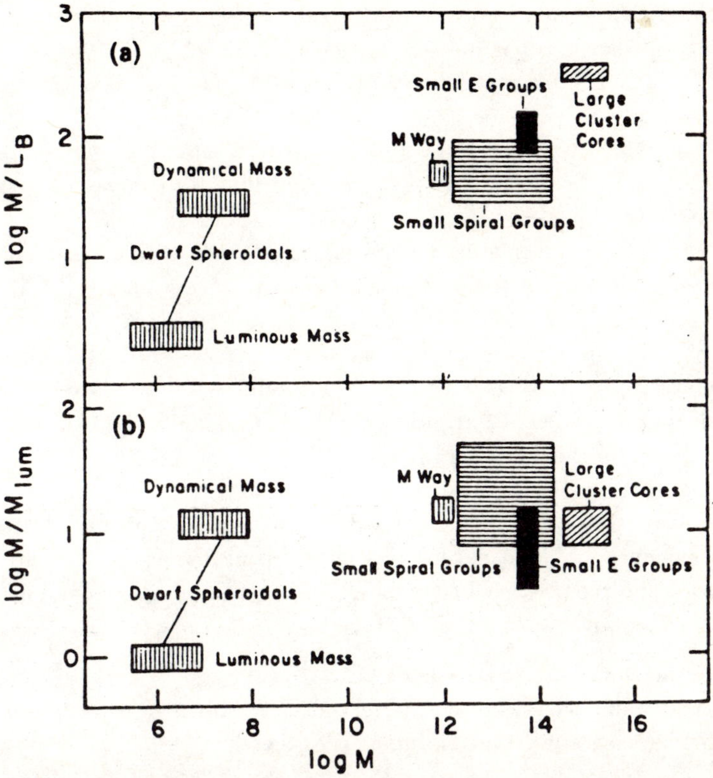

Figure 2.4. (*a*) The apparent increase with scale of the mass-to-light ratio. This increase is due to two distinct trends: (i) ordinary stars and gas are a decreasing proportion of the mass of the larger systems, and (ii) in the larger systems even the "ordinary" (star + gas) component has a higher M/L, because they consist primarily of elliptical galaxies with few young stars, and contain much hot gas revealed only by its X-ray emission. (*b*) The same data replotted, with effect (ii) subtracted out. One finds that the physically more fundamental ratio of "ordinary" matter to dark matter is *independent* of scale in all virialized systems larger than galaxies and has a value consistent with $\Omega \cong 0.1$–0.2.

This figure is adapted from Faber (1984) and Blumenthal et al. (1984); fuller details of the data on which it is based can be found in the latter paper. (The masses of dwarf spheroidal galaxies are uncertain and controversial; the diagrams show these systems plotted twice, depending on whether they do or do not contain dark matter.)

hot gas in clusters which probe the depth of the gravitational potential well.

There is, of course, no reason why dark matter should all be the same stuff—conceivable different forms dominate on the scales of individual galaxies, clusters, and superclusters. An important question is the dependence of M/L on scale. The data are summarized in Figure 2.4*a*, adapted from Faber (1984) and Blumenthal et al. (1984), which displays the often-cited trend for M/L to increase with scale. However, a more relevant quantity than M/L is the ratio of "ordinary" matter to hidden matter. The ordinary matter, which is certainly composed of baryons, comprises the stars (which themselves have an M/L depending on galaxy type) and also the diffuse gas. The graph of $M_{\rm lum}/M_{\rm tot}$ does not display the same monotonic rise (Figure 2.4*b*—see legend for further explanation); on the contrary, it suggests that on scales $M_{\rm tot} \gtrsim 10^{12}\ M_\odot$ the ratio $M_{\rm lum}/M_{\rm tot}$ may remain steady at a value ~0.1. The fact that this quantity is no larger for a rich cluster than for a single galaxy-plus-halo, even though the M/L for the cluster is larger, is due to the different stellar populations in the ellipticals in rich clusters plus the large contribution of X-ray-emitting gas in clusters to $M_{\rm lum}$.

This evidence, taken at face value, suggests that Ω (defined as the ratio of the actual mean density to the cosmological critical density $\rho_{\rm crit} = ((8/3)\pi G t^2)^{-1})$ is in the range 0.1–0.2 but that only 10% of this ($\Omega_b = 0.01$–0.02) is definitely baryonic. However, there is no dynamical evidence for $\Omega = 1$: there are no bound systems with $M/L = 1000\ h_{50}$ solar units, which would be the universal value if $\Omega = 1$ (h_{50} denotes Hubble's constant H_0 in units of 50 km s^{-1} Mpc^{-1}). I shall return to this question in §2.4.

This factor of 10 discrepancy between the amount of "luminous" mass and the amount reliably inferred from the dynamics of groups and clusters is the prime evidence for dark matter; this matter could, as far as the astronomical data are concerned, either be composed of baryons in some nonluminous form or be nonbaryonic.

2.2.2 Is the Dark Matter Baryonic?

A constraint on the mean baryonic density comes from primordial nucleosynthesis (Yang et al. 1984 and references therein):

$$0.04 \lesssim \Omega_b h_{50}^2 \lesssim 0.15. \tag{3}$$

This restriction comes primarily from the measured D and ^{3}He abundance; values outside this range cannot be excluded but require modification of the standard homogeneous hot big bang or some alternative (noncosmological) mechanism for producing light elements such as deuterium. The inequalities (3) imply that, for a low Hubble constant, *some* dark mass *must* be baryonic, and everything that is dynamically inferred could be. Contrariwise, a standard hot big bang with a high Hubble constant requires nonbaryonic matter even to account for galactic halos and the virial equilibrium of clusters.

The most clear-cut way to settle the nature of the hidden mass would of course be to detect the objects that make it up. Baryonic systems—stars or their remnants—are already severely constrained by the fact that the dark mass is so inconspicuous. More remarkably, there are genuine prospects that elementary particles of the kinds that could dominate the halo mass may be individually detectable by terrestrial experiments.

If $H_0 = 50$ km s^{-1} Mpc^{-1}, all the dynamically inferred dark matter could be baryonic. The astronomical constraints on this option have been recently discussed in detail by Carr and collaborators, and are summarized in Figure 2.5 (from Carr, Bond, and Arnett 1984). The import of this diagram is that stars or their remnants can only contribute $\Omega \geq 0.1$ if they are *either* predominantly

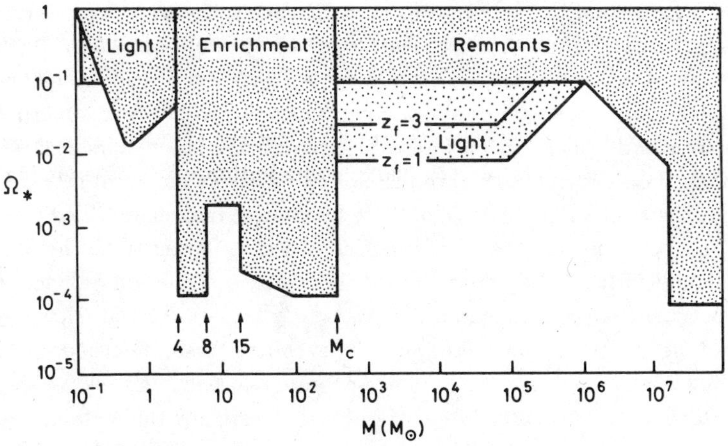

Figure 2.5. This diagram, from Carr, Bond, and Arnett (1984), shows various constraints on the fraction Ω of the critical cosmological density that can be contributed by first-generation stars (or their remnants) in different mass ranges. The objects are assumed to form at a redshift z_f. Dynamical friction arguments constrain the number of greater than $10^6\ M_\odot$ black holes in galactic halos. The requirement not to overproduce heavy elements constrains the number of remnants of ordinary heavy stars which end their lives by exploding as supernovae. The possible options are "Jupiters" of below $0.1\ M_\odot$, or the remnants of very massive objects (VMOs) in the mass range from a few hundred to $10^6\ M_\odot$, which could have formed at a large z_f. These objects do not necessarily eject any material processed beyond helium, and leave black hole remnants.

"Jupiters" (stars of below $0.1\ M_\odot$) or else black holes which are the remnants of very massive objects (VMOs) with masses between a few hundred solar masses and $10^6\ M_\odot$.

The main constraint on the masses of low mass stars in galactic halos comes from limits to the observed optical and infrared emission. A very faint optical halo in M87 has been traced out to 300 kpc (Arp and Bertola 1969). In the edge-on spiral NGC 4565 there are limits on the near-infrared emission corresponding to 76 solar units in the I band (Hegyi and Gerber 1977) and 38 solar units in the K band (Boughn, Saulson, and Seldner 1981). The constraints thereby imposed on the slope of the initial mass function (IMF) have been discussed by Peebles (1985) and Hegyi and Olive (1985). If the entire halo mass were contributed by stars with an IMF of the form

$$\frac{dn}{dm} \propto m^{-(1+x)} \tag{4}$$

down to some minimum mass $M_{\rm min}$, then if $M_{\rm min}$ exceeds 0.007, $x \geq 1.9$. (Salpeter's classic [1955] study derived $x = 1.35$ for our galactic disk.) In fact the infrared colors are not specially helpful in pinning down x, since for any $x < 2$ they are dominated by red giants. Note that the power law example (4) is only illustrative; the same amount of mass could equally be concealed by a population with a log Gaussian distribution peaking below $0.1\ M_\odot$.

One thing is clear: one cannot invoke a smooth extension of the IMF that is observed below $1\ M_\odot$, which actually seems to flatten off below $\sim 0.3\ M_\odot$. The objects constituting the dark matter must in some sense be a "special creation." However, this is perhaps not a cogent objection to the idea. After all, the IMF derived by Miller and Scalo (1979) and Scalo (1985) for the solar neighborhood does not look like a single power law but rather resembles two superposed log Gaussian distributions. In an interesting recent discussion Larson (1985) suggests that these represent the products of two distinct modes of star formation and that the relative importance of these two modes may have varied over galactic history. Conceivably the star formation process relevant to the halo involved a third mode with a different characteristic mass. If the halo objects formed at an early pregalactic epoch and subsequently clustered nondissipatively into galaxies and clusters, then there would be even less reason to suspect that their IMF should resemble that of stars forming here and now. Indeed, we should remain open-minded even about the IMF of stars forming within galaxies. Cooling flows in clusters of galaxies (Fabian, Nulsen, and Canizares 1984), where the gas pressure is ~100 times higher than in our galactic disk, have been interpreted as implying star formation with a very steep IMF. When stars in the halo formed, conditions may have resembled those in cooling flows more than they resembled those in our Galaxy now. Perhaps the Salpeter-Miller-Scalo function pertains only in environments of a typically low pressure.

Heavy elements are expelled from massive stars in their terminal phases unless they are so massive that they end their lives by collapsing to black holes after the pair-production instability (Truran and Cameron 1971; Carr, Bond, and Arnett 1984 and references therein). Collapse rather than explosion is thought to occur for core masses above $\sim 200\ M_{\odot}$. If the hidden mass were VMOs, then the requirement that heavy elements not be overproduced therefore requires a very flat IMF with a value of x that is actually negative. Moreover, there must be a cutoff above $\sim 10^6\ M_{\odot}$, at least for objects within individual halos, because dynamical friction would have increased the velocity dispersion of disk stars to an excessive degree. Within the context of VMO theories, we have little evidence on whether the preferred mass is closer to 10^3 or to $10^6\ M_{\odot}$. The upper mass limit could be pushed downward if we had a firmer understanding of what luminosity would result from accretion onto black holes passing through the galactic disk (Ipser and Price 1977; McDowell 1985; Ostriker and Lacey 1985).

Halos could—consistently with present data—be made up of compact objects whose masses range from $10^6\ M_{\odot}$, down to "Jupiters," about 10^8 times smaller. One way of discriminating between these options is by searching for manifestations of gravitational lensing. The probability of seeing lensing due to an object in our own halo is only of order 10^{-6}. However, it is, ironically, much easier to detect objects in the halos of galaxies halfway out to the Hubble radius. As was first clearly realized by Refsdal (1970), the probability that a compact source at a redshift $z > 1$ is significantly lensed by objects along its line of sight is of order Ω_{lens}, independent of the individual lens masses involved. However, the angular separation θ of the lens images is a diagnostic of the masses. For a path length of order the Hubble radius

$$\theta \cong 10^{-6}\left(\frac{M_{\rm lens}}{M_\odot}\right)^{1/2} \quad \text{arcsec.} \tag{5}$$

For $M_{\rm lens} > 10^5\ M_\odot$, very long baseline radio interferometers provide adequate resolution. For $M_{\rm lens} < 0.1\ M_\odot$ ("Jupiters"), the angular scale is less than 10^{-6} arcsec. This cannot be directly resolved by any technique until optical interferometers are deployed in space. There is nevertheless a genuine prospect of detecting lensing of this kind because of the variability that would ensue if the lens were to move transversely (Gott 1981; Young 1981). It takes only a few years for an object at the Hubble radius moving at 10^3 km s^{-1} to traverse an angle 10^{-6} arcsec. Another possibility, emphasized by Canizares (1982), is that "minilensing" might be detectable because it would affect the optical continuum of quasars but not the spectral lines, since the latter come from a more extended region with intrinsic angular size exceeding θ. If there were a firm observational limit to the scatter in the equivalent widths of the lines from quasar to quasar (i.e., in the line/continuum ratio), this would constrain the value of Ω contributed by small compact objects.

2.2.3 Nonbaryonic Dark Matter

Provided we know the mass and annihilation cross section of an elementary particle, we can in principle calculate how many of them survive from the big bang, and hence its present contribution to Ω. Progress in experimental particle physics *may* therefore reveal a particle which *must* contribute significantly to Ω, unless we abandon the hot big bang theory entirely. No such definite candidate is known at present; in the case of some known particles such as neutrinos, the masses are not well enough determined experimentally; and there are many possible species whose existence is still conjectural. The idea of nonbaryonic dark matter is nonetheless attractive, especially because the clustering properties of such matter would mimic the inferred mass distributions in galaxies and clusters in a gratifying way (see §2.3).

Many ways have been recently proposed for detecting, or at least constraining, candidate particles. I shall follow Press's terminology and refer to these particles generically as "cosmions." If cosmions were unstable and photons were among the decay products, there may be observational evidence even for a decay time scale as long as 10^{24} s. This is because limits to the hard radiation background amount to only 10^{-8} of the critical density. Antiprotons observed in the cosmic radiation may even be decay products of cosmions (Silk and Srednicki 1984).

There has recently been a spate of interesting suggestions about how cosmions might reveal their presence relatively close at hand. They could have cross sections σ of order 10^{-36} cm^2 for interactions with nucleons. The "optical depth" of the Sun is of order $\sigma/10^{-36}$ cm^{-2}. As discussed by Press (this volume), a cosmion scattering elastically off a nucleon in the Sun would lose energy via the recoil and could thereby become trapped. Over the lifetime of the Sun, an accumulated isothermal core of cosmions could build up a mass of 10^{-12} $M_\odot$ if annihilations did not occur. However, annihilations would restrict this buildup, unless one adopts a rather artificial model in which the cross section for annihilation is far below that for scattering. However, even though annihilations may prevent a dense enough core building up to affect the standard solar neutrino problem, high energy neutrinos from these annihilations may reveal their presence in the underwater detectors developed to search for proton decay. Already, cosmions with mass exceeding 6 GeV can be excluded. Analogous limits come from considering annihilations in the Earth rather than the Sun, as discussed by Silk, Olive, and Srednicki (1985) and Krause, Srednicki, and Wilczek (1985).

Very recently, Goodman and Witten (1985) and Drukier, Freese, and Spergel (1985) discussed direct detection of cosmions in the laboratory. This can best be done using a so-called super-CD, an array of superconducting grains maintained just below the transition temperature. The heat deposited by a single cosmion could raise the temperature of one of these grains above the critical value, thereby allowing magnetic flux to penetrate in a manner that could be detected. If the cosmions were, for instance, scalar neutrinos of mass greater than 5 GeV, the halo density would yield up to 10^4 counts per day per kilogram of detector. The count rate is proportional to the 7th power of the cosmion velocity. By adjusting the threshold, one could thereby, if such effects were detected, determine the velocity distribution of the halo particles and see if the halo were rotating.

Witten (1984) suggested that grains or nuggets of "strange matter," containing up, down, and strange quarks, may survive stably from the quark hadron transition at $t = 10^{-4}$ s. Such objects, in some sense intermediate between elementary particles and lumps of astrophysical size, would count as nonbaryonic matter in the context of nucleosynthesis. Recent work (Applegate and Hogan 1985; Alcock and Farhi 1985) suggests that neutrino heating would destroy nuggets unless they had a mass of planetary order, and it is unclear that any larger than this would even form, since this would involve coordination over a scale larger than the particle horizon at the relevant epoch. I find the "demise" of Witten's nuggets disappointing for two reasons. First, they might, as De Rujula and Glashow (1984) have suggested, have been detectable: interesting constraints could be set from the results of monopole searches, proton decay experiments, the number of meteor showers, and limits on the frequency of small scale seismic events. A second appealing feature of the nugget concept is that it leads naturally to a universe where the respective contributions of ordinary and dark matter to Ω do not differ by more than an order of magnitude. If the dark matter were in, for instance, axions, some "fine tuning" would have to be invoked to prevent these contributions from differing by many powers of 10.

2.3. Formation of Protogalaxies

So there are at least three serious candidates for the dark matter in galactic halos and clusters: low mass stars; black hole remnants of very massive objects; or nonbaryonic matter, in the form of supersymmetric particles or axions. I would myself lay even odds between these three options at the moment. However, it is gratifying that we can expect the odds to change quite rapidly, owing either to improved observational and experimental searches, or to progress in particle physics.

A separate line of attack that might shorten the list of candidates entails exploring the consequences of each hypothesis for the earliest stages of the cosmogonic scheme—the processes whereby small primordial perturbations evolve into protogalaxies and clusters. The key parameter here is the spectrum of the density fluctuations at the era of recombination, which is determined by initial conditions and by damping processes during the fireball phase of the expansion ($t < 10^6$ yr). The mass scale of the first bound systems (i.e., the scale on which the growing perturbations after recombination first become nonlinear) roughly corresponds to the scale at which the fluctuation spectrum peaks. Figure 2.6 depicts three rival cosmogonic schemes (see legend for further details). The disparity between them is highlighted if we envisage what the universe would have been like, according to each, at the epoch corresponding to $z \simeq 10$. In hierarchical "bottom-up" schemes a lot could have happened already. In others the universe is then still amorphous neutral hydrogen, with all parts expanding in nearly undeviated Hubble flow. The cosmological "dark age" starts when microwave background photons shift longward of the visible band, when the universe is 10^6 yr old: it ends when the first bound systems (or their constituent stars) light up. We do not know whether this dark age is brief or lasts for a billion years. We are more confused and ignorant about this phase of cosmic history than many seem to be about the first 10^{-35} s.

A related cosmogonic issue is this. To what extent is the present large scale structure a rather direct consequence of initial fluctuations, imprinted at 10^{-35} s? Or does it, contrariwise, result from secondary perturbations generated or seeded by the first bound systems? Hogan (1985) has suggested the phrase "paleogeny versus neogeny" to denote this dichotomy.

2.3.1 The "Cold Dark Matter" Cosmogony

The three schemes depicted in Figure 2.6 have all been reviewed elsewhere. Here I shall focus on the specific cosmogonic scheme which postulates that the hidden mass is so-called cold dark matter. In other words this material is envisaged as weakly interacting nonbaryonic stuff whose thermal motions are too low for free streaming to damp out fluctuations on any significant scale. (In this respect it should be contrasted with a model where light [$\sim$10 eV] neutrinos dominate: these would be "hot," in the sense that scales less than $10^{15}\ M_\odot$ are homogenized by free streaming, leading to a cosmogony where superclusters form first and then have to fragment into individual galaxies.)

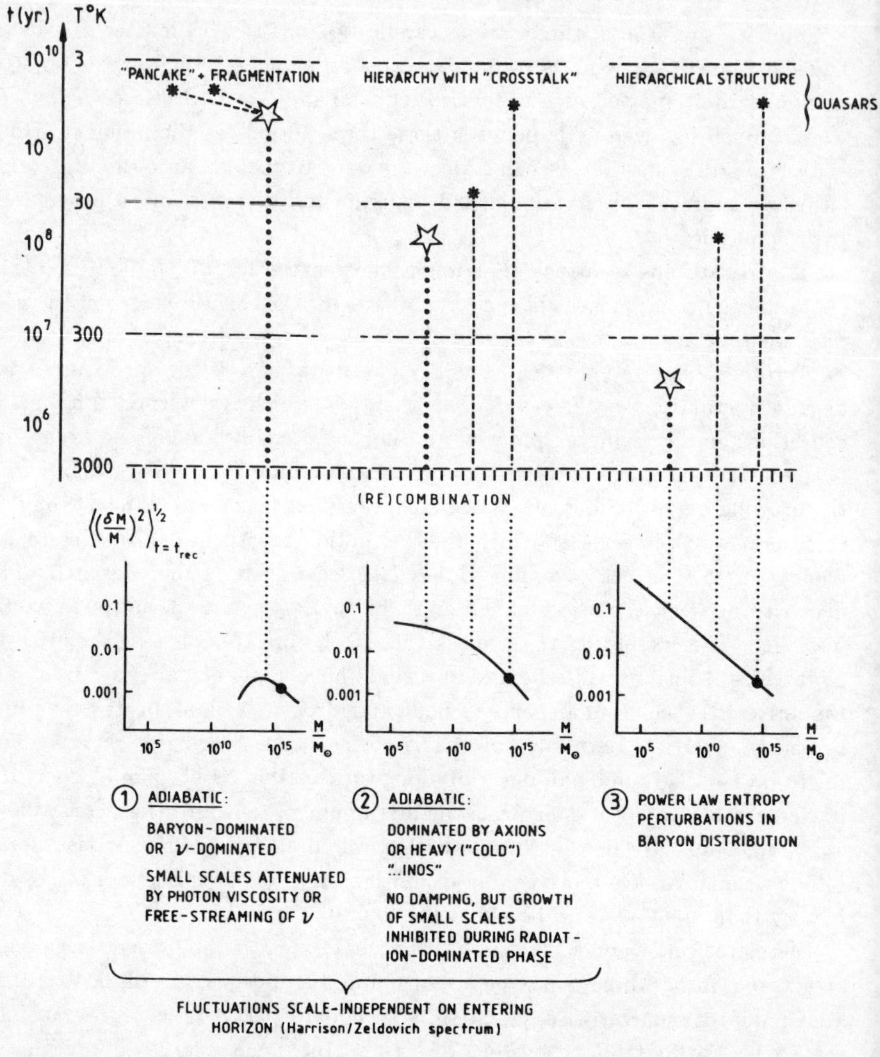

Figure 2.6. The cosmogonic processes after (re)combination depend on the spectrum of fluctuations which have survived damping processes, etc., at earlier times. After $t_{\rm rec}$, linear growth proceeds roughly according to the law $(\delta\rho/\rho) \propto t^{2/3}$, and the first gravitationally bound systems to form will have the mass for which the density contrasts are largest at $t_{\rm rec}$. In case 1, (super)clusters are the first systems to condense, and they form quite recently. In case 2, baryonic systems of greater than $\sim 10^6\ M_\odot$ condense in potential wells produced by "cold dark matter" or "inos," which are presumed to be slow moving, so that they are not homogenized on small scales as neutrinos are. (This specific case is discussed more fully in the text.) The third case shows the spectrum that might arise from primordial entropy perturbations. In cases 2 and 3, subgalactic systems would form *before* galaxies; if these subgalactic systems provided an energy input, they could generate "secondary" perturbations on larger scales which could swamp the genuinely primordial ones.

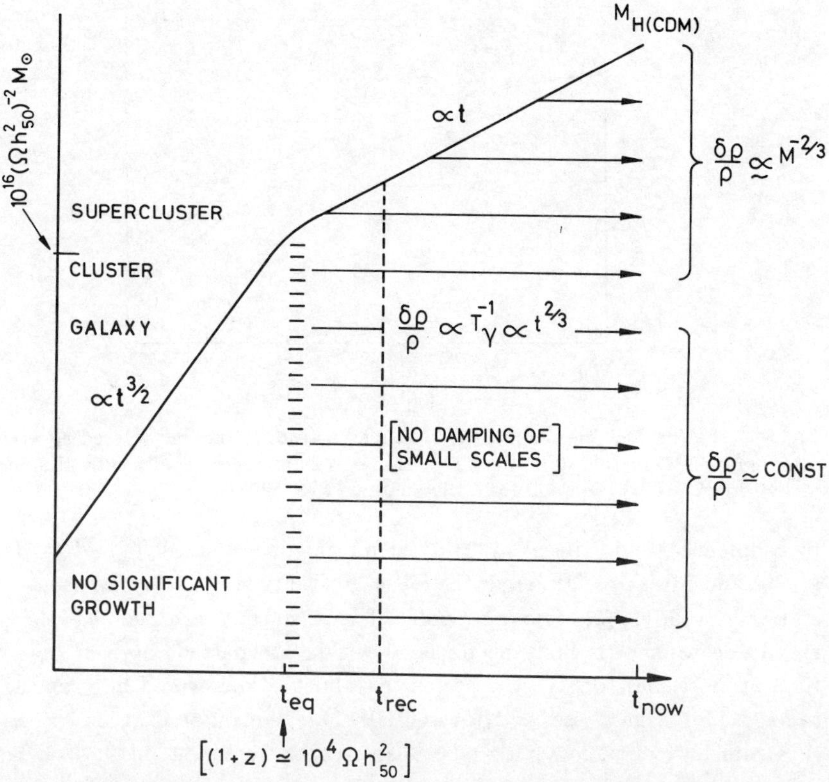

Figure 2.7. The growth of adiabatic fluctuations in a universe dominated by "cold dark matter." The mass of cold dark matter within the particle horizon is shown as a function of time on a log-log plot. For $t > t_{\rm eq}$ (corresponding to the redshift indicated on the figure), all scales grow at the same rate. Before $t_{\rm eq}$, when the expansion is radiation-dominated, there is essentially no growth, because the growth time scale much exceeds the expansion time scale. If the fluctuations on all scales enter the horizon with equal amplitude (the Harrison-Zel'dovich hypothesis), then the present-day spectrum would have the approximate form written on the right-hand side. (The accurately calculated spectrum is shown in Figure 2.8.) The cold dark matter perturbations start to grow at $t_{\rm eq}$, whereas radiation pressure inhibits growth of baryonic fluctuations on relevant scales until the (later) recombination time $t_{\rm rec}$. Cold dark matter fluctuations thus have a "head start" (baryons being able to fall into the resultant potential wells after $t_{\rm rec}$); this permits an acceptable cosmogonic scheme with lower Harrison-Zel'dovich amplitude, and smaller microwave background fluctuations, than in a baryon-dominated universe.

It is relatively straightforward to calculate the growth factor for density perturbations in cold dark matter between the era when they enter the particle horizon and the present. All scales grow roughly as the scale factor (i.e., $\delta\rho/\rho \propto R \propto T^{-1} \propto t^{2/3}$) after the epoch when the universe becomes dynamically dominated by dark matter (see Figure 2.7). However, at earlier times the universe is dominated by photons and the expansion time scale is proportional to $\rho_{\rm rad}^{1/2}$; perturbations in the dark matter alone with a growth time $\sim(G\rho_{\rm CDM})^{-1/2}$ do not then have time to grow significantly. This effect, termed "stagspansion"

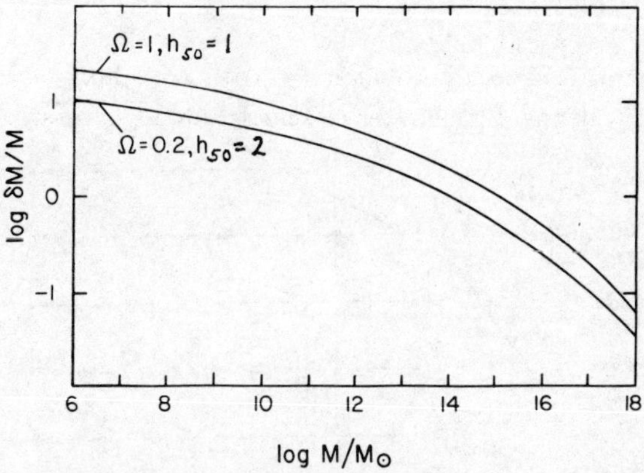

Figure 2.8. The root-mean-square fluctuations within a randomly placed sphere containing mass M for cold dark matter, for two specific models calculated by Blumenthal and Primack (1983) (adapted from Blumenthal et al. 1984).

by Blumenthal and Primack (1983), was first calculated by Guyot and Zel'dovich (1970) and Meszaros (1974). If the initial fluctuations had the Harrison (1970) and Zel'dovich (1972) scale-independent form—so that each scale has the same amplitude when it is first encompassed within the particle horizon (i.e., at a time proportional to M)—the present-day fluctuations would be proportional to $M^{-2/3}$ for large masses and be essentially independent of M at low mass. The spectrum flattens at low masses because all scales entering the horizon before the epoch of equal densities undergo essentially the same growth. The resultant spectrum of density perturbations actually has a rather gradual "rollover" in the transition region (Figure 2.8). There is nevertheless a characteristic mass imprinted by cosmology, which is smaller than the mass now within a "Hubble volume" by a factor $(\Omega_{\rm rad}/\Omega_{\rm DM})^{3/2}$, where $\Omega_{\rm rad}$ ($\sim 10^{-4}$) is the present contribution of primordial radiation to Ω.

The amplitude of the fluctuations cannot yet be reliably predicted theoretically. The vertical scaling in Figure 2.8 can therefore be normalized to ensure that $<(\delta\rho/\rho)^2>^{1/2}$ is now ~ 1 on a scale $\sim 10\ h_{50}$ Mpc. (This is inferred from the clustering properties of galaxies; the normalization would need some adjustment if, as discussed further in §2.4, the galaxies do not trace the overall mass distribution.)

Because the spectrum in Figure 2.8 is so nearly flat for small M, the typical fluctuation of $\sim 10^6\ M_\odot$ would collapse no earlier than the epoch corresponding to $z \cong 10$. The buildup of structure is hierarchical, in the sense that smaller scales tend to form earlier. However, because of the flat spectrum, there would be complicated "cross talk" between many different scales (see the middle panel of Figure 2.6): the 3 σ peaks in the density distribution on $10^{11}\ M_\odot$ scales would have the same amplitude as typical $(1-\sigma) \sim 10^6\ M_\odot$ peaks and would therefore collapse at the same time. It is consequently hard to analyze (either analytically or numerically) even the purely dynamical and nondissipative aspects of the clustering. However, those studies that have been done (e.g., Davis et al. 1985; Frenk et al. 1985) are encouraging in that, when the amplitude of the fluctuations is normalized so as to match the data on galaxy clustering, the finer scale disposition of the dark matter closely reproduces the sizes and profiles of galactic halos.

2.3.2 The First Stars

Pressure gradients in the baryons would prevent bound systems of $\ll 10^5\ M_\odot$ from collapsing, even after recombination. If there were no primordial perturbations in the baryon-photon ratio (i.e., the initial fluctuations were adiabatic), the first bound baryonic systems would be associated with high σ peaks in the dark matter distribution on mass scales 10^5–$10^6\ M_\odot$. These would collapse at a redshift z of order 10 ν, where ν is the number of standard deviations above the mean. Everything is straightforward while all perturbations remain linear, but once the first bound baryonic systems have formed—once "first light" occurs—what happens next could be crucially influenced by feedback from these systems. This influence depends on the mass spectrum of the first-generation stellar objects. (Although I have introduced this issue in the context of the cold dark matter picture, our poor understanding of early star formation is a stumbling block impeding quantitative study of any galaxy formation schemes particularly the hierarchical ones where the first stars form at pregalactic epochs.

We cannot reliably decide whether a collapsing baryonic cloud of $\sim 10^5\ M_\odot$ will turn into a single very massive star or fragment into 10^7 stars of ultralow mass. In the latter case the first stars would not inject any significant energy into the remaining gas. On the other hand, if the first stars were ordinary massive stars or supermassive objects, their ultraviolet radiation would photoionize the medium. Indeed, only $\sim 10^{-4}$ of the initial mass need turn into such stars in order to generate enough UV radiation to photoionize everything else (Hartquist and Cameron 1975). An additional possibility is that heavy elements generated by pregalactic supernovae may be spread through the intergalactic medium at an early stage. Details of these feedback processes are discussed by Couchman and Rees (1985). The baryonic Jeans mass would be raised from $\sim 10^5\ M_\odot$ to $\sim 10^8\ M_\odot$, with a consequent pause in the cosmogonic process until systems exceeding this larger mass undergo collapse; moreover, no more than 10^{-4} of the baryons (i.e., the greater than 3 σ peaks) need condense into massive stars before reheating chokes off any further collapse of $\sim 10^5\ M_\odot$ systems.

Gas would subsequently condense into dark matter potential wells in the mass range 10^8 to $10^{12}\ M_\odot$. The lower limit is set by the Jeans mass after ionization, and the upper is the maximum scale where dissipative cooling occurs at $z > 10$ (see Figure 2.3; Compton cooling would allow larger and hotter clouds to collapse at larger z). The luminosity function of galaxies depends primarily on two things.

1. We need to know the mass distribution of isolated virialized systems of dark matter. This is a difficult problem because of the cross talk between various

scales, and its solution awaits N-body simulations with larger dynamic range than have so far been carried out. At the moment it is unclear whether well-defined substructure survives or whether bound systems on scales of, say, 10^8 $M_\odot$ are rapidly engulfed in a larger system before baryons have much chance to condense within them.

2. The luminosity of a galaxy resulting from infall into a given potential well depends on what fraction of infalling gas is retained in each virialized clump of dark matter (likely to be larger for deeper potential wells, i.e., those of large mass and/or those which evolve from high σ peaks) and on the kind of stars it turns into.

The above two issues will need to be settled before we can reliably estimate the luminosity function of galaxies, even when our starting point is the specific assumption of a cold dark matter spectrum that evolved from Harrison-Zel'dovich initial fluctuations.

A comment may be worthwhile about the cosmogonic relevance of globular clusters. A wide class of schemes yield $\sim 10^6\ M_\odot$, the post-recombination Jeans mass, as a characteristic scale for the first bound baryonic systems. Many authors, following Dicke and Peebles (1968), have suggested that globular clusters may be survivors of these first-generation objects. This is an interesting possibility. However, there is no necessary reason for this identification, because a similar characteristic mass ($\sim 10^6\ M_\odot$) emerges in a quite different context (Fall and Rees 1985).

A gaseous collapsing protogalaxy which was initially not quite homogeneous would develop a two-phase structure: clouds at $T_c \simeq 10^4$ degrees would be embedded in a hot medium at the virial temperature $T_{\rm virial}$ (which is several times 10^6 K for a big galaxy with $M_G \gtrsim 10^{11}\ M_\odot$). Pressure balance between the two phases implies $n_c T_c = n_{\rm H} T_{\rm virial}$, and the characteristic Jeans mass for the cool clouds is then

$$M_{\rm J} \propto n^{-1/2} T^{3/2} \cong \left(\frac{T_c}{T_{\rm virial}}\right)^2 M_G. \tag{6}$$

The above expression, which is modified by a further factor $(\rho_b/\rho_{\rm total})^{-1/2}$ if there is a nonbaryonic component of mass in the protogalaxy, suggests that the largest natural scale of substructure in a protogalaxy is of order $10^6\ M_\odot$. The fact that the gas cools very efficiently down toward 10^4 degrees suggests that all intermediate scales of substructure between the galaxy itself and $10^6\ M_\odot$ may be missed out. Because of this argument there is no firm motivation for relating globular clusters to pregalactic phenomena.

2.3.3 Some Tests

The following are among the observations that would help to decide among the three cosmogonic schemes illustrated in Figure 2.6, and to test the "cold dark matter" model in particular.

a) The upper limits on the microwave background fluctuations on small angular scale constrain the amplitude of $\delta\rho/\rho$ at the recombination epoch and are already sensitive enough to exclude some models. Uson and Wilkinson (1984), for instance, find $<(\Delta T/T)^2>^{1/2} < 2.1 \times 10^{-5}$ for angles $\sim 3.5'$. A further improvement, even by a factor no more than 2, could be crucial. Measurements on larger angular scales are directly probing the metric fluctuation and could test the consistency of a Harrison-Zel'dovich spectrum where this amplitude is scale-independent.

b) It would be valuable to improve data on clustering correlations (and non-Hubble velocities) on scales greater than 10 Mpc. Because the cold dark matter fluctuation spectrum falls off steeper than a "white noise" spectrum at long wavelengths, the correlation function in this model should go negative for scales greater than $\sim 70\ h_{50}^2$ Mpc.

c) One would like to know whether there is dark matter in dwarf galaxies. The observational evidence here is uncertain, because the stellar velocities are below 10 km s^{-1}. However, such dark matter certainly could not be light neutrinos, because they would need a higher phase space density than they had in the big bang (Tremaine and Gunn 1979; Madsen and Epstein 1984), whereas *cold* nonbaryonic matter would be an acceptable possibility.

d) Most crucial of all would of course be firm evidence for particles that could constitute "cold dark matter." Experimental hopes for axion searches, and for the detection of supersymmetric particles, are by no means fanciful.

2.4. A Flat ($\Omega = 1$) Universe?

If the universe had resembled a Friedmann model since the Planck time t_P, a comoving scale initially equal to the Planck length would by now (when $t \cong t_0 \cong 10^{60}\ t_P$) have grown by $\sim 10^{30}$, because the scale factor varies as $t^{1/2}$ for a radiation-dominated expression. But the present Hubble radius is $\sim 10^{60}$ Planck lengths. The Robertson-Walker curvature radius, which is certainly at least comparable to the present Hubble radius, is thus $\sim 10^{30}$ larger than the "natural" scale if the universe were "set up" at a time of order t_P. This so-called flatness problem introduces a large dimensionless number:

$$\mathcal{R} = \left(\frac{\text{Robertson}-\text{Walker curvature radius}}{\text{Planck scale expanded to present epoch } t_0}\right)$$
$$= \left(\frac{t_0}{t_P}\right)^{1/2} \left(\frac{t_{eq}}{t_0}\right)^{1/6} \frac{1}{|(\Omega - 1)|^{1/2}}. \tag{7}$$

The extra factor $(t_{eq}/t_0)^{1/6}$, where t_{eq} is the time at which the expansion switches from being radiation-dominated to being matter-dominated, arises because for $t > t_{eq}$ the scale factor grows as $t^{2/3}$ rather than as $t^{1/2}$.

The value of $\mathcal{R}$ is thus at least $\sim 10^{30}$; the vastness of this number seems to imply some kind of fine tuning in the initial conditions.

One reaction to this requirement might be an anthropic one: the universe would not have offered a hospitable environment for galaxies and stars if $\mathcal{R}$ were not so huge. But it would be more satisfactory if there were some fundamental reason. Such reasons are offered by the concepts of inflationary and quantum cosmology. Such ideas indeed suggest that the universe could readily have inflated by *far more than* 30 powers of 10, yielding a value of Ω indistinguishable from unity: to have anything else demands a coincidence in that it requires the curvature scale to be just of order the present Hubble distance. If it involved an improbable effort for the universe to "heave itself up" to this size, one could perhaps invoke anthropic arguments that the Robertson-Walker curvature scale would be no larger than our present Hubble radius. But the inflation is so readily achieved that it would probably overshoot by many powers of e. (In any case, if the universe had inflated by just the amount needed to yield Ω of, say, 0.1, it would be unnatural to find such small amplitude anisotropies in the microwave background—this would require the preinflation geometry of our observable universe to have been rather special, with a curvature which was essentially constant [to 1 part in $\sim 10^4$] over scales of order the curvature radius.)

At an IAU symposium held in Krakow in 1973, the mean density of the universe was a topic of discussion. During one session the chairman, Professor J. A. Wheeler, conducted a poll of the audience to seek the favored value of Ω. One gratifying outcome was that a majority of participants accepted that Ω was unknown. However, many shared Professor Wheeler's aesthetic preference for a closed universe with Ω well in excess of unity. I suspect that a similar poll taken today would reveal "reasoned prejudice" in favor of $\Omega = 1$, instilled in us by the appeal of inflationary cosmology.

If the universe were indeed flat, what could make it so? Recall that the evidence summarized in Figure 2.4 suggests that Ω is only 0.1–0.2. Moreover, our infall toward Virgo (relative to the Hubble flow) is only ~ 250 km s^{-1}. This tells us, essentially, the amount of excess mass within a sphere centered on the Virgo cluster and whose surface lies near the Local Group. The galaxies within this sphere are ~ 3 times more close-packed than in a typical volume of space, and the relatively low infall velocity is then inconsistent with $\Omega = 1$, unless for some reason the galaxy distribution is more clumped than the mass in general. There are two ways of nevertheless reconciling the observations with a flat universe, both of which require that the dominant hidden mass must be more smoothly distributed on large scales than are the galaxies, or at least the conspicuous galaxies included in surveys. (Note that, if the conventional hot big bang model is correct, the nucleosynthesis constraint [4] favors a nonbaryonic form for the dynamically dominant constituent of an $\Omega = 1$ universe.)

i) The universe may be dynamically dominated by ultrahot weakly interacting particles which do not cluster. One difficulty here is that if such particles had always been present they would have inhibited gravitational clustering altogether as well as yielding an unacceptably fast expansion time scale at the era of nucleosynthesis. This problem is eased if the hot particles represent decay products of massive particles with a lifetime $\sim 10^9$ yr. A nonzero cosmological constant (Λ-term) is an alternative hypothesis whose consequences are similar (but to postulate a value of Λ such that it is dynamically competitive with matter at the present epoch introduces the kind of unappealing fine tuning that inflationary cosmology seeks to avoid).

ii) Some kind of *biasing* in the galactic distribution might render galaxies more clumped than the overall mass distribution even on scales as large as 20 Mpc. Were this so, voids would not be as empty as they look and the local "Virgo Supercluster" would not be a threefold enhancement in the *total* density. It is unlikely, especially in the otherwise attractive cold dark matter cosmology discussed in §2.3, that material (even baryonic material alone) could be pushed over distances exceeding 20 Mpc. So could the efficiency with which baryons transform into luminous galaxies be patchy? If so, the ratio of baryons to cold dark matter could be constant on all scales larger than one or two megaparsecs (up to which we expect some segregation due to cooling flows, etc.).

The latter possibility involves a simple consistency requirement. If clusters such as Coma embody a fair sample of the contents of the universe—i.e., if their ratio of baryonic to total mass equals $\Omega_b/\Omega_{\rm total}$—then the fraction of baryons in clusters cannot exceed the value of Ω naively estimated from Coma-like systems, i.e., 0.1 or 0.2. So if

$$(M/L)_{\rm universe} \cong 1000\, h_{50}\Omega_{\rm total}\ , \tag{8}$$

then $\Omega_{\rm total} = 1$ and $\Omega_b = 0.1$ are compatible with $(M/L)_{\rm cluster} = 100$ provided that M/L for galaxies is less than 10; these illustrative "round numbers" do indeed seem marginally consistent. This idea suggests that up to 90% of baryons may remain as diffuse gas in voids, or else in faint or low surface brightness galaxies; the *mean* (M/L) for *baryonic* matter would then be $\sim$100. In this connection, one wonders whether there might, in some voids, be dark halos with no luminous galaxies in them. Such objects might account for double quasars with no sign of a galaxy to act as a gravitational lens.

Physical mechanisms for bringing about this biasing have been discussed by Rees (1985), Silk (1985), and others. A feature common to several such mechanisms is that the first galaxies to form would exert negative feedback on the formation of later ones. This has the advantage that the resulting galaxies would then automatically display enhanced clustering, for reasons described in detail by Kaiser (1984). No mechanism has yet been worked out in convincing detail. However, the idea of biasing is not just an ad hoc contrivance, introduced to shore up the philosophically attractive $\Omega = 1$ model against apparently conflicting evidence. It would be astonishing if no such mechanism were important—if *no* large scale environmental effects influenced galaxy formation and if light did indeed trace mass on all scales greater than 1 Mpc. Any convincing determination of Ω must await much further data on galactic morphology and evolution, on the content of voids, and on the nature of the hidden mass. In the meantime the virial evidence does not seem a severe embarrassment for advocates of $\Omega = 1$.

2.5. Concluding Comments

The problems of large-scale cosmogony—how galaxies evolved from amorphous cosmic beginnings—are so intermeshed that we will not really solve any until the whole picture comes into better focus. For instance, we cannot test theories of galaxy formation and evolution until we understand star formation (and the possible role of active galactic nuclei) as well as the initial fluctuations. It is not that the *physics* is all specially recondite—indeed, most of the physics that astrophysicists need is in "Landau and Lifshitz," and the most intractable problems of all are complex manifestations of Newtonian gravity and dissipative gas dynamics. But when we ask where the fluctuations come from or what the hidden mass is, we realize that even the most ordinary galaxies pose questions that may transcend the physics we understand.

Finally, a sociological note. Astrophysical cosmology is a field where much of the best work around the world has been spearheaded by the over-60s. So it augers well for future progress that Al Cameron is now bolstering their ranks.

References

Alcock, C., and Farhi, E. 1985, *Phys. Rev.,* **D,** in press.

Applegate, J., and Hogan, C. J. 1985, preprint.

Arp, H., and Bertola, F. 1969, *Ap. Letters,* **4,** 23.

Binney, J. J. 1977, *Ap. J.,* **215,** 483.

Blumenthal, G.; Faber, S. M.; Primack, J. R.; and Rees, M. J. 1985, *Nature,* **311,** 517.

Blumenthal, G., and Primack, J. 1983, in *Fourth Workshop on Grand Unification,* ed. H. A. Weldon et al., p. 256 (Boston: Birkhauser).

Boughn, S. P., Saulson, P. R., and Seldner, M. 1981, *Ap. J. (Letters),* **250,** L15.

Canizares, C. R. 1982, *Ap. J.,* **263,** 508.

Carr, B. J., Bond, J. R., and Arnett, W. D. 1984, *Ap. J.,* **277,** 445.

Couchman, H. M. P., and Rees, M. J. 1985, *M.N.R.A.S.,* in press.

Davis, M.; Efstathiou, G.; Frenk, C. S.; and White, S. D. M. 1985, *Ap. J.,* in press.

De Rujula, A., and Glashow, S. L. 1984, *Nature,* **312,** 734.

Dicke, R. H., and Peebles, P. J. E. 1968, *Ap. J.,* **154,** 891.

Drukier, A. K., Freese, K., and Spergel, D. N. 1985, preprint.

Faber, S. M. 1984, in *Proc. ESO/CERN Conference on Cosmology and Particle Physics,* ed. G. Setti and L. Van Hove, p. 187 (CERN Publications).

Fabian, A. C., Nulsen, P. E. J., and Canizares, C. R. 1984, *Nature,* **310,** 733.

Fall, S. M., and Rees, M. J. 1985, *Ap. J.,* **298,** 18.

Frenk, C.; Davis, M.; Efstathiou, G.; and White, S. D. M. 1985, *Ap. J.,* submitted.

Goodman, M. W., and Witten, E. 1985, *Phys. Rev.,* **D31,** 3059.

Gott, J. R. 1981, *Ap. J.,* **243,** 140.

Guyot, M., and Zel'dovich, Y. B. 1970, *Astr. Ap.,* **9,** 227.

Harrison, E. R. 1970, *Phys. Rev.,* **D1,** 2726.

Hartquist, T. W., and Cameron, A. G. W. 1977, *Ap. Space Sci.,* **48,** 145.

Hegyi, D. J., and Gerber, G. L. 1977, *Ap. J. (Letters),* **218,** L7.

Hegyi, D. J., and Olive, K. A. 1985, *Ap. J.*, in press.

Hogan, C. J. 1985, in *Inner Space—Outer Space,* ed. D. N. Schramm et al. (Chicago UP).

Ipser, J. R., and Price, R. H. 1977, *Ap. J.,* **216,** 578.

Kaiser, N. 1984, *Ap. J. (Letters),* **284,** L9.

Kormendy, J., and Knapp, G., eds. 1986, *IAU Symposium 117* (Dordrecht: Reidel).

Krause, L., Srednicki, M., and Wilczek, F. 1985, preprint.

Larson, R. B. 1985, *M.N.R.A.S.,* in press.

Madsen, J., and Epstein, R. I. 1984, *Ap. J.,* **282,** 11.

McDowell, J. 1985, *M.N.R.A.S.,* in press.

Meszaros, P. 1974, *Astr. Ap.,* **37,** 225.

Miller, G. E., and Scalo, J. M. 1979, *Ap. J. Suppl.,* **41,** 513.

Ostriker, J. P., and Lacey, C. 1985, *Ap. J.,* in press.

Ostriker, J. P., and Meisels, A. 1985, *Ap. J.,* in press.

Peebles, P. J. E. 1985, in *Theoretical Aspects of Astrophysics and Cosmology,* ed. J. Sanz (Singapore: World Scientific Publishers).

Rees, M. J. 1985, *M.N.R.A.S.,* **213,** 75P.

Rees, M. J., and Ostriker, J. P. 1977, *M.N.R.A.S.,* **179,** 541.

Refsdal, S. 1970, *Ap. J.,* **159,** 357.

Salpeter, E. E. 1955, *Ap. J.,* **121,** 161.

Scalo, J. M. 1985, *Fund. Cosm. Phys.,* in press.

Silk, J. I. 1977, *Ap. J.,* **211,** 638.

——. 1985, *Ap. J.,* **297,** 1.

Silk, J. I., Olive, K., and Srednicki, M. 1985, preprint.

Silk, J. I., and Srednicki, M. 1984, *Phys. Rev. Letters,* **53,** 624.

Tremaine, S., and Gunn, J. E. 1979, *Phys. Rev. Letters,* **42,** 407.

Truran, J. W., and Cameron, A. G. W. 1971, *Ap. Space Sci.,* **14,** 179.

Uson, J., and Wilkinson, D. T. 1984, *Nature,* **312,** 427.

Witten, E. 1984, *Phys. Rev.,* **D30,** 272.

Yang, J.; Turner, M. S.; Steigman, G.; Schramm, D. N.; and Olive, K. A. 1984, *Ap. J.,* **281,** 493.

Young, P. J. 1981, *Ap. J.,* **244,** 756.

Zel'dovich, Y. B. 1972, *M.N.R.A.S.,* **160,** 1P.

Zwicky, F. 1933, *Helv. Phys. Acta,* **6,** 110.

3

Chemical Evolution of Galaxies

B. E. J. Pagel

Abstract

The broad distribution of metallicities in Baade's window suggests that the solution to the classic G dwarf problem lies in special factors related to the formation of the galactic disk rather than in a universal metallicity dependence of the IMF. Stellar abundances of various elements indicate that C, iron peak elements, Sr, Zr, and Eu mostly vary rather little relative to iron, whereas O, Si, Ca, and a few other elements vary less than iron and Y and Ba vary more. Stellar data suggest a primary origin for N, but there is probably substantial secondary N at the center of the Galaxy.

Oxygen abundances in irregular and spiral galaxies show some correlation with gas fraction and a better one with total mass and surface density in the disk, respectively. They do not fit the Simple Model, but could be explained by (i) a variable IMF, (ii) outflows of enriched gas, or (iii) inflow of unprocessed gas, provided that the effects of these processes are correlated with the local surface density.

3.1. Introduction

Al Cameron has made many fundamental contributions to the problem of the chemical evolution of galaxies including his work on nucleosynthesis in stars, his studies of stellar evolution and the origin of the solar system, his systematic compilations of abundances, and his calculations of specific models of galactic chemical evolution as such, notably the modified Simple Model of Truran and Cameron (1971), which still compares quite favorably with the constraints imposed by observations in the solar neighborhood.

Since 1971 there have been major advances in the theory of stellar evolution and nucleosynthesis and in the observational data, notably, on the one hand, abundance distributions for stars in our Galaxy and, on the other, the determination of abundance gradients and relations between composition and structural properties for galaxies in general. Galactic chemical evolution models have been calculated, with varying degrees of sophistication, to account for abundance gradients and the data for the solar neighborhood, but despite this we are still a long way from having any galactic chemical model that can be regarded as definitive, because there are too many free parameters: uncertainty in the effects of mixing, mass loss, and binary star evolution; the choice between variable and invariable initial mass functions; past rates of star formation; and the effects of inflow, outflow, and cross-flow of material with very little direct evidence for or against. The existence of all these variables complicates the problem of checking nucleosynthesis predictions against observed abundances, e.g., in calculating yields or distinguishing between primary and secondary production of nitrogen. In the present state of uncertainty there is still a lot to be said for the Simple Model, or simple modifications thereof, at least as a frame of reference (cf. Edmunds and Pagel 1984*a*; Mould 1984), because although the Simple Model is often wrong it is not always clear which aspect of it is wrong. I shall therefore refer to the Simple Model quite often in this paper.

3.2. The G Dwarf Problem Revisited

I shall begin with a few comments on the classic G dwarf problem (van den Bergh 1962; Schmidt 1963; Pagel and Patchett 1975; Lynden-Bell 1975), which is the presence of a narrow and nearly lognormal distribution of metallicities among long-lived stars in the solar neighborhood compared to a broader and exponential distribution e^{-z} predicted by the Simple Model with instantaneous recycling, where z is the heavy-element abundance in units of the yield. Bond (1981) has claimed that a similar problem exists in the galactic halo because he does not find stars with [Fe/H] < -3 in his sample although a few such stars do exist.

Proposed solutions to this problem can be divided into two classes: effects resulting from some universal law, such as a dependence of the initial mass function (IMF) on metallicity, and effects resulting from the particular conditions of formation of the galactic disk, such as formation of the disk by gradual accretion, as in the models of Larson (1976) and Lynden-Bell (1975), or Ostriker and Thuan's (1975) model of initial enrichment by ejecta from massive stars in the halo. Some other suggestions, like metal-enhanced star formation (Talbot and Arnett 1974), could belong in either category.

If the correct explanation for the solar neighborhood is a universal law, then there should be a G dwarf problem everywhere, and this is not the case in fact. For one thing, if there is a problem in the halo, it is obviously at quite a different level from the problem in the solar neighborhood, but the halo is quite likely to have been affected by special factors such as loss of enriched gas in dissipative processes (Hartwick 1976). A more relevant sample for comparison with the solar neighborhood comes from the interesting observations by Whitford and Rich (1983) of red giants in Baade's window near the center of the Galaxy. In Figure 3.1, I show a cumulative plot, similar to Maarten Schmidt's G dwarf plot, of metallicity against number of stars, after removing high luminosity stars that could bias the sample because M giants were not measured. There are only 14 stars and they are not G dwarfs, but, as far as they go, the data are in strikingly good agreement with the Simple Model and this seems to be evidence against any universal law that only massive stars can be formed at a metallicity less than a tenth of solar and in favor of explanations of the solar neighborhood

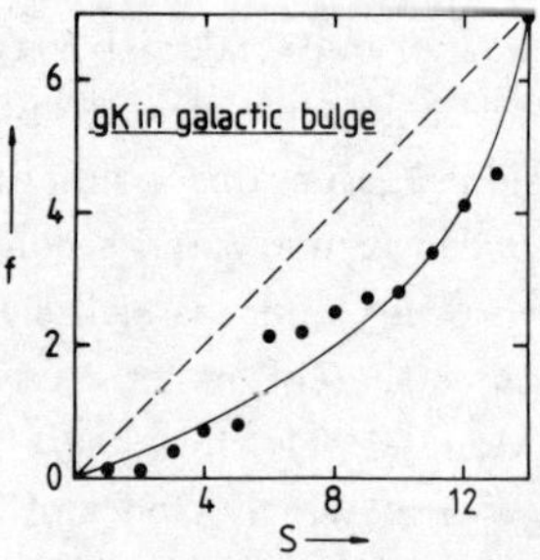

Figure 3.1. Cumulative distribution of metallicity f, relative to the Sun, for GK giants observed in Baade's window by Whitford and Rich (1983). The curve represents the prediction of the Simple Model with a yield corresponding to $f = 2.5$.

distribution in terms of special circumstances related to the formation of the disk. This is not to say that the IMF does not vary with metallicity in the sense suggested by Schmidt, as there is evidence from H II regions that at least the upper part of it does (Shields and Tinsley 1976; Terlevich and Melnick 1983), and this could be relevant to the situation in the halo; but I do not think it is the main factor in the disk.

A related question is whether clean, unprocessed material exists anywhere in the universe or whether there is some prompt initial enrichment by stars or more massive objects either on a pregalactic scale (which was the original concept of Population III) or on a galactic scale. One candidate for clean material is provided by the intergalactic clouds that produce the Lyman-α forest of absorption lines in high redshift QSOs (Sargent et al. 1980), but even if one discounts the claims to have detected O VI or Si III in these systems (Norris et al. 1983; Chaffee et al. 1985) the limits on abundance are model-dependent and not very stringent. Stronger constraints on Population III are provided by the two most metal-deficient stars to have been analyzed: the dwarf G64-12 with [Fe/H] $= -3.5$ (Carney and Peterson 1981) and the giant CD $-38°245$ with [Fe/H] $= -4.5$ (Bessell and Norris 1984), which are pretty clean by anybody's definition. The only marked peculiarity of CD $-38°245$ relative to the general run of Population II stars is an overabundance of nitrogen relative to iron by an order of magnitude or more (no data are available for oxygen) so that there is little room for pregalactic metal enrichment but a possible role for a subset of massive stars that produce mainly primary nitrogen (Woosley and Weaver 1982*a*).

3.3. Metallicity, Age, and Differential Enrichment

A related problem is the relation between the ages of stars and their chemical composition, which in the past has often been regarded as being definable by a single parameter equivalent to metallicity. But we have to be careful here because abundant elements like oxygen do not vary in lockstep with iron, which is the chief cause of the overall strength of stellar absorption lines.

Ignoring this problem for the moment, the Simple Model predicts that stellar abundances should increase with the time of formation in some way: e.g., the exponential model of Truran and Cameron (1971) predicts that elements C–Fe should increase linearly with time. The age-metallicity relation of stars of the disk has been studied by Mayor (1976) and Twarog (1980). Twarog's data do indeed indicate a linear increase in Fe/H with time up to about half the age of the solar system, but there are signs that it flattens off at later times, as young stars have essentially the same metallicity as the Sun. Twarog, assuming a constant rate of star formation, attributed the flattening to continual inflow of unprocessed material, but there is no necessity for this, as it can equally arise from a reduction in the rate of gas consumption or as a result of the emission of secondary gas from old stars—an effect that is neglected in the instantaneous recycling approximation.

The crucial test here consists in deciding how much of the present-day interstellar gas has been astrated, which can be deduced from the abundances of D and ^{6}Li. If we accept current estimates of the primordial solar system abundance of deuterium as being $D/H = 2 \times 10^{-5}$ (Yang et al. 1984) and also believe the claim of Vidal-Madjar et al. (1983) that the present ratio D/H is only 5×10^{-6}, then it follows that most of the interstellar gas we have now is indeed secondary material that has been recycled through stars (cf. Güsten and Mezger 1982). This interpretation is still quite controversial, but it receives some support from the low ^{6}Li/^{7}Li ratio reported by Ferlet and Dennefeld (1983) in interstellar gas toward ς Oph and δ Sco.

In any case the enrichment of disk and halo stars seems to follow a fairly coherent pattern, both in iron and in other elements with the exception of nitrogen. Differential trends long suspected among the light elements (Wallerstein 1962; Pagel 1964; Conti et al. 1967; Gasson and Pagel 1966) have been confirmed and extended in more accurate modern work (Clegg, Lambert, and Tomkin 1981; Peterson 1981; Luck and Bond 1981, 1985; Arpigny and Magain 1983; Tomkin and Lambert 1984; Edvardsson, Gustafsson, and Nissen 1984; Tomkin, Lambert, and Balachandran 1985; Spite and Spite 1985) although there are still some outstanding discrepancies, e.g., for Al and Mg (Spite and Spite 1985).

Among the lighter elements carbon varies almost in lockstep with iron, but oxygen is quite different with $[O/Fe] \simeq -0.5\,[Fe/H]$ in disk stars ($[Fe/H] > -1.0$) and $[O/Fe] \simeq 0.7$ in halo stars. Thus iron in the disk stars displays secondary-like behavior relative to oxygen. Mg, Al, Ca, and Ti behave qualitatively like oxygen but with smaller amplitude while Sc, iron group elements, and zinc follow iron apart from reports of an overabundance of Ni in extreme halo giants (Luck and Bond 1983, 1985) which are still controversial (Spite and Spite 1985). The trend shown by oxygen has been attributed to the presence of a distinct IMF, with relatively more massive stars, in the halo phase (Twarog and Wheeler 1982), so that the yield of oxygen relative to iron has gone down in the course of time, but the trend could also be understood on the basis of a constant IMF with a time delay between oxygen production by more massive stars and iron production mainly by less massive stars (cf. Tinsley 1979; Matteucci and Tornambé 1985), depending on the masses and time scales involved. Superposed on these trends there is a marked odd-even effect in Na/Mg and Al/Mg which is shown by Tomkin, Lambert, and Balachandran (1985) to be in quantitative accord with the dependence of neutron supply in explosive nucleosynthesis on CNO abundance predicted by Arnett (1971) and by Woosley and Weaver (1982*b*). Mg isotopes show a similar effect in the subdwarf Groombridge 1830 ([Fe/H] $\simeq -1.3$), though not in stars with a smaller deficiency (Tomkin and Lambert 1980).

For C, N, and O we have data on galactic and extragalactic H II regions as well as nearby stars which bear on the controversy as to whether nitrogen is a primary or a secondary element or both, and the overall situation is highly confusing. In halo stars nitrogen can be slightly underabundant or strongly overabundant relative to iron (Bessell and Norris 1982, 1984; Laird 1985) suggesting primary production, and Tomkin and Lambert (1984) find a close relation between nitrogen and carbon in their sample of stars. These results agree with the suggestion of Edmunds and Pagel (1978) that nitrogen is a primary product of relatively low mass progenitors. We made this suggestion because a plot of N/O versus O/H in H II regions (Pagel and Edmunds 1981; Edmunds and Pagel 1984*a*), while showing a secondary-like trend overall—i.e., N/O increases statistically with O/H—shows little trend in N/O with O/H either in any one galaxy such as our own (Shaver et al. 1983) or among irregular galaxies taken as a group. Alloin et al. (1979) noted the same trend and explained it on the basis of a detailed model of galactic chemical evolution with part primary and part secondary nitrogen combined with variations in the slope of the IMF, whereas Serrano and Peimbert (1983) argue in favor of purely secondary nitrogen combined with a primary yield that increases with metallicity and explain the scatter as a consequence of variable inflow and time-lag effects. Alternatively, White and Audouze (1983) suggest that N/O gradients in spiral disks are flattened out by inflow with stochastic mixing. On the other hand, Tomkin and Lambert (1984) have pointed out that the overall trend in N/O versus O/H in stars is much like that in H II regions, but the trend in stars as shown by them merely reflects the behavior of oxygen relative to other primary elements like iron!

There have been two other developments on this problem that point in opposite directions. The first is the measurement of carbon abundances in H II regions from ultraviolet spectra taken with the *IUE* (Dufour, Shields, and Talbot 1982; Dufour, Schiffer, and Shields 1984). Dufour (1984) finds that carbon falls off with diminishing oxygen abundance even faster than nitrogen does, although a plot of C/N against O/H (Figure 3.2) indicates that the difference between the trends of C and N is fairly marginal. In either case, the relation favors primary nitrogen if carbon is primary.

The second development is the measurement of isotope ratios in molecular

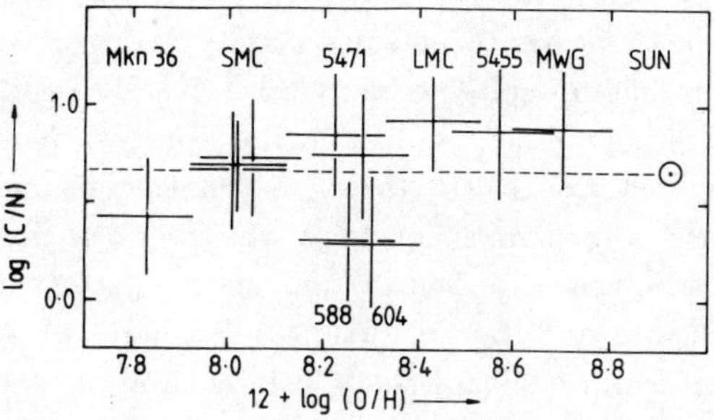

Figure 3.2. Plot of log (C/N) against log (O/H) based on the data of Dufour (1984).

clouds near the Sun and at the center of the Galaxy. Measurements of $^{13}C/^{12}C$ from formaldehyde (Henkel, Wilson, and Bieging 1982) suggest that the ratio in the local medium is like that in the solar system but that there is a distinct gradient with galactocentric distance of about 0.05 dex kpc^{-1} as might be expected for secondary ^{13}C. Furthermore, Güsten and Ungerechts (1985) have recently measured $^{14}N/^{15}N$ from ammonia inversion transitions in two local clouds (DR 21(OH) and S140), where it is somewhat higher than in the solar system, and at the galactic center, where it is higher again by another factor of 2, and they interpret their results in terms of a chemical evolution model in which ^{15}N is primary but ^{14}N is 87% secondary in the solar system. Furthermore, although N/O measured from N^+/O^+ in galactic H II regions has not shown a significant gradient (Shaver et al. 1983), there are measurements of N^{++}/O^{++} from fine structure transitions in the far-infrared (Lester et al. 1983) which are somewhat discrepant. Theoretically both N^+/O^+ and N^{++}/O^{++} should be close to the true abundance ratio N/O, but the infrared determinations are larger by a factor of 3 (which may be due to uncertainties in atomic data) and show signs of a gradient at 6 kpc from the galactic center where the optical observations do not reach. Thus there are signs that carbon-13 and nitrogen-14 show secondary behavior between here and the galactic center.

In the galactic center itself, the secondary model predicts that O/H and N/O should both be enhanced by the same factor as $^{14}N/^{15}N$, i.e., about 3. This is in good agreement with determinations of O/H (Aitken, Griffiths, and Jones 1976). We do not know N/O at the center of our Galaxy, but our estimates for H II regions near the center of other galaxies (Pagel and Edmunds 1981; Edmunds and Pagel 1984*a*) suggest only a modest enhancement of the order of 0.15 dex. Nuclei of early-type galaxies with emission lines are usually "Liners" (Heckman 1980) with strong lines of [N II], and in M81 Peimbert and Torres-Peimbert (1981) inferred a considerable overabundance of nitrogen by fitting a shock model. However, it now seems more likely that Liners, like Seyfert galaxies, are photoionized by a hard spectrum like a power law (Ferland and Netzer 1983; Halpern and Steiner 1983; Keel 1983; Péquignot 1984; Diaz, Pagel, and Wilson 1985; Binette 1985; Diaz, Pagel, and Terlevich 1985), and in this case it is hard to tell whether N/O is solar or up to 3 times greater.

Thus the question of primary versus secondary nitrogen is in a state of great confusion. I think there has to be some primary nitrogen to begin with, but that secondary production quite probably takes over at a certain stage.

Another interesting problem is posed by the *s*-process elements. These ought to show secondary behavior relative to iron, if the neutron exposure per iron seed nucleus in the progenitor star is constant, which may or may not be the case according to the various neutron sources and neutron poisons in the helium shell-flashing phase (cf. Iben and Renzini 1983). In disk stars, barium varies more or less in lockstep with iron (Edvardsson, Gustafsson, and Nissen 1984), while in halo stars barium shows the expected secondary-like behavior for $[Fe/H] < -1.5$ (Barbuy, Spite, and Spite 1985 and references therein), while yttrium varies logarithmically about half as much. On the other hand, Sr and Zr vary little with respect to iron, e.g., in the much studied bright metal-deficient red giant HD 122563 (Pagel 1965; Sneden and Parthasarathy 1984), and the classical *r*-process elements Eu and Gd can be either mildly overdeficient relative to iron, as in the case of 122563, or mildly overabundant, as in the case of HD 110184 (Sneden and Pilachowski 1985), where in accordance with a suggestion by Truran (1981) it seems that Ba and heavier elements are due to the *r*-process rather than the *s*-process. Other halo stars with relative overabundances of Eu have been found by Spite and Spite (1978) and by Luck and Bond (1981), and according to Luck and Bond (1985) Nd behaves generally rather like Eu. To complicate matters yet further, Griffin et al. (1982) discovered a still more metal-deficient giant, HD 115444, with $[Fe/H] = -2.9$, which has relative enhancements of Zr, Ba, La, *and* Eu. While the authors have succeeded in finding *r*- and *s*-process distributions that fit the various abundance patterns observed, the significance of all these results in the context of galactic chemical evolution models is far from clear.

3.4. The Relation between Abundance and Galactic Parameters

The third topic I wish to discuss is the interpretation of abundance gradients and the relation between metallicity, gas fraction, and mass or local surface density.

The Simple Model with instantaneous recycling predicts the well-known relation between abundance in the gas phase and gas fraction

$$z = \ln(m/g), \tag{1}$$

where z is the abundance of primary elements by mass in the interstellar gas in units of the yield, m is the total mass of the system, and g is the mass of gas which remains. A relation of this kind was believed some years ago to hold for irregular and blue compact galaxies, which are well-mixed systems without marked abundance gradients (e.g., Lequeux et al. 1979; Talent 1980)—or at least for a high proportion of them (Kinman and Davidson 1981; Webster et al. 1983). The existence of such a relation has been challenged recently, however (Hunter, Gallagher, and Rautenkrantz 1982; Matteucci and Chiosi 1983; Gallagher and Hunter 1984), with the implication that inflows and/or outflows dominate the

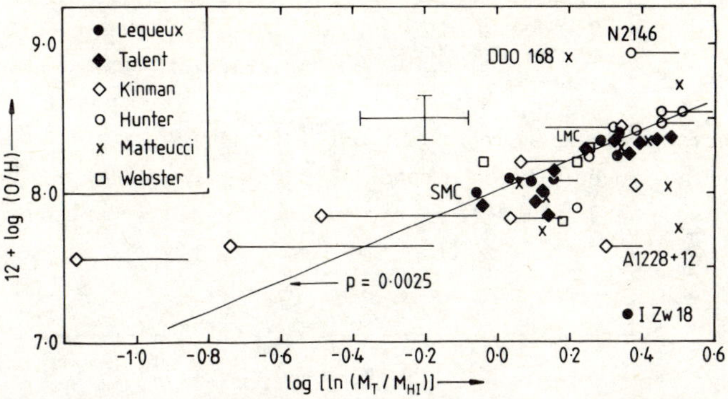

Figure 3.3. Plot of log (O/H) in the gas phase against the double logarithm of the gas fraction in irregular and blue compact galaxies. The slope of the line is fixed by equation (1).

situation. There are difficulties with the determination of total mass, and Hunter et al. and Gallagher and Hunter use the luminosity with a mass-to-light ratio rather than a direct dynamical determination. If we use dynamical estimates of mass, we obtain the relation shown in Figure 3.3 in which the abundances and gas fractions have been taken in most cases from the original authors, but horizontal lines mark the differences between their gas fractions and those calculated by Matteucci and Chiosi, which give some idea of the uncertainties. The straight line with slope fixed by equation (1) seems to give a reasonably good fit to the majority of points, although, as Matteucci and Chiosi pointed out, a number fall below, notably the most deficient galaxy known, IZw 18 (Lequeux and Viallefond 1980), and this could represent departures from the Simple Model caused by inflows or outflows as discussed by Matteucci and Chiosi.

A more remarkable property of irregular and blue compact galaxies is the relation between abundance and dynamically estimated mass (Figure 3.4), which is better than the relation with gas fraction (though still violated by IZw 18) and very similar to the mass-metallicity relation given by Mould (1984) for elliptical galaxies where one looks at the average of a stellar population, but some care is needed here because the mass scales are not necessarily comparable and both neglect the possible presence of dark halos. The usual interpretation for irregular galaxies is that larger ones consume their gas more rapidly while the trend in ellipticals has been explained by loss of material through galactic winds (e.g., Larson 1974; Vigroux, Chièze, and Lazareff 1981), possibly after a series of mergers of smaller gas-rich systems in which the star formation efficiency increases with the total mass (Tinsley and Larson 1979). This last effect could perhaps explain the similarity in slope between the two mass-metallicity relations, but in the escape and merger models one needs a yield of the order of 0.02, which is very different from the yield of 0.0025 that we infer by applying the Simple Model to irregular galaxies.

Still more ambiguities of interpretation arise in the discussion of abundance

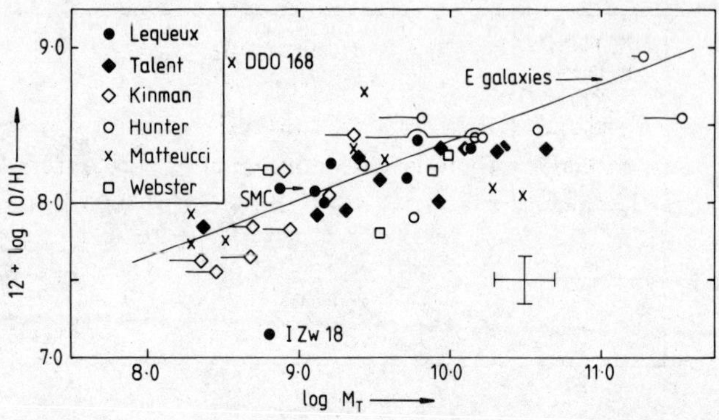

Figure 3.4. Plot of log (O/H) in the gas phase of irregular and blue compact galaxies against the logarithm of the dynamically estimated mass. The straight line is the relation given for the mean stellar abundance in elliptical galaxies by Mould (1984).

gradients. Color gradients are observed in some elliptical galaxies (Wirth 1981; Boroson, Thompson, and Shectman 1983; Davis et al. 1985), but they are usually fairly small and the interpretation is complicated by the presence of dust lanes. Larger gradients are found in the bulges of spirals (Wirth 1981; van der Kruit and Searle 1982*a*, *b*), suggesting the operation of dissipative processes during their formation. However, it seems that the isochromes are not flatter than the isophotes, as they should be on the basis of this picture (Larson 1976), and so van der Kruit and Searle (1982*b*) suggested an alternative picture in which the abundance gradient is primordial and related to the mass-to-light ratio; the spheroid would in this case have undergone a dissipationless collapse in the early stages. Alternatively, Davis et al. (1985) refer to N-body calculations by R. G. Carlberg, in which the collapse is slowed down by pressure support, which may give a better fit to the data.

In our own Galaxy, the existence of an abundance gradient in the outer halo beyond 10 kpc is a subject of controversy (Freeman and Norris 1981; Carney 1984). Near the galactic plane and at right angles to it, however, there is a strong gradient of about 0.2 dex in the first kpc (Blaauw and Garmany 1976; Trefzger 1981; Hartkopf and Yoss 1982), followed by a flatter trend which can be interpreted as a bimodal distribution with fixed parameters in which the metal-poor population becomes more dominant relative to the metal-rich component with increasing height. Jones and Wyse (1983) identify the metal-poor component with the "thick disks" observed in edge-on S0–Sbc galaxies (Tsikoudi 1979; Burstein 1979; van der Kruit and Searle 1982*a*, *b*) and argue that thick gaseous disks are the first objects to form when protogalactic gas collapses in the potential of a nonspherical dark halo, followed by the formation of Population II stars, of the central bulge, and finally of the ordinary thin disk. Gilmore (1984), on the other hand, using star count data, identifies the thick disk as a third component, of only moderate metal deficiency (more or less the same as Intermediate Population II), between metal-rich disk stars and extremely metal-poor stars of the halo; this third component formed in a dissipationless collapse similar to the classical picture of Eggen, Lynden-Bell, and Sandage (1962).

Another interesting aspect of abundance gradients is the presence of radial gradients in the disks of spirals, to which attention was first drawn by Searle (1971). In our own Galaxy there is a wide range of evidence from observations of stars and H II regions for a gradient of the order of 0.05 dex kpc^{-1} in the abundances of both oxygen and iron (Pagel and Edmunds 1981; Shaver et al. 1983), and oxygen and nitrogen gradients have been determined in a number of nearly spiral galaxies with H II regions using either a direct determination based on measured electron temperatures (Smith 1975; Shields and Searle 1978) or a more indirect so-called empirical method based on model fits to the regular behavior of ratios of strong lines with radial distance (Alloin et al. 1979; Pagel et al. 1979; Pagel, Edmunds, and Smith 1980; Pagel and Edmunds 1981; McCall 1982; Edmunds and Pagel 1984*b*; McCall, Rybski, and Shields 1985). The second method is useful for metal-rich or faint H II regions in which weak temperature-sensitive lines are unobservable and appears to be accurate to ±0.2 dex, although more calibration data are still needed at abundances exceeding solar.

Oxygen abundances have been plotted by several authors against radial distance in units of the isophotal radius and the resulting diagrams are quite confusing, but they do indicate a broad correlation with morphological type in the sense that later-type galaxies have lower abundances near the outer isophote. McCall has suggested what is probably a more informative method of looking at the data, which is to plot them against the local surface density of mass in the disk, which can be obtained from the rotation curve after deconvolving the disk from the bulge by means of surface photometry, and against the gas fraction, but the gas fraction is more difficult to estimate because neutral hydrogen only provides a lower limit and there is an additional contribution from molecular hydrogen that has to be inferred from observations of CO, and the conversion from CO line intensity to H_2 column density is quite uncertain (Li, Riley, and Wolfendale 1983; Bhat et al. 1985). Edmunds and Pagel (1984*b*) have collected oxygen abundances in H II regions from their own observations and from the literature, using a slightly revised calibration of their empirical method, and compared them with McCall's estimates of disk surface density and gas fraction, taking only neutral atomic hydrogen into account. When this is done, it turns out that the Scd galaxies define a fairly tight relation between oxygen abundance and either total disk surface density or gas fraction, and that early-type (Sab–Sc) spirals define similar relations but with more scatter.

The relation with gas fraction does not fit the Simple Model with a constant yield: the observed abundance gradients are too steep, even when H_2 molecules are neglected. Three kinds of model have been put forward to account for steep abundance gradients in spiral disks. (1) Vertical inflow models based on the ideas of dynamical collapse (e.g., Tinsley and Larson 1978; Chiosi 1980) or steady inflows of unprocessed material can either flatten or enhance the abundance gradient according to the way in which their effect varies with radial distance, and for a spatially constant inflow rate the gradient is enhanced because the outer regions are affected more. In this way Diaz and Tosi (1984) and Tosi and Diaz (1985) have fitted the abundance gradients in several galaxies using a unique initial mass function but have problems in fitting the high abundances found in M51. (2) Combined vertical and inward radial flows due to low angular momentum of the inflowing gas provide models that can give an excellent fit to the abundance gradient in the Milky Way (Mayor and Vigroux 1981; Lacey and Fall 1985). This model has not been tried out for other galaxies. (3) Variable yield models have been proposed by Peimbert and Serrano (1982), who suggest a yield increasing with metallicity, and by Güsten and Mezger (1982), who propose a bimodal IMF in our own Galaxy with only stars above 2 $M_{\odot}$ being formed in spiral arms and an increase in the spiral arm fraction toward the center of the Galaxy; this has a similar effect in enhancing the abundance gradient and at the same time is required in order to reconcile the rate of star formation deduced from Lyman continuum photon production rates derived from galactic free-free emission with the present-day gas fraction. A similar lower truncation of the IMF has been deduced by Rieke et al. (1980) in the case of the starburst galaxy M82.

Variable yield models are quite promising, although it is not too clear what is the driving factor. Peimbert and Serrano's relation with metallicity fits our data very well, but this does not mean that it is the metallicity that drives the yield. It could just as well be the other way around, and one obtains an equally good fit to the data by assuming that the yield increases linearly with the surface density in the disk. This variation either may represent that of the true yield, governed primarily by the IMF, or may just be a variation in the effective yield due to loss of some of the ejected material from supernovae in the form of galactic winds and fountains, with more of it retained in regions of high surface density and presumably higher escape velocity, which would link the effect to the mass-metallicity relation in elliptical galaxies. Whatever the explanation finally turns out to be, and whether it involves inflow, outflow, or variable initial mass function, the effective yield seems to be closely related to local conditions.

References

Aitken, D. K., Griffiths, J., and Jones, B. 1976, *M.N.R.A.S.*, **176**, 73P.

Alloin, D.; Collin-Souffrin, S.; Joly, M.; and Vigroux, L. 1979, *Astr. Ap.*, **78**, 200.

Arnett, W. D. 1971, *Ap. J.*, **166**, 153.

Arpigny, C., and Magain, P. 1983, *Astr. Ap.*, **127**, L7.

Barbuy, B., Spite, F., and Spite, M. 1985, *Astr. Ap.*, **144**, 343.

Bessell, M. S., and Norris, J. 1982, *Ap. J. (Letters)*, **263**, L29.

——. 1984, *Ap. J.*, **285**, 622.

Bhat, C. L.; Issa, M. R.; Houston, B. P.; Mayer, C. J.; and Wolfendale, A. W.

1985, *Nature,* **314**, 11.

Binette, L. 1985, *Astr. Ap.,* **143**, 334.

Blaauw, A., and Garmany, C. D. 1976, in *Third European Astronomical Meeting,* ed. A. Kharadse, p. 351 (Abastumani Obs.).

Bond, H. E. 1981, *Ap. J.,* **248**, 606.

Boroson, T. A., Thompson, I. B., and Shectman, S. A. 1983, *Astr. J.,* **88**, 1707.

Burstein, D. 1979, *Ap. J.,* **234**, 829.

Carney, B. W. 1984, *Pub. A.S.P.,* **96**, 841.

Carney, B. W., and Peterson, R. C. 1981, *Ap. J.,* **245**, 238.

Chaffee, F. H.; Foltz, C. B.; Röser, H. J.; Weymann, R. J.; and Latham, D. W. 1985, *Ap. J.,* **292**, 362.

Chiosi, C. 1980, *Astr. Ap.,* **83**, 206.

Clegg, R., Lambert, D. L., and Tomkin, J. 1981, *Ap. J.,* **250**, 262.

Conti, P. S.; Greenstein, J. L.; Spinrad, H. E.; Wallerstein, G.; and Vardya, M. S. 1967, *Ap. J.,* **148**, 105.

Davis, L. E.; Cawson, M.; Davies, R. L.; and Illingworth, G. 1985, *Astr. J.,* **90**, 169.

Diaz, A. I., Pagel, B. E. J., and Terlevich, E. 1985, *M.N.R.A.S.,* **214**, 41P.

Diaz, A. I., Pagel, B. E. J., and Wilson, I. R. G. 1985, *M.N.R.A.S.,* **212**, 737.

Diaz, A. I., and Tosi, M. 1984, *M.N.R.A.S.,* **208**, 365.

Dufour, R. J. 1984, *Future of Ultra-violet Astronomy Based on Six Years of IUE Research* (NASA).

Dufour, R. J., Schiffer, F. H., and Shields, G. A. 1984, *Future of Ultra-violet Astronomy Based on Six Years of IUE Research* (NASA).

Dufour, R. J., Shields, G. A., and Talbot, R. J. 1982, *Ap. J.,* **252**, 461.

Edmunds, M. G., and Pagel, B. E. J. 1978, *M.N.R.A.S.,* **185**, 77P.

——. 1984*a,* in *Stellar Nucleosynthesis,* ed. C. Chiosi and A. Renzini, p. 341 (Dordrecht: Reidel).

——. 1984*b, M.N.R.A.S.,* **211**, 507.

Edvardsson, B., Gustafsson, B., and Nissen, P. E. 1984, *ESO Messenger,* no. 38, p. 33.

Eggen, O. J., Lynden-Bell, D., and Sandage, A. R. 1962, *Ap. J.,* **136**, 748.

Ferland, G. J., and Netzer, H. 1983, *Ap. J.,* **264**, 605.

Ferlet, R., and Dennefeld, M. 1983, in *ESO Workshop on Primordial Helium,* ed. P. A. Shaver, D. Kunth, and K. Kjär, p. 373.

Freeman, K. C., and Norris, J. 1981, *Ann. Rev. Astr. Ap.,* **19**, 319.

Gallagher, J. A., and Hunter, D. A. 1984, *Ann. Rev. Astr. Ap.,* **22**, 37.

Gasson, R. E. M., and Pagel, B. E. J. 1966, *Observatory,* **86**, 196.

Gilmore, G. 1984, *M.N.R.A.S.,* **207**, 223.

Griffin, R.; Griffin, R.; Gustafsson, B.; and Vieira, T. 1982, *M.N.R.A.S.,* **198**, 637.

Güsten, R., and Mezger, P. G. 1982, *Vistas in Astr.,* **26**, 159.

Güsten, R., and Ungerechts, H. 1985, *Astr. Ap.,* in press.

Halpern, J. P., and Steiner, J. E. 1983, *Ap. J. (Letters)*, **269**, L37.
Hartkopf, W., and Yoss, K. 1982, *Astr. J.*, **87**, 1679.
Hartwick, F. D. A. 1976, *Ap. J.*, **209**, 418.
Heckman, T. M. 1980, *Astr. Ap.*, **87**, 152.
Henkel, C., Wilson, T. L., and Bieging, J. 1982, *Astr. Ap.*, **109**, 344.
Hunter, D. A., Gallagher, J. S., and Rautenkrantz, D. 1982, *Ap. J. Suppl.*, **49**, 53.
Iben, I., and Renzini, A. 1983, *Ann. Rev. Astr. Ap.*, **21**, 271.
Jones, B. J. T., and Wyse, R. F. G. 1983, *Astr. Ap.*, **120**, 165.
Keel, W. C. 1983, *Ap. J.*, **269**, 466.
Kinman, T. D., and Davidson, K. 1981, *Ap. J.*, **243**, 127.
Lacey, C. G., and Fall, S. M. 1985, *Ap. J.*, in press.
Laird, J. B. 1985, *Ap. J.*, **288**, 556.
Larson, R. B. 1974, *M.N.R.A.S.*, **166**, 585.
——. 1976, *M.N.R.A.S.*, **176**, 31.
Lequeux, J.; Peimbert, M.; Rayo, J.; Serrano, A.; and Torres-Peimbert, S. 1979, *Astr. Ap.*, **80**, 155.
Lequeux, J., and Viallefond, F. 1980, *Astr. Ap.*, **91**, 269.
Lester, D. F.; Dinerstein, H. L.; Werner, M. W.; Watson, D. M.; and Genzel, R. L. 1983, *Ap. J.*, **271**, 618.
Li, T. P., Riley, P. A., and Wolfendale, A. 1983, *M.N.R.A.S.*, **203**, 87.
Luck, R. E., and Bond, H. E. 1981, *Ap. J.*, **244**, 919.
——. 1983, *Ap. J. (Letters)*, **271**, L75.
——. 1985, *Ap. J.*, in press.
Lynden-Bell, D. 1975, *Vistas in Astr.*, **19**, 299.
Matteucci, F., and Chiosi, C. 1983, *Astr. Ap.*, **123**, 121.
Matteucci, F., and Tornambé, A. 1985, *Astr. Ap.*, **142**, 13.
Mayor, M. 1976, *Astr. Ap.*, **48**, 301.
Mayor, M., and Vigroux, L. 1981, *Astr. Ap.*, **98**, 1.
McCall, M. L. 1982, thesis, University of Texas at Austin.
McCall, M. L., Rybski, P. M., and Shields, G. 1985, *Ap. J. Suppl.*, **57**, 1.
Mould, J. R. 1984, *Pub. A.S.P.*, **96**, 773.
Norris, J., Hartwick, F. D. A., and Peterson, B. W. 1983, *Ap. J.*, **273**, 450.
Ostriker, J. B., and Thuan, T. X. 1975, *Ap. J.*, **202**, 353.
Pagel, B. E. J. 1964, *R. Obs. Bull.*, no. 87.
——. 1965, *R. Obs. Bull.*, no. 104.
Pagel, B. E. J., and Edmunds, M. G. 1981, *Ann. Rev. Astr. Ap.*, **19**, 77.
Pagel, B. E. J.; Edmunds, M. G.; Blackwell, D. E.; Chun, M. S.; and Smith, G. 1979, *M.N.R.A.S.*, **189**, 95.
Pagel, B. E. J., Edmunds, M. G., and Smith, G. 1980, *M.N.R.A.S.*, **193**, 219.
Pagel, B. E. J., and Patchett, B. E. 1975, *M.N.R.A.S.*, **172**, 13.
Peimbert, M., and Serrano, A. 1982, *M.N.R.A.S.*, **198**, 563.
Peimbert, M., and Torres-Peimbert, S. 1981, *Ap. J.*, **245**, 845.

Péquignot, D. 1984, *Astr. Ap.*, **131**, 159.

Peterson, R. C. 1981, *Ap. J.*, **244**, 989.

Rieke, G. H.; Lebovsky, A. J.; Thompson, R. I.; Low, F. J.; and Tokunaga, A. T. 1980, *Ap. J.*, **238**, 24.

Sargent, W. L. W.; Young, P. J.; Boksenberg, A.; and Tytler, D. 1980, *Ap. J.*, **42**, 41.

Schmidt, M. 1963, *Ap. J.*, **137**, 758.

Searle, L. 1971, *Ap. J.*, **168**, 327.

Serrano, A., and Peimbert, M. 1983, *Rev. Mex. Astr. Astrofis.*, **8**, 117.

Shaver, P. A.; McGee, R. A.; Danks, A. C.; and Pottasch, S. R. 1983, *M.N.R.A.S.*, **204**, 53.

Shields, G. A., and Searle, L. 1978, *Ap. J.*, **222**, 821.

Shields, G., and Tinsley, B. M. 1976, *Ap. J.*, **203**, 66.

Smith, H. E. 1975, *Ap. J.*, **199**, 591.

Sneden, C., and Parthasarathy, M. 1984, *Ap. J.*, **267**, 757.

Sneden, C., and Pilachowski, C. 1985, *Ap. J. (Letters)*, **288**, L55.

Spite, M., and Spite, F. 1978, *Astr. Ap.*, **67**, 23.

——. 1985, *Ann. Rev. Astr. Ap.*, **23**, 225.

Talbot, R. J., and Arnett, W. D. 1974, *Ap. J.*, **190**, 605.

Talent, D. L. 1980, thesis, Rice University.

Terlevich, R., and Melnick, J. 1983, ESO preprint, no. 264.

Tinsley, B. M. 1979, *Ap. J.*, **229**, 1046.

Tinsley, B. M., and Larson, R. B. 1978, *Ap. J.*, **221**, 554.

——. 1979, *M.N.R.A.S.*, **186**, 503.

Tomkin, J., and Lambert, D. L. 1980, *Ap. J.*, **235**, 925.

——. 1984, *Ap. J.*, **279**, 220.

Tomkin, J., Lambert, D. L., and Balachandran, S. 1985, *Ap. J.*, **290**, 289.

Tosi, M., and Diaz, A. I. 1985, *M.N.R.A.S.*, **217**, 571..

Trefzger, C. F. 1981, *Astr. Ap.*, **95**, 184.

Truran, J. W. 1981, *Astr. Ap.*, **97**, 391.

Truran, J. W., and Cameron, A. G. W. 1971, *Ap. Space Sci.*, **14**, 179.

Tsikoudi, V. 1979, *Ap. J.*, **234**, 842.

Twarog, B. 1980, *Ap. J.*, **242**, 242.

Twarog, B., and Wheeler, J. C. 1982, *Ap. J.*, **261**, 636.

van den Bergh, S. 1962, *Astr. J.*, **67**, 486.

van der Kruit, P. C., and Searle, L. 1982*a*, *Astr. Ap.*, **110**, 61.

——. 1982*b*, *Astr. Ap.*, **110**, 79.

Vidal-Madjar, A.; Laurent, C.; Gry, C.; Bruston, P.; Ferlet, R.; and York, D. 1983, *Astr. Ap.*, **120**, 58.

Vigroux, L., Chièze, J. P., and Lazareff, B. 1981, *Astr. Ap.*, **98**, 119.

Wallerstein, G. 1962, *Ap. J. Suppl.*, **6**, 407.

Webster, B. L.; Longmore, A. J.; Hawarden, T. G.; and Mebold, U. 1983, *M.N.R.A.S.*, **205**, 643.

White, S., and Audouze, J. 1983, *M.N.R.A.S.*, **203**, 603.
Whitford, A. E., and Rich, R. M. 1983, *Ap. J.*, **274**, 723.
Wirth, A. 1981, *Astr. J.*, **86**, 981.
Woosley, S. E., and Weaver, T. A. 1982*a*, in *Supernovae: A Survey of Current Research*, ed. M. J. Rees and R. J. Stoneham, p. 79 (Dordrecht: Reidel).
——. 1982*b*, in *Essays in Nuclear Astrophysics*, ed. C. A. Barnes, D. D. Clayton, and D. N. Schramm, p. 377 (Cambridge University Press).
Yang, J.; Turner, M. S.; Steigman, G.; Schramm, D. N.; and Olive, K. A. 1984, *Ap. J.*, **281**, 493.

Part II

Nucleosynthesis

4

The Age of the Observable Universe in the Inflationary Cosmology

William A. Fowler and Chris C. Meisl

Aonaibh ri cheile
Motto of Clan Cameron

Abstract

We work in the framework of a flat universe following the early inflationary phase of the expanding universe. Current values for t_0, the age of the observable universe, and for the reciprocal of the Hubble constant, $T_0 = H_0^{-1}$, are reviewed and compared. We review in considerable detail and update calculations on the determination of t_0 through nucleocosmochronology. The uncertainties in $t_0/T_0 = H_0 t_0$ are such that no strong conclusions can be reached. The current value of the dimensionless parameter for mass-energy density, Ω_0, can be unity. This is greater than the current dimensionless baryon density, $\Omega_b \approx 0.2$, so there must be missing mass in the form of exotic particles such as "massive" neutrinos, axions, photinos, or other "inos." On the other hand Ω_0 can also be equal to Ω_b, in which case the missing mass is in the vacuum and we live at an epoch when the dimensionless cosmological "constant," $\lambda_0 \approx 0.8$, is the same order of magnitude as Ω_0. We come to the happy conclusion that there is more work for all involved in cosmological chronology.

A motto of the Clan Cameron is *Aonaibh ri cheile,* Gaelic for *unite.* Alastair Graham Walter Cameron has united physics, astronomy, chemistry, and the geosciences in his lifelong diligent and immensely productive research on nucleosynthesis, abundance of the elements, origin of the solar system, and more. We salute him on his 60th birthday and are honored to add this contribution to his *Festschrift.*

4.1. Introduction

The extraordinary success of the new inflationary cosmology (Guth 1981; Guth and Steinhardt 1984; Steinhardt 1984) in solving the cosmological horizon problem, and thus explaining the large scale homogeneity and isotropy of the observable universe, prompts further investigation of other consequences of the inflationary model. One of the most important consequences is that the universe is flat as a result of the exponential expansion during the early inflationary phase of the expanding universe. The curvature of spacetime was reduced to zero during this phase.

It is the purpose of this account to discuss the age of the observable universe within the framework of a flat universe. The adjective "observable" has a special meaning in the inflationary cosmology. According to Guth and Steinhardt (1984) our observable universe may be only an expanding bubble in an otherwise steady-state universe. We are content to study our observable universe and let others elaborate on matters of larger scale.

In general the universal expansion rate is governed by the Friedmann equation

$$H^2 = \left(\frac{1}{a}\frac{da}{dt}\right)^2 = \frac{8\pi G}{3}(\rho + \rho_v) - k\frac{c^2}{a^2}, \tag{1}$$

which reduces, when the curvature parameter k equals zero, to

$$H^2 = \frac{8\pi G}{3}(\rho + \rho_v), \quad k = 0. \tag{2}$$

In these equations H is the Hubble expansion parameter, $a(t)$ is the arbitrary distance measure, G is the gravitational constant, ρ is the mass-energy density in g cm^{-3} of matter and radiation, and ρ_v is the vacuum mass-energy density for which the corresponding pressure is $p_v = -\rho_v c^2$ with c the velocity of light. The matter and radiation density is commonly expressed in terms of the dimensionless parameter

$$\Omega = 8\pi G\rho/3H^2, \tag{3}$$

and the vacuum density in terms of the dimensionless cosmological "constant"

$$\lambda = 8\pi G\rho_v/3H^2, \tag{4}$$

which is related to the usual cosmological constant Λ by

$$\Lambda = 8\pi G\rho_v = 3H^2\lambda, \tag{5}$$

although in some literature Λ is replaced by Λc^2. It will be clear for the flat universe, $k = 0$, that

$$\Omega + \lambda = 1 \tag{6}$$

at all times. Thus $\Omega_0 + \lambda_0 = 1$ at the present time t_0, the age of the observable universe. If one wishes, a_0 can be taken as the present radius of the observable universe.

We specifically seek to compare determinations of t_0 with independent determinations of the Hubble time, $T_0 = H_0^{-1}$, where H_0, the Hubble constant, is the present value of the Hubble parameter. For H_0 in units of km s^{-1} Mpc^{-1} one finds $T_0 = 976.8\ H_0^{-1}$ Gyr (sidereal). From the ratio t_0/T_0 or the product $H_0 t_0$ we will then determine Ω_0 and λ_0, always under the condition $\Omega_0 + \lambda_0 = 1$. This has been done by Turner, Steigman, and Krauss (1984) and Bludman (1984) in great generality, but we wish to present our own analysis of the current observational, experimental, and theoretical situation.

We note that $\Omega/\lambda = \rho/\rho_v$, where ρ_v is a constant independent of time after inflation. Thus in the early universe after inflation when ρ was extremely large, $\Omega/\lambda \gg 1$ or $\Omega \approx 1$ when $a \gtrsim 0$. It has been noted by a number of authors (e.g., Peebles 1984*a*) that it would be an extraordinary coincidence for λ_0 to be comparable to Ω_0 at just the present epoch. Moreover, if Ω differed slightly from unity at one time, it will differ greatly a short time later (Steinhardt 1984). Thus there is a prejudice, which we share, for $\Omega_0 \approx 1$, $\lambda_0 \approx 0$, but this remains to be seen. We also note ρ cannot be negative, so $\Omega \geq 0$ and $\lambda \leq 1$. However, λ can be negative, so Ω can be greater than unity. These statements also apply to Ω_0 and λ_0. For future reference, recall that the cosmological deceleration parameter is given by

$$q \equiv -\frac{\ddot{a}/a}{(\dot{a}/a)^2} = \frac{1}{2}\,\Omega - \lambda \Longrightarrow \frac{3}{2}\,\Omega - 1 = \frac{1}{2} - \frac{3}{2}\,\lambda \tag{7}$$

for $\Omega + \lambda = 1$ after all matter is nonrelativistic and radiation can be neglected. Note that $q = \Omega/2$ for $\lambda = 0$.

4.2. Evolution in the Inflationary Universe

The time variation or evolution of cosmological parameters in a flat universe has been studied for many years (e.g., Bondi 1952) and very recently by Bludman (1984) and Peebles (1984*a*). In all cases the solutions we are interested in apply to the time after the matter mass-energy density became greater than the radiation energy density. Since this occurred when the universe was about 5×10^5 yr old, we are safe in neglecting this short interval of time. Thus, for all but a short time period, it is assumed that $\rho/\rho_0 = (a_0/a)^3$. It is not a subject for discussion in this account, but we gratuitously note that big bang nucleosynthesis was completed by $t = 300$ s when the universe was radiation-dominated and ρ was proportional to a^{-4}.

Translating Bondi's notation into ours, the variation with time of the distance measure is given by

$$\text{I. } (a/a_0)^3 = \left(\frac{\Omega_0}{\Omega_0 - 1}\right) \sin^2[3(\Omega_0 - 1)^{1/2} H_0 t/2], \tag{8}$$

$$\text{for } \lambda_0 < 0,\ \Omega_0 > 1,\ q_0 > \tfrac{1}{2},$$

$$\approx \frac{9}{4}\Omega_0 {H_0}^2 t^2 \quad \text{for small } t$$

$$\text{II. } (a/a_0)^3 = \frac{9}{4}\, H_0^2 t^2 = 6\pi G \rho_0 t^2, \tag{9}$$

$$\text{for } \lambda_0 = 0,\ \Omega_0 = 1,\ q_0 = \tfrac{1}{2}.$$

$$\text{III. } (a/a_0)^3 = \left(\frac{\Omega_0}{1 - \Omega_0}\right) \sinh^2[3(1 - \Omega_0)^{1/2} H_0 t/2], \tag{10}$$

$$\text{for } 1 \geq \lambda_0 > 0,\ 0 \leq \Omega_0 < 1,\ -1 \leq q_0 < \tfrac{1}{2},$$

$$\approx \frac{9}{4}\Omega_0 {H_0}^2 t^2 \quad \text{for small } t.$$

Note that the distance measure can be arbitrarily chosen since a appears in a/a_0, where a_0 is the present value of a.

Case I represents a closed universe for which the distance measure reaches a maximum given by

$$a_{\max}/a_0 = \left(\frac{\Omega_0}{\Omega_0 - 1}\right)^{1/3} \tag{8a}$$

at

$$t = \frac{\pi}{3(\Omega_0 - 1)^{1/2} H_0} \tag{8b}$$

and returns to zero at

$$t = \frac{2\pi}{3(\Omega_0 - 1)^{1/2} H_0}. \tag{8c}$$

Bludman (1984) discusses this final crunch in detail. The canonical condition for case I, $q_0 > 1/2$, follows from equation (7).

Case II represents a flat, Euclidean universe in which the distance measure expands forever with $a \propto t^{2/3}$, but the expansion rate $\dot{a} \propto t^{-1/3}$ approaches zero. Again equation (7) yields the canonical condition for case II, $q_0 = 1/2$.

Case III represents an open universe with the distance measure eventually expanding exponentially with time according to

$$(a/a_0) \propto \exp[(1 - \Omega_0)^{1/2} H_0 t] = \exp[\lambda_0^{1/2} H_0 t]. \tag{10a}$$

Once again equation (7) yields the canonical condition for case III, $q_0 < 1/2$.

It is now appropriate to turn to the evolution of Ω, λ, and H. From equation (6) in Bludman (1984) one finds, for $\Omega_0 + \lambda_0 = 1$, that

$$\frac{\Omega}{\Omega_0} = \frac{1}{\Omega_0 + (1 - \Omega_0)(a/a_0)^3}. \tag{11}$$

Equation (11) can be rewritten as

$$\frac{\Omega_0}{\Omega} = \frac{1}{\Omega + (1 - \Omega)(a_0/a)^3}, \tag{11a}$$

which shows the extraordinary sensitivity of the present value of Ω, namely, Ω_0, to small deviations from unity of the value after inflation when $a_0 \gg a$. It seems most reasonable to take $\Omega_0 = \Omega = 1$ and have done with it, but we shall see. Equation (11) also leads directly to Bludman's equation (5):

$$\left(\frac{\lambda}{\lambda_0}\right)^{-1} = \left(\frac{H}{H_0}\right)^2 = (a_0/a)^3(1 - \lambda_0) + \lambda_0 = (a_0/a)^3\Omega_0 + 1 - \Omega_0. \tag{12}$$

Substitution of equations (8), (9), and (10) into (12) then yields equation (4) in Peebles (1984a). We write the equation in terms of parameters at the current

Table 4.1
Dependence of $t_0/T_0 = H_0t_0$ on λ_0, Ω_0, and q_0

λ_0	Ω_0	q_0	H_0t_0 $\lambda_0+\Omega_0=1$	H_0t_0 $\lambda_0=0$†	λ_0	Ω_0	q_0	H_0t_0 $\lambda_0+\Omega_0=1$	H_0t_0 $\lambda_0=0$†
1.00	0.00	−1.000	∞	—	0.30	.70	0.050	0.749	0.898
0.99	0.01	−0.985	2.006	—	0.20	.80	0.200	0.717	0.779
0.95	0.05	−0.925	1.490	—	0.10	.90	0.350	0.690	0.713
0.90	0.10	−0.850	1.278	—	0.00	1.00	0.500	0.667	0.667
0.80	0.20	−0.700	1.076	—	−0.33	1.33	1.000	0.605	0.571
0.74	0.26	−0.606	1.000*	—	−1.00	2.00	2.000	0.524	0.473
0.70	0.30	−0.550	0.964	—	−2.00	3.00	3.500	0.450	0.397
0.60	0.40	−0.400	0.888	—	−3.00	4.00	5.000	0.403	0.351
0.50	0.50	−0.250	0.831	—	−4.00	5.00	6.500	0.369	0.320
0.40	0.60	−0.100	0.786	—	−9.00	10.00	14.000	0.278	0.239
0.33	0.67	0.000	0.760	1.000	−∞	∞	∞	0.000	0.000

*For $H_0t_0 = 1$, $\lambda_0 = 0.737125$, $\Omega_0 = 0.262875$, $q_0 = -0.605688$.
†For $\lambda_0 = 0$, $k = 0$ or ± 1, $q_0 = \Omega_0/2 \geq 0$, do not use λ_0 and Ω_0 in this table; use only q_0.

epoch,

$$\text{IA.}\quad \frac{t_0}{T_0} = H_0t_0 = \frac{2}{3(\Omega_0 - 1)^{1/2}} \tan^{-1}(\Omega_0 - 1)^{1/2}$$

$$= \frac{2}{3|\lambda_0|^{1/2}} \tan^{-1}|\lambda_0|^{1/2} \qquad (13)$$

$$\text{for } \lambda_0 < 0,\ \Omega_0 > 1,\ q_0 > \tfrac{1}{2};$$

$$\text{IIA.}\quad \frac{t_0}{T_0} = H_0t_0 = \frac{2}{3}, \quad \text{for } \lambda_0 = 0,\ \Omega_0 = 1,\ q_0 = \tfrac{1}{2}; \qquad (14)$$

$$\text{IIIA.}\quad \frac{t_0}{T_0} = H_0t_0 = \frac{2}{3(1 - \Omega_0)^{1/2}} \tanh^{-1}(1 - \Omega_0)^{1/2}$$

$$= \frac{2}{3\lambda_0^{1/2}} \tanh^{-1}\lambda_0^{1/2} \qquad (15)$$

$$\text{for } 1 \geq \lambda_0 > 0,\ 0 \leq \Omega_0 < 1,\ -1 \leq q_0 < \tfrac{1}{2}.$$

As Peebles (1984*a*) makes clear, equations (13), (14), and (15) can be written in terms of $t/T = Ht$, Ω, and λ at any time, t, not just t_0. However, we are primarily concerned with the present values of these quantities for which equations (13), (14), and (15) hold. In Table 4.1 values for H_0t_0 are tabulated as a function of λ_0, Ω_0, and q_0. We will return to this table after the discussion, to follow, of independent determinations of H_0, the present value of the Hubble expansion parameter, and t_0, the present age of the observable universe.

In Table 4.1 we also give, for comparison, values of H_0t_0 for the canonical *big bang* cosmology with zero cosmological constant, $\lambda_0 = 0$, and arbitrary spacetime curvature, $k = 0$ or ± 1. In this case $q_0 = \Omega_0/2 \geq 0$, and the columns giving λ_0 and Ω_0 in the table are to be disregarded. The values for H_0t_0 for this cosmology are given in columns (5) and (10) of the table.

4.3. The Hubble Constant

It is appropriate to begin with the personal reminiscences of one of us

(W.A.F.).—"In my lifetime no 'constant' has changed in value so markedly as the Hubble constant. When I was a graduate student at Caltech in 1934, the great Edwin Hubble, himself, told me at breakfast in the Athenaeum, the faculty club, that his constant was $H_0 = 530$ km s^{-1} Mpc^{-1} corresponding to $T_0 = 1.84$ Gyr. He later published this value in, among other places, *The Realm of the Nebulae,* p. 170, Hubble, 1936. His value for T_0 taken to be the approximate age of the universe seemed incredible at the time but even then was under challenge by geologists who were convinced that there existed terrestrial rocks some 3 Gyr in age."

Hubble's analysis of his redshift observations is in principle still correct, but his expression $H_0 = (c/r)(\Delta\lambda/\lambda_0)$, where $\Delta\lambda$ is the redshift in light from a galaxy at proper distance r, has always been plagued by uncertainties in the distance scale for r. The wavelength of the light from a source at zero velocity with respect to the observer is λ_0 (not to be confused with λ_0 above). Baade, Humason, Mayall, Sandage, and Tammann have corrected Hubble's distance scale, always using observational techniques and analysis which improved with the years, so that Sandage and Tammann (1984) claim $H_0 = 50 \pm 7$ km s^{-1} Mpc^{-1} on the basis of their latest observations, or

$$T_0 = 19.5 \pm 2.7 \text{ Gyr}. \tag{16}$$

That is not all of the story. There is considerable controversy even today about the value of the Hubble constant as determined by classical redshift-distance observations. The value $H_0 = 100 \pm 10$ km s^{-1} Mpc^{-1}, $T_0 = 9.8 \pm 1$ Gyr has been claimed by de Vaucouleurs (1979) and $H_0 = 95(+15, -12)$ km s^{-1} Mpc^{-1}, $T_0 = 10.3 \pm 1.4$ Gyr by van den Bergh (1975). See Hodge (1981) for a review of determinations of H_0 over the period 1974–1980. There are controversies about the distance scale and about the velocity component of the local group of galaxies toward the Virgo cluster which is superimposed on the cosmological Hubble flow and about the velocity component of the local supercluster toward the supercluster in Hydra-Centaurus (see reports in *Science,* **227,** 622 (1985)). Our Galaxy, the Milky Way, is a member of the local group, which in turn is a member of the local supercluster. For the extent of the disagreement about the Virgocentric velocity see Davis and Peebles (1983).

A promising new technique as a distance indicator is the Tully-Fisher (T-F) (1977) relation for spiral galaxies. It relates the velocity width of the 21 cm line emission of neutral hydrogen from galaxies (i.e., essentially twice their maximum rotational velocity) to their absolute magnitude. Comparison of absolute magnitude with the observed apparent magnitude yields the distance to the galaxy. With the distance the observed redshift yields the Hubble constant. Calibration of the T-F relation is discussed in detail by van den Bergh (1984) using observations in blue light and in the infrared. Van den Bergh concludes that $H_0 = 68 \pm 17$ km s^{-1} Mpc^{-1}, $T_0 = 14.4 \pm 3.6$ Gyr from the T-F relation using observations in blue light and $H_0 = 82 \pm 18$ km s^{-1} Mpc^{-1}, $T_0 = 11.9 \pm 2.6$ Gyr from infrared observations. Van den Bergh (1984) suggests that these values can be averaged to yield $H_0 = 75 \pm 17$ (internal error) km s^{-1} Mpc^{-1}, or

$$T_0 = 13.0 \pm 3.1 \text{ Gyr}. \tag{17}$$

To summarize, H_0 varies from 43 through 75 to 110 km s^{-1} Mpc^{-1} and T_0 from 22 through 13 down to 9 Gyr. What a pity!

There is some hope, once again in the application of nuclear physics to cosmology. Little doubt remains that the light curve of Type I supernovae (Population II stars) is powered by the radioactive decay of ^{56}Ni through ^{56}Co to ^{56}Fe. Arnett, Branch, and Wheeler (1985) have investigated exploding carbon-oxygen white dwarf models for Type I supernovae. These models yield excellent agreement with the observations on the light curves in many details. The model calculations indicate that the mass of ^{56}Ni produced in the degenerate carbon-oxygen deflagration falls in the range $0.4 \leq M_{\text{Ni}}/M_\odot \leq 1.4$, where the limits are quite firm with $M_{\text{Ni}}/M_\odot = 0.6$ the best value.

Arnett, Branch, and Wheeler (1985) proceed to calculate the maximum absolute luminosity of their model Type I supernovae and find $L_{\text{max}} = 2.2 \times 10^{43}(M_{\text{Ni}}/M_\odot)$ ergs s^{-1}. In detail they find that this maximum occurs 17 days after the initial event. This is in good agreement with those cases where the initial event or the rise to the maximum luminosity has been observed.

The apparent luminosity is related to the absolute luminosity by $L_{\text{app}} \propto r^{-2}L_{\text{abs}} \propto r^{-2}M_{\text{Ni}}/M_\odot$, where r is the distance to the supernova. Thus measurement of the apparent luminosity at maximum permits determination of r and thus of the Hubble constant, $H_0 = v/r = cz/r$, if the redshift z of the galaxy in which the supernova occurs is known. Observations of six Type I supernovae in clusters of galaxies with known redshifts greater than $cz = 3000$ km s^{-1} are used to determine H_0. With redshifts this large, departures from the ideal Hubble flow, discussed previously in this section, introduce only small uncertainties. The result is $H_0 = 46(M_\odot/M_{\text{Ni}})^{1/2}$ km s^{-1} Mpc^{-1} so that from the $M_{\text{Ni}}/M_\odot$ values given above one finds a best value $H_0 = 59.4$ km s^{-1} Mpc^{-1} and a firm range $39 \leq H_0 \leq 73$ in the same units. These results yield the Hubble time expressed as $T_0 = 21.2(M_{\text{Ni}}/M_\odot)^{1/2}$ Gyr with a best value $T_0 = 16.4$ Gyr and a firm range $13 \leq T_0 \leq 25$ Gyr.

Careful study of the paper by Arnett, Branch, and Wheeler (1985) and of a more detailed paper by Branch (1984) convinces us that their range of values will be substantially reduced by further analysis but that their best value will not be appreciably changed. In particular it is to be noted that their upper limit, 25 Gyr, results from the value $M_{\text{Ni}} = 1.4\ M_\odot$, the Chandrasekhar maximum mass for a white dwarf. It is highly unlikely that all of the white dwarf mass is converted into ^{56}Ni, and Branch (1984) suggests an upper limit equal to $M_{\text{Ni}} = 1.0\ M_\odot$. In addition, Branch (1984) discusses the use of Type I supernovae as standard candles and the Baade-Wesselink method as well as the method based on ^{56}Ni radioactivity in Type I supernovae. He concludes, "It is likely that $H_0 \approx 50$–60 km s^{-1} Mpc^{-1} and the ^{56}Ni-method provides a rigorous lower limit, $H_0 \gtrsim 40$, and a firm upper limit, $H_0 \lesssim 70$ km s^{-1} Mpc^{-1}." We interpret this to mean $H_0 = 55 \pm 5$ km s^{-1} Mpc^{-1} for the value and standard deviation in H_0. Accordingly one has

$$T_0 = 17.8 \pm 1.6 \text{ Gyr}. \tag{18}$$

4.4. Ages of Globular Clusters and Galaxies

It is generally thought that galaxies could form only after matter decoupled from radiation in the universe at approximately 3×10^5 yr. The time for galaxy formation, t_g, is usually expressed in terms of the redshift parameter at formation, z, where $1 + z = a_0/a$. For Case II, the flat, Euclidean universe, one finds, as an example, from equation (9),

$$t_g = \frac{2/3}{H_0(1+z)^{3/2}\Omega_0^{1/2}}. \tag{19}$$

For $h_0 = H_0/100$, H_0 in km s^{-1} Mpc^{-1} one has

$$t_g = \frac{6.5}{h_0(1+z)^{3/2}\Omega_0^{1/2}} \text{ Gyr} \approx 1 \text{ Gyr}. \tag{19a}$$

From $z = \Delta\lambda/\lambda_0$, with $\Delta\lambda$ the redshift and λ_0 the laboratory wavelength, we note that z is the redshift for light reaching us now which left a galaxy when the universal distance measure was $a = a_0(1+z)^{-1}$. Parenthetically, it is useful that we can formally use this expression for the future with z approaching -1 in the distant future for cases II and III and with z returning to ∞ in the distant future for case I.

Peebles (1984*b*) states that the formation of large galaxies was nearly complete by $z \sim 4$. We chose $z = 4.5$, $\Omega_0 = 1$, and $h_0 = 1/2$ so that equation (19a) yields 1 Gyr for the time of galaxy formation. Note that this time becomes 2 Gyr for $\Omega_0 = 0.2$. Thus at least 1 Gyr must be added to the age of galaxies to obtain t_0, the age of the observable universe.

The oldest systems in galaxies are thought to be the globular clusters so that the ages of the globular clusters are taken to be the ages of the respective galaxies. The age of globular clusters depends on a number of quantities which must be determined by astronomical observations on the clusters. Iben and Renzini (1984) give the following equation:

$$\begin{aligned}\log t_9 = &-0.352 - 1.88(Y_I - 0.3) - 1.44(Y_{\rm HB} - Y_I) \\ &- 0.088(3 + \log Z) + 0.41\Delta M_{\rm TO}^{\rm RR}(\text{bol}),\end{aligned} \tag{20}$$

where t_9 is the globular cluster age in Gyr (10^9 yr), Y_I is the initial He abundance by mass on the main sequence, $Y_{\rm HB}$ is the He abundance by mass on the horizontal branch of stellar evolution, Z is the initial cluster heavy-element abundance by mass, and $\Delta M_{\rm TO}^{\rm RR}$(bol) is the difference in bolometric magnitude between the RR Lyrae variables in the cluster and the turnoff point for main-sequence stars in the cluster.

Iben and Renzini (1984) take M92 as representative of globular clusters in our Galaxy for which $\log Z = -3.89$ and $\Delta M^{\rm RR}_{\rm TO}(\rm bol) = 3.35$. They also argue that $Y_{\rm HB} = Y_I + 0.02$ on the basis of the first dredge-up episode on the red giant branch. Thus equation (7) reduces to

$$\log t_9 = 1.071 - 1.88(Y_I - 0.3). \tag{21}$$

This equation essentially expresses the fact that the larger the initial He abundance the less time required for the globular cluster's stars to turn off the main sequence when they have exhausted the hydrogen in their central core.

Y_I is frequently taken equal to Y_P, the primordial abundance produced in big bang nucleosynthesis. Since Wagoner, Fowler, and Hoyle (1967) many attempts have been made to determine Y_P from the abundances of ^{2}D, ^{3}He, ^{4}He, and ^{7}Li produced in the big bang. We agree with Iben and Renzini (1984) that these attempts yield $Y_P = 0.23 \pm 0.04$ with a rather large uncertainty. In addition, pregalactic and protogalactic nucleosynthesis can increase Y_I over Y_P. Thus it is preferable to turn to observations such as those of Kunth and Sargent (1983), who find what they term Y_P and which we term $Y_I = 0.245 \pm 0.003$ from high precision spectroscopy on He/H in H II regions in 12 emission-line galaxies. Kunth and Sargent (1983) call their Y_P the primordial or pregalactic helium abundance and show that other estimates which lead to a lower value of Y_P do not stand up to critical study, usually because of lack of attention to errors. For contrary results, see, for example, Janes and Demarque (1983).

We accept Kunth and Sargent (1983) and find from equation (21) that a best value for the age of M92 and thus for the age of our Galaxy and to a good approximation for other galaxies is

$$t_G = t(\rm M92) = 14.9 \pm 2.4 \ Gyr, \tag{22}$$

where the uncertainty is reduced from the ± 3.5 Gyr given by Iben and Renzini (1984) because the uncertainty in Y_P given by Kunth and Sargent (1983) is so small. However, see Paczyński (1984), who prefers a conservative estimate of 14 ± 4 Gyr.

From equations (19a) and (22) the result for the age of the universe as determined from globular cluster ages is

$$t_0(GC) = 15.9 \pm 2.4 \ \rm Gyr. \tag{23}$$

4.5. Nucleocosmochronology

Nucleocosmochronology uses the production and decay of radioactive nuclei as chronometers to determine the duration of nucleosynthesis in the Galaxy before the formation of the solar system. Addition of the age of the solar system to the duration time period yields the age of the Galaxy and then, through equation (19a), the age of the universe. Here we will neglect small additions (~0.1 Gyr) for the time of formation and evolution of the stars which first synthesized the chronometers in the Galaxy and for a possible interval of free decay between the end of nucleosynthesis and the formation of the solar system.

The most suitable chronometers are ^{232}Th(^{208}Pb) with mean lifetime $\bar{\tau}$ = 20.34 Gyr, ^{235}U(^{207}Pb) with $\bar{\tau}$ = 1.015 Gyr and ^{238}U(^{206}Pb) with $\bar{\tau}$ = 6.446 Gyr. The mean lifetimes have been calculated from the half-lives given in Lederer and Shirley (1978). The final decay products, all isotopes of Pb, are given in parentheses. Other chronometers, their decay products, and their mean lifetimes include ^{26}Al(^{26}Mg)1.033 Myr, ^{87}Rb(^{87}Sr)68.1 Gyr, ^{129}I(^{129}Xe)22.6 Myr, ^{187}Re(^{187}Os)65.8 Gyr, and ^{244}Pu(^{240}U and fission products ~0.1%)0.116 Gyr. Of these, we believe that the three short-lived chronometers can tell us little about the long duration of nucleosynthesis in the Galaxy and that there are many observational and theoretical problems involved in the use of the two long-lived ones.

This history of nucleocosmochronology and the current situation are discussed in Fowler (1984). The most recent calculations disagree with older calculations. Using ^{129}I, ^{232}Th, ^{235}U, ^{238}U, and ^{244}Pu, Thielemann, Metzinger, and Klapdor (1983) find $t_0 = 21.8(+2, -4)$ Gyr. Thielemann (1984) omitted ^{129}I and revised this to $t_0 = 18.9(+2, -4)$ Gyr. Even this latter value is in poor accord with the value given by Fowler (1977), $t_0 = 12(+3, -2)$ Gyr, using the same chronometers, also with the exception of ^{129}I, even though the values overlap when the uncertainties are taken into account.

Here we will discuss three models for nucleosynthesis in the Galaxy using observational input on the relative abundances, N_A, of ^{232}Th, ^{235}U, and ^{238}U plus theoretical calculations of their relative production, P_A, in the r-process involving rapid neutron capture and β-decay. ^{238}U is taken as the standard of comparison, and the mass number, A, will be designated by the last digit in the number when used as a subscript.

The input data reduce to

$$K_{28} = \frac{N_2/N_8}{P_2/P_8} \tag{24}$$

and

$$K_{58} = \frac{N_5/N_8}{P_5/P_8}, \tag{25}$$

where the quantities N_A are evaluated at the origin of the solar system 4.6 Gyr ago. The quantities P_A are taken as the average over all r-process events, presumably supernovae, which occurred in the Galaxy from its formation to that of the solar system. It will be clear that, early in the Galaxy before appreciable disintegration took place, $(N_2/N_8)_{\rm GF} = P_2/P_8$ and $(N_5/N_8)_{\rm GF} = P_5/P_8$, where the subscript GF designates Galaxy formation. It is taken that the r-process occurs in fairly massive stars, so the time for the formation and evolution of these stars is neglected (< 0.1 Gyr).

Since we restrict ourselves to two input data, our models can only involve two parameters to be evaluated by the use of equations (26), (27), and (28) given

Table 4.2
Abundance (N) and Production (P) Input Data
(N and P are ratios for ^{232}Th and ^{235}U relative to ^{238}U)

Case	A	N_0	Source	N	P	Source	$K = N/P$
A	232	4.00	WAF	2.457	1.80	WAF	1.365
	235	7.25(−3)	LED	0.330	1.42	WAF	0.232
B	232	3.75	FKT	2.303	1.39	TMK	1.657
	235	7.25(−3)	LED	0.330	1.24	TMK	0.266
C	232	3.75	FKT	2.303	1.80	WAF	1.280
	235	7.25(−3)	LED	0.330	1.42	WAF	0.232
D	232	4.00	WAF	2.457	1.784	CCT	1.377
	235	7.25(−3)	LED	0.330	1.516	CCT	0.217
E	232	3.75	FKT	2.303	1.784	CCT	1.291
	235	7.25(−3)	LED	0.330	1.516	CCT	0.217

Note: N calculated from N_0 for differential free decay over 4.6 Gyr. WAF: Fowler (1977); FKT: Thielemann (1984); LED: Lederer and Shirley (1978); CCT: Cowan, Cameron, and Truran (1985); and TMK: Thielemann, Metzinger, and Klapdor (1983).

below. For the derivation of these, see, for example, Fowler (1972). We give only one expression for each of our models. To obtain the necessary two equations, set K_{AS} (S for standard in general) equal to K_{28} and K_{58}. All models have a duration of nucleosynthesis equal to Δ in Gyr. It is the important parameter to be determined by solving the two equations for K_{28} and K_{58}.

Model I. Exponential synthesis (exp $-t/\tau_R$) with time constant, τ_R:

$$K_{AS} = \left[\frac{\tau_A(\tau_S - \tau_R)}{\tau_S(\tau_A - \tau_R}\right]\left[\frac{\exp(-\Delta/\tau_A) - \exp(-\Delta/\tau_R)}{\exp(-\Delta/\tau_S) - \exp(-\Delta/\tau_R)}\right]. \tag{26}$$

The mean lifetimes are written as τ_A and τ_S.

Model II. Early spike S_E plus uniform synthesis ($\tau_R = \infty$):

$$K_{AS} = \frac{S_E \exp(-\Delta/\tau_A) + (1 - S_E)(\tau_A/\Delta)[1 - \exp(-\Delta/\tau_A)]}{S_E \exp(-\Delta/\tau_S) + (1 - S_E)(\tau_S/\Delta)[1 - \exp(-\Delta/\tau_S)]}. \tag{27}$$

S_E is the fraction of synthesis in an initial "delta-function" contribution, and $1 - S_E$ is the fraction in uniform synthesis over the duration Δ. The initial fraction may have been "protogalactic" if it represents short-term r-process production in massive stars before the formation of the disk of the Galaxy in which the solar system is located.

Model III. Uniform synthesis ($\tau_R = \infty$) plus late spike S_L:

$$K_{AS} = \frac{(1 - S_L)(\tau_A/\Delta)[1 - \exp(-\Delta/\tau_A)] + S_L}{(1 - S_L)(\tau_S/\Delta)[1 - \exp(-\Delta/\tau_S)] + S_L}. \tag{28}$$

S_L is the fraction of synthesis in a final "delta-function" contribution, and $1 - S_L$ is the fraction in uniform synthesis. The final fraction may have been produced in a single supernova or in a short period burst of supernova activity. In any case the supernova or supernovae were near the solar nebula and may have triggered its collapse to form the solar system. These intriguing possibilities are discussed in Cameron and Truran (1977).

Table 4.2 shows the abundance (N_A/N_8) and production ($P = P_A/P_8$) input data for five cases, A to E, for which we have made calculations. The sources of the input data will be described in the following paragraphs.

Fowler (1977) took $(N_2/N_8) = 4.0 \pm 0.2$ on the basis of data given in Table 6 of his paper. Thielemann (1984) took $(N_2/N_8) = 3.75 \pm 0.00$ on the basis of measurements by Anders and Ebihara (1982). Correction for the differential lifetime in the free decay over the age of the solar system yields $(N_2/N_8) = (N_2/N_8)_0 \exp(-4.6/9.437) = 0.6142(N_2/N_8)_0$. In all cases we have used $(N_5/N_8)_0 = 7.253 \times 10^{-3}$ from Lederer and Shirley (1978). Correction for the differential lifetime in the free decay over the age of the solar system yields $(N_5/N_8) = (N_5/N_8)_0 \exp(+4.6/1.205) = 0.3297$.

Fowler (1977) took $(P_2/P_8) = 1.80 \pm 0.10$ and $(P_5/P_8) = 1.42 \pm 0.19$ on the basis of calculations of the relative abundances of the progenitors of ^{232}Th, ^{235}U, and ^{238}U produced in the *r*-process. These calculations incorporated modifications to those of Seeger, Fowler, and Clayton (1965). A substantial decrease ($\times$ 0.8) in P_5/P_8 was made to take into account the greater β-delayed neutron emission for odd-*A* nuclei taking part in the transition from the *r*-process path to the valley of stability.

Thielemann, Metzinger, and Klapdor (1983) introduced β-delayed fission as well as neutron emission into the *r*-process. Their results depend critically on the shape and heights of fission barriers determined by Howard and Möller (1980). They find $P_2/P_8 = 1.39$ and $P_5/P_8 = 1.24$ but do not explicitly state the uncertainties in these values. However, Thielemann (1984) has considered values as high as 1.50 and 1.30, respectively. It seems reasonable to take $P_2/P_8 = 1.39 \pm 0.1$ and $P_5/P_8 = 1.24 \pm 0.1$. A cursory study on our part of the Thielemann, Metzinger, and Klapdor (1983) calculations reveals that an effective cutoff at $A = 246$ in *r*-process calculations without β-delayed fission yields relative production rates after progenitor decay very close to theirs. This simplification will be used in what follows. Cowan, Cameron, and Truran (1985) have produced dynamical *r*-process calculations in which the classical waiting point method is abandoned and neutron-capture and β-decay rates are explicitly employed. In a major tour de force they also use the latest nuclear rates in their calculations. They show that, for a sufficient flux of neutrons produced by the $^{13}C(\alpha,n)^{16}O$ during helium burning, *r*-process nuclei in solar proportions can be produced. The amount of ^{13}C required is found to be exceedingly high, but this is characteristic of all *r*-process scenarios and will eventually be solved. Cowan (1984) has provided us with their detailed numerical calculations. Their odd-*A* to even-*A* production rates are quite high, and Cowan (1984) has provided an explanation for this. Nevertheless, in keeping with Fowler (1977), we have multiplied their P_5/P_8 by 0.8. They do not include β-delayed fission, so in calculating P_2/P_8 and P_5/P_8 we cut off the summation over progenitors (see Table 5, Fowler 1977) at $A = 246$. All the above results are summarized under N and P and the ratio $K = N/P$ in Table 4.2. Note that K_{28} varies from 1.280 to 1.657 and K_{58} from 0.217 to 0.266. The high values are characteristic only of case B.

4.6. Discussion of Results

Our results are displayed in Table 4.3. Cases A and D, model I, confirm

Table 4.3
Results of Nucleocosmochronological Calculations
(all ages in Gyr. S_E and S_L in fraction of total synthesis)

	Model I		Model II		Model III	
Case	τ_R	Δ	S_E	Δ	S_L	Δ
A	17.92	6.16	0.28	5.04	—	—
B	−3.92*	9.64*	—	—	0.047	13.27
C	2.37	3.87	0.57	3.08	—	—
D	9.42	6.01	0.38	4.78	—	—
E	2.08	3.89	0.62	3.10	—	—

*Including early spike $S_E = 0.49$ calculated by Thielemann (1984) for ^{187}Os(radiogenetic)/$^{187}Re = 0.13$ and for an effective ^{287}Re half-life equal to 35 Gyr.

the calculations of Fowler (1977) for exponential nucleosynthesis. We conclude that *the increase in Δ due to β-delayed fission found by Thielemann (1984) is counterbalanced by the explicit use of neutron-capture and β-decay rates by Cowan, Cameron, and Truran (1985).* The duration of nucleosynthesis is $\Delta =$ 6.1 Gyr yielding a galactic age of 10.7 Gyr and the age of the universe $t_0 = 11.7$ Gyr. For model II, cases A and D, $\Delta \approx 5$ Gyr and t_0 is correspondingly smaller. However, even for case A, the early spike $S_E = 28\%$ of the total nucleosynthesis in the Galaxy. This value is somewhat larger than the average heavy-element abundance (~20%) found in early disk stars relative to that in the solar system (Trimble 1982). Calculations for model III did not converge, and this is to be expected. Only rarely will both model II and model III give convergent solutions except in the trivial case, $S_E = S_L = 0$. Using the assumed uncertainties and using the dependence of Δ on K_{28} and K_{58} given in Table 3 of Fowler (1972), one finds a standard deviation of Δ equal to 2.5 Gyr so that

$$t_0 = 11.7 \pm 2.5 \text{ Gyr}. \tag{29}$$

We must concede the possibility that the disk of the Galaxy in which nucleosynthesis takes place may have formed several billion years after the nucleus of the Galaxy and the globular clusters were formed. Compare equations (23) and (29)! It is important to note that the value for Δ is essentially independent of the choice of input data for P by Fowler (1977) or Cowan, Cameron, and Truran (1985). However, comparison of case C, model I, with case A, model I, shows that Δ decreases rapidly with a decrease in (N_2/N_8) when (P_2/P_8) is held constant. This behavior is adequately covered by the standard deviations stated above.

For case B, that of Thielemann, Metzinger, and Klapdor (1983) as modified by Thielemann (1984), we were only able to find a convergent solution for model III with a late spike, $S_L = 4.7\%$ and $\Delta = 13.3$ Gyr, in exact agreement with the Thielemann (1984) modification of Thielemann, Metzinger, and Klapdor (1983). However, Thielemann (1984) kindly calculated results for a combined early spike plus exponential model. He found $\Delta = 9.64$ Gyr, $\tau_R = -3.92$ Gyr, and $S_E = 49\%$. The increasing rate of exponential synthesis compensated for the large early spike. This value falls just within the range of uncertainty in Δ expressed as +2, −4 Gyr by Thielemann (1984). Case E, model I, in comparison with case A, model I, shows the marked decrease in Δ when both K_{28} and K_{58} are substantially decreased. See Table 3 of Fowler (1972) for early calculations which showed the same effect. The result $\Delta = 3.89$ Gyr for case E, model I, is essentially the same as $\Delta = 3.87$ Gyr for case C, model I, and both give an age of the universe at the lower range shown in equation (29). Finally, we conclude that, given his input data, there is no reason to change the results of Thielemann (1984), who found $\Delta + \delta = 13.3 \pm 4.0$ Gyr, where δ is a small interval between the end of galactic synthesis and the formation of the solar system. This finally yields

$$t_0 = 18.9 \pm 4.0 \text{ Gyr}. \tag{30}$$

Equation (30) can be combined with equation (29) to yield $\bar{t}_0 = 13.7 \pm 3.$ Gyr with internal error 2.1 Gyr, external error 3.3 Gyr, and $\chi_2 = 2.4$.

Before leaving nucleocosmochronology it is necessary to comment on the ^{187}Re/^{187}Os chronology (Clayton 1964). For the situation late in 1983 the reader is referred to Fowler (1984). Since that time the cross section at 30 keV for $^{189}\mathrm{Os}^g(n,\gamma)^{190}\mathrm{Os}$ has been measured by Macklin and Winters (1984) to be 1.168 barns. The superscript g indicates the ground state of ^{189}Os which has spin and parity $3/2^-$ and Nilsson numbers 512. We assume that the excited state of $^{187}\mathrm{Os}^*$ at 9.75 keV with the same spin, parity, and Nilsson numbers has the same neutron-capture cross section as $^{189}\mathrm{Os}^g$. We use equation (5) of Clayton (1964) with current input parameters and modified by equation (1) of Woosley and Fowler (1979) to find that radiogenic ^{187}Os is *now* 12.9% of its parent ^{187}Re. Given our assumption this result is quite firm, the best available, and, moreover, independent of the lifetime of ^{187}Re, one key to the ^{187}Re/^{187}Os chronology. The remaining problem in the chronology is the lifetime of ^{187}Re under astrophysical circumstances. The lifetime for the 2.64 keV β-decay $^{187}\mathrm{Re}(e^-\nu)^{187}\mathrm{Os}$ is very sensitive to these circumstances, and the problem has been discussed in detail by Yokoi, Takahashi, and Arnould (1983). They point out that the reverse process, electron capture by ^{187}Os, can also occur in astrophysical environments. They conclude, and we concur, that the ^{187}Re/^{187}Os cannot yet be regarded as a truly reliable chronometer. We recommend, as Thielemann (1984) has done, using the Th/U chronometers to determine the effective lifetime for $^{187}\mathrm{Re} \Longleftrightarrow {}^{187}\mathrm{Os}$. Much can be learned about the overall astrophysical environment to which the products of nucleosynthesis are subjected in the stars and interstellar medium of the Galaxy.

4.7. Conclusion

In Table 4.4 we list our perhaps arbitrary choices of recent (~1984) determinations of the age of the universe t_0 and of the reciprocal of the Hubble constant T_0. Because of the large internal and external uncertainties in t_0 and T_0 it is difficult to draw conclusions. Nevertheless we make several attempts. Readers may draw their own conclusions.

For a minimum value for t_0/T_0 we divide the minimum t_0 by the maximum

Table 4.4
Age of the Universe (t_0) and Reciprocal of the Hubble Constant (T_0)

Method	Source	Result
Nucleocosmochronology	This work	$t_0 = 11.7 \pm 2.5$ Gyr
Age of Globular Clusters	This work using Iben & Renzini 1984 Kunth & Sargent 1983	$t_0 = 15.9 \pm 2.4$ Gyr
Nucleocosmochronology	Thielemann 1984	$t_0 = 18.9 \pm 4.0$ Gyr
Tully-Fisher Relation	van den Bergh 1984	$T_0 = 13.1 \pm 2.2$ Gyr
Redshift & SN I Luminosity	Arnett et al. 1984 Branch 1984	$T_0 = 17.8 \pm 1.6$ Gyr
Redshift Distance	Sandage & Tammann 1984	$T_0 = 19.5 \pm 2.7$ Gyr
Possibilities		
Missing Mass in Exotic Particles	$\Omega = 1 > \Omega_b \approx 0.2,\ \lambda_0 = 0$	$t_0/T_0 = 2/3$
Missing Mass in the Vacuum	$\Omega_0 \approx \Omega_b \approx 0.2,\ \lambda_0 \approx 0.8$	$t_0/T_0 \approx 1$

T_0 to obtain

$$(H_0 t_0)_{\min} = (t_0/T_0)_{\min} = 0.60 \pm 0.15. \tag{31}$$

This corresponds from Table 4.1 to the range in values for λ_0 given by

$$-2.0 \le (\lambda_0)_{\min} \le 0.30. \tag{32}$$

Note that $\lambda_0 = 0$, for which $H_0 t_0 = 2/3$ falls in this range. In this case $\Omega_0 = 1$, which is in excess of the universal baryon density $\Omega_b \approx 0.2$ from big bang nucleosynthesis and observations on luminous mass in galaxies (Yang et al. 1984). This leads to the so-called missing mass problem with exotic candidates for the missing mass such as "massive" neutrinos, axions, photinos, and other "inos."

For a maximum value for t_0/T_0 we divide the maximum t_0 by the minimum T_0 to obtain

$$(H_0 t_0)_{\max} = (t_0/T_0)_{\max} = 1.44 \pm 0.39. \tag{33}$$

This corresponds from Table 4.1 to the range in values for λ_0 given by

$$0.80 \le (\lambda_0)_{\max} \le 0.98. \tag{34}$$

Note that $\lambda_0 = 0.8$, for which $H_0 t_0 \approx 1$, falls within this range. In this case $\Omega_0 \approx \Omega_b \approx 0.2$ and the missing mass is in the vacuum with $\rho_v = 3\lambda_0/8\pi G T_0^2 \approx 10^{-29}$ g cm^{-3} for $\lambda_0 = 0.8$, $\Omega_0 = 0.2$, and $T_0 = 13.1$ Gyr $= 4.1 \times 10^{17}$ s. In the language of the inflationary model this means that most, but not all, of the false vacuum density before inflation was converted into radiation density during inflation. The small remaining constant vacuum density now dominates since radiation and relativistic particle densities have decreased as a^{-4} and nonrelativistic matter density as a^{-3}. As Peebles (1984*a*) has emphasized, it "does not seem plausible" that λ_0 and Ω_0 are of the same order of magnitude just at the present epoch. However, there is some comfort that the missing mass is in the vacuum and exotic particles can be exorcized!

Going further than many would care to go, we calculate $<t_0>$ and $<T_0>$ to obtain

$$<t_0> = 14.7 \pm 1.9, \tag{35}$$

and

$$<T_0> = 16.8 \pm 1.7, \tag{36}$$

where external errors are quoted so that

$$<H_0t_0> = <t_0/T_0> \equiv <t_0>/<T_0> = 0.87 \pm 0.14. \tag{37}$$

This corresponds from Table 4.1 to the range in values for λ_0 given by

$$-0.45 \leq \lambda_0 \leq 0.84, \tag{38}$$

using a spread of 2 standard deviations from equation (37).

Again, readers may draw their own conclusions. For our part we list two interesting possibilities at the bottom of Table 4.4. We conclude that the value $H_0t_0 = t_0T_0 = 2/3$ predicted by the inflationary model of an early exponential expansion followed by the canonical Friedmann-Robertson-Hubble expansion with $\lambda_0 = 0$ (zero cosmological constant) cannot be ruled out by current determinations for t_0 and T_0. In this case the missing mass, proportional to $1 - \Omega_b \approx 0.8$, must be in the form of exotic particles so far undetected or so far not completely understood. We also conclude that the value $H_0t_0 = t_0T_0 \approx 1$ cannot be ruled out. This value is consistent with $\Omega_0 \approx \Omega_b \approx 0.2$, $\lambda_0 \approx 0.8$, and the missing mass, $1 - \Omega_b \approx 0.8$, is in the vacuum. We emphasize that neither of these conclusions is strongly supported by the values of t_0 and T_0 in Table 4.4. However, we have a preference for $t_0 = 11.7 \pm 2.5$ Gyr and $T_0 = 17.8 \pm 1.6$ Gyr, which yields $t_0/T_0 = 0.66 \pm 0.15$. This is close to $t_0/T_0 = 2/3$, for which $\Omega = \Omega_0 = 1$.

In regard to the determination of the age of the observable universe, t_0, through nucleocosmochronology we emphasize that *detailed calculations for the r-process must be made incorporating rates of neutron capture and rates of β-decay, β-delayed emission, and β-delayed fission.* This will be quite a job! We come to the happy conclusion that there is more work for all involved in cosmological chronology.

Acknowledgments

We are grateful to F.-K. Thielemann for informative discussion and correspondence and for communicating results of his calculations prior to publication. We also thank J. J. Cowan for details on the production of the progenitors of ^{232}Th, ^{235}U, and ^{238}U resulting from the calculations described in Cowan, Cameron, and Truran (1985). This work was supported in part by the National Science Foundation, grant (PHY82-15500).

References

Anders, E., and Ebihara, M. 1982, *Geochim. Cosmochim. Acta,* **46,** 2363.

Arnett, W. D., Branch, D., and Wheeler, J. C. 1985, *Nature,* **314,** 337.

Bludman, S. A. 1984, *Nature,* **308,** 319.

Bondi, H. 1952, *Cosmology* (Cambridge University Press).

Branch, D. 1984, in *Lecture Notes in Physics,* ed. N. Bartel (Springer-Verlag). Also see Wagoner, R. V. 1982, in *Supernovae: A Survey of Current Research,* ed. M. J. Rees and R. J. Stoneham, p. 253 (Dordrecht: Reidel).

Burbidge, E. M.; Burbidge, G. R.; Fowler, W. A.; and Hoyle, F. 1957, *Rev. Mod. Phys.*, **29**, 547.

Cameron, A. G. W., and Truran, J. W. 1977, *Icarus*, **30**, 477.

Clark, D. H., and Stephenson, F. R. 1977, *The Historical Supernovae* (Pergamon).

Clayton, D. D. 1964, *Ap. J.*, **139**, 637.

Cowan, J. J. 1984, private communication.

Cowan, J. J., Cameron, A. G. W., and Truran, J. W. 1984, in *Stellar Nucleosynthesis,* ed. C. Chiosi and A. Renzini (Dordrecht: Reidel).

——. 1985, *Ap. J.*, **294**, 656.

Davis, M., and Peebles, P. J. E. 1983, *Ann. Rev. Astr. Ap.*, **21**, 109.

de Vaucouleurs, G. 1979, *Ap. J.*, **233**, 433.

Fowler, W. A. 1972, in *Cosmology, Fusion and Other Matters,* ed. F. Reines, p. 67 (Boulder: Colorado Associated University).

——. 1977, in *Proceedings of the Welch Foundation Conferences on Chemical Research XXI, Cosmochemistry,* ed. W. D. Milligan, p. 61 (Houston: Robert A. Welch Foundation).

——. 1984, *Rev. Mod. Phys.*, **56**, part 1, 149.

Guth, A. H. 1981, *Phys. Rev.*, **D23**, 347.

Guth, A. H., and Steinhardt, P. J. 1984, *Sci. Am.*, **250**, 116.

Hodge, P. W. 1981, *Ann. Rev. Astr. Ap.*, **19**, 357.

Howard, W. M., and Möller, P. 1980, *Atomic Data Nucl. Data Tables,* **25**, 219.

Hubble, E. 1936, *The Realm of the Nebulae* (New Haven: Yale University Press) (reprint 1958, Dover Press).

Iben, I., Jr., and Renzini, A. 1984, *Phys. Reports,* **105**, 331.

Janes, K., and Demarque, P. 1983, *Ap. J.*, **264**, 206.

Kunth, D., and Sargent, W. L. W. 1983, *Ap. J.*, **273**, 81.

Lederer, C. M., and Shirley, V. S. 1978, *Table of Isotopes,* 7th ed. (Wiley).

Macklin, R. L., and Winters, R. R. 1984, private communication.

Paczyński, B. 1984, *Ap. J.*, **284**, 670.

Peebles, P. J. E. 1984*a*, *Ap. J.*, **284**, 439.

——. 1984*b*, *Science,* **224**, 1385.

Sandage, A., and Tammann, G. A. 1984, *Nature,* **307**, 326.

Seeger, P. A., Fowler, W. A., and Clayton, D. D. 1965, *Ap. J. Suppl.*, **11**, 121.

Steinhardt, P. J. 1984, *Comments Nucl. Part. Phys.*, **A12**, 273.

Thielemann, F.-K. 1984, in *Stellar Nucleosynthesis,* ed. C. Chiosi and A. Renzini (Dordrecht: Reidel), and private communication.

Thielemann, F.-K., Metzinger, J., and Klapdor, H. V. 1983, *Zs. Phys.*, **A309**, 301.

Trimble, V. 1982, in *Extraterrestrials: Where Are They?* ed. M. H. Hart and B. Zuckerman (Pergamon).

Tully, R. B., and Fisher, J. R. 1977, *Astr. Ap.*, **54**, 661.

Turner, M. S., Steigman, G., and Krauss, L. M. 1984, *Phys. Rev. Letters,* **52**,

2090.

van den Bergh, A. 1975, *Galaxies and the Universe,* p. 509 (Chicago: University of Chicago Press).

——. 1984, *Quart. J.R.A.S.,* **25,** 137, and private communication.

Wagoner, R. V., Fowler, W. A., and Hoyle, F. 1967, *Ap. J.,* **148,** 3.

Woosley, S. E., and Fowler, W. A. 1979, *Ap. J.,* **233,** 411.

Yang, J.; Turner, M. S.; Steigman, G.; Schramm, D. N.; and Olive, K. A. 1984, *Ap. J.,* **281,** 493.

Yokoi, K., Takahashi, K., and Arnould, M. 1983, *Astr. Ap.,* **117,** 65.

5

Isotopic Anomalies and SUNOCON Survival during Galactic Evolution

Donald D. Clayton

Abstract

The equations for the chemical evolution of the Galaxy are enlarged to include the fate of refractory supernova condensates (SUNOCONs), which are of potentially great significance as carriers of isotopic anomalies of the refractory elements. I first demonstrate that the differential equations for their concentration are, in the linear model, formally identical to those of the abundances of radioactive nuclei, where the decay rate is determined by the combined effects of incorporation into stars (astration) and interstellar destruction (sputtering). Analytic solutions in the face of gaseous infall onto the galactic disk are presented as a potentially useful "standard model" of chemical evolution, in which the SUNOCON abundances can be explicitly studied. I find that roughly 40% of interstellar Al resided in SUNOCONs at the time of formation of the solar system. The largest uncertainty is in the sputtering destruction rate, which I argue to be at least 10 times smaller than the corresponding rate for unshielded dust components. Finally, I present possible applications to the problems of the observed excesses of ^{16}O, of ^{26}Mg, and of ^{50}Ti in Al-rich meteoritic inclusions.

5.1. Introduction

It has been commonplace in the astrophysical literature for the phrase "chemical evolution of the Galaxy" to refer in fact to the evolution of the bulk abundances of the elements. The evolution of actual chemical structures, on the other hand, has more often been the province of the astrophysicists' search for the origins of observed molecules or dust components. But it is clear that in principle the chemical evolution must involve the evolution of chemical objects. Addressing this for the moment in a personal vein, I note that A. G. W. Cameron, whose 60th birthday we celebrate by this book, has constantly perceived the origin of the solar system, and solid bodies within it, as an organic event in the physical and chemical evolution of the Galaxy. His leadership played a big role in my own thinking about the fate of cosmic dust. It was exactly 10 years ago, at the conference "Frontiers of Astronomy" in Venice in honor of the 60th birthday of another pioneer in cosmic dust, Fred Hoyle, when I first described the need to include the supernova condensates (SUNOCONs) in the equations for the chemical evolution of the Galaxy. Only about seven months before that conference I had realized (Clayton 1975*a*, *b*) that many isotopic anomalies in the meteorites might depend upon the evolutionary fate of that component of the interstellar dust. The connection has never been transparent, because both the meteorites and the Galaxy are complicated and poorly understood. Despite a certain element of prematurity I laid out the basic working framework and its confrontation with then established dogma among meteoritic scientists at the "Protostars and Planets" meeting in Tucson. In those years there were few adherents of the viewpoint I nonetheless published (Clayton 1978). Many of those arguments have now become generally accepted, however, and nothing has played a larger role in their acceptance than A. G. W. Cameron's opinion that they, and not an inhomogeneous admixture from a neighboring supernova trigger to the origin of the solar system (Cameron and Truran 1977), are the true causes of the bulk of the isotopic anomalies (e.g., Consolmagno and Cameron 1980). It is therefore fitting that the exact analytic models that I now present, as well as my speculations about their possible roles in the history of galactic dust and in meteoritics, be presented in this volume in tribute to A. G. W. Cameron's 60th birthday. His influence on these scientific areas and on my own contributions to them has been paramount.

Before turning to the mathematical models it is advisable to motivate them with a simple physical description. Consider to that end the ^{16}O richness of the CaAl-rich inclusions of carbonaceous meteorites. These inclusions had initially a 5% excess of ^{16}O (R. N. Clayton et al. 1977), although some of the minerals within them (e.g., melilite) have lost that initial ^{16}O richness by diffusive exchange with normal ambient oxygen. We may speculate, considering that the CaAl-rich inclusions are dominated by their huge Al abundance, that the ^{16}O richness has something to do with their Al richness. Although that speculation is not required by known facts, it is also supported by a plausible theoretical argument. Clayton (1977*a*, *b*) argued at some length that aluminum begins its galactic existence in ^{16}O-rich solids because the carbon-burning supernova shells that created the aluminum have also bathed it in pure ^{16}O during the expansion of the supernova interior and because the Al cannot escape in gaseous form owing to the highly refractory sequences $Al + O \rightarrow Al_2O_3$ and $Al_2O_3 + MgO \rightarrow MgAl_2O_4$ (spinel). Figure 5.1 illustrates this sequence schematically. It may or may not be a coincidence that spinel, $MgAl_2O_4$, remains today the most ^{16}O-rich mineral to be found within the CaAl-rich inclusions, but I continue to believe, eight years after this initial explanation was advanced, that a genetic line does exist even though the structural alteration may have been great.

But if this speculation be correct, it immediately confronts the problem of galactic evolution. Dust cannot survive forever in its initial form. Much of the interstellar aluminum was synthesized during early galactic epochs and would have been particularly vulnerable to destruction by incorporation into later forming stars (astration) or by nonthermal destruction within the interstellar medium

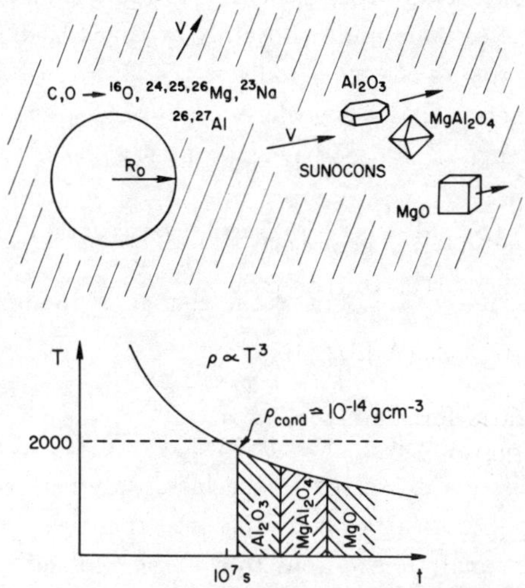

Figure 5.1. Supernova condensates (SUNOCONs) condense thermally while the interior cools by expanding (Hoyle and Wickramasinghe 1970). Shown here is a shell that explosively burned carbon, creating Mg and Al isotopes in a bath of pure ^{16}O. The first condensate is corundum, Al_2O_3, which is partially or totally converted to spinel, $MgAl_2O_4$, later in the expansion (Clayton 1977*b*). This condensation begins after a few months when the remnant is only about 100 AU in size, long before the gases can have mixed with circumstellar matter. This freezes the correlation of ^{16}O with Al. Figure from Clayton (1982*a*).

itself (sputtering). Both processes erase the remembered correlation of Al with ^{16}O, because even if, as is likely, the Al recondenses thermally upon reinjection from stars following astration of the initial Al dust, the new Al stellar condensates (STARDUST) will not be ^{16}O-rich; and similarly for recondensation (sticking) of sputtered Al atoms. Thus a clear question of relevance is, *What fraction of the SUNOCONs escaped destruction during galactic evolution?* And actually the question must be formulated more rigorously than that: viz, if the fraction of all aluminum that began its interstellar life as an Al-rich SUNOCON is $\beta^0_{\rm Al}$, what is the fraction of *interstellar* aluminum at time t, $\beta_{\rm Al}(t)$, that remains in its initial SUNOCON? Since only that Al fraction carries pure ^{16}O, the ^{16}O richness of the bulk of interstellar Al is also equal to $\beta_{\rm Al}(t)$. With this example in mind we now address the mathematical and physical problem.

5.2. A Specific Mathematical Problem

Consider a one-zone model of the chemical evolution of the Galaxy, allowing, however, its mass to increase as matter falls onto the galactic disk. Concentrate on one particular kind of SUNOCON, and let $S_i (i = 1, 2 \ \dots)$ represent the mass fraction of interstellar matter contained in that SUNOCON. Then

$$M_i = S_i M_G. \tag{1}$$

where M_G is the total mass of the interstellar material. This concept is in exact analogy to the usual mass of element Z, $M(Z) \equiv X_Z M_G$. One may equally well concentrate only on a specific element, for example Al, and let S_{Al} be the mass fraction of interstellar aluminum contained within SUNOCONs, in which case equation (1) still applies. In what follows, therefore, I drop the index i and let S be the mass fraction under consideration. We may then construct the terms for the rate of change of that SUNOCON mass:

$$\frac{d}{dt}(SM_G) = \text{creation term} - \text{destruction terms}. \tag{2}$$

In so doing we must keep account of the change of the mass of the interstellar material

$$\frac{dM_G}{dt} = -\psi(1-R) + f, \tag{3}$$

where ψ is the rate at which mass is turned into stars, R is the return fraction given back, herein expressed in the instantaneous cycling approximation, and $f(t)$ is the rate at which mass falls onto the disk. With the objective of constructing exact analytic solutions to a posed problem, I will follow Clayton (1984*b*, 1985*a*, *b*), who showed how this can be done for the elemental abundances within the context of linear models of star formation, within which the star formation rate ψ is proportional to the mass of the interstellar matter. Defining the linear rate coefficient ω by

$$\psi(1-R) \equiv \omega M_G, \tag{4}$$

the total ISM mass can be written

$$M_G(t) = M_{G0}\exp(-\omega t + \theta(t)), \tag{5}$$

where

$$\theta(t) \equiv \int_0^t \frac{f(t')}{M_G(t')}dt'. \tag{6}$$

Now turn to the creation and destruction terms for the SUNOCON in question. It is sensible to assume that incorporation into stars destroys that SUNOCON:

$$\text{astration rate} = S\psi = S\omega M_G/(1-R) \tag{7}$$

is one destruction term for equation (2). The destruction of SUNOCONs also occurs in the ISM itself by sputtering and grain-grain fragmentation in the wake of interstellar shock waves. If the average SUNOCON lifetime (when averaged over separate phases of the ISM) is designated by $\tau_S \equiv \lambda_s^{-1}$, then a second destruction term for equation (2) is

$$\text{sputtering rate} = \lambda_s S M_G.$$

As implied, λ_s is a rather complicated average over all ISM phases, but in this global treatment I take λ_s to be the appropriate average over those phases. And although λ_s is not necessarily a constant, I take it to be so on the grounds that the supernova rate, which does the sputtering, is proportional to M_G via the stellar birthrate, but it must also be absorbed by M_G. Thus the supernova rate per unit mass of ISM is constant by construction. Since it is not implausible that λ_s be constant in such a picture, I take it to be so. It also allows an analytic mathematical solution. The sputtering rate λ_s does depend upon the type of SUNOCON, however. Aluminum is less gaseous (more depleted) than is magnesium specifically because it has greater shielding against sputtering. Thus $\lambda_s = \lambda_s(S_i) = \lambda_{si}$ in a full description.

The creation term in equation (2) is easy only because it is SUNOCONs that are being described. They are, by definition (Clayton 1978), condensed during the expansion and cooling of the supernova interior, so their creation rate is proportional to the stellar birthrate. It is convenient to use the "yield" concept common to studies of the evolution of galactic abundances. Let y_S be the yield of S-type SUNOCONs per unit mass of matter locked into stars. Then the creation term in equation (2) is

$$\text{creation} = y_S(1-R)\psi = y_S\omega M_G. \tag{8}$$

It will frequently be convenient in what follows to think of

$$y_S \equiv y_Z\beta_Z^0 \tag{9}$$

as the product of the yield of element Z and the fraction β_Z^0 of that element that initially condenses into SUNOCONs. For aluminum we will have $\beta_{\text{Al}}^0 \simeq 1$. Finally, there may in principle be a nonnegligible concentration S_f of SUNOCONs in the matter falling onto the disk of the Galaxy.

Writing equation (2) explicitly now:

$$\frac{d}{dt}(SM_G) = y_S\omega M_G - \lambda_s SM_G - S\omega M_G/(1-R) + S_f f \tag{10}$$

describes, along with equation (3), the evolution of a specific SUNOCON. It can be cast into a more useful form by taking the chain rule derivative on the left-hand side and substituting equation (3) for dM_G/dt. This operation yields

$$\frac{dS}{dt} = y_S\omega - \frac{\omega R}{1-R}S - \lambda_s S - (S - S_f)\frac{f(t)}{M_G(t)}, \tag{11}$$

or, defining the effective destruction rate

$$\lambda \equiv \lambda_s + \frac{\omega R}{1-R}, \tag{12}$$

we have more simply

$$\frac{dS}{dt} = y_S\omega - \lambda S - (S - S_f)\frac{f}{M_G}. \tag{13}$$

It is interesting that this equation is identical to that of the concentration Z of a radioactive nucleus (see Clayton 1984b, equation [10]):

$$\frac{dZ}{dt} = y_Z\omega - \lambda Z - (Z - Z_f)\frac{f}{M_G} \tag{14}$$

except that the radioactive decay rate λ is for the SUNOCON replaced by the sum of the sputtering rate λ_s and the astration rate $\omega R/(1-R)$, as displayed in equation (12). This means that all of the results obtained for radioactivity concentrations (e.g., Clayton 1984*b*, §§ *g* and *h*) can be taken over directly to SUNOCON concentrations. The physical problem is the evaluation of equation (12). In particular, I define a "remainder" in analogy to the radioactive remainder; it is the ratio of the SUNOCON concentration $S(t)$ to the concentration of the stable element $Z(t)$ from which the SUNOCON is built. Call this SUNOCON remainder r_S:

$$r_S \equiv S(t)/Z(t). \tag{15}$$

The good physical example is aluminum, for which $S_{\rm Al}(t)$ is the mass fraction of the ISM in the form of Al in SUNOCONs, whereas $Z_{\rm Al}(t)$ is the mass fraction of Al in the ISM. The initial value of this SUNOCON remainder has simple physical significance which can be seen by assuming that $S_f = 0$ and $Z_f = 0$ in the matter falling onto the Galaxy. Then at short times the ratio of equations (13) and (14) gives

$$\frac{dS}{dZ} = \frac{dS/dt}{dZ/dt} \underset{t \to 0}{\to} \frac{y_S}{y_Z} = \beta_Z^0, \tag{16}$$

where the last equality comes from equation (9) for the fraction β_Z^0 of the element that initially condenses into SUNOCONs. The value of β_Z^0 may be as high as unity for a superrefractory element like aluminum that condenses totally during supernova expansion. Thus the SUNOCON remainder begins at the value $r_S = \beta_Z^0$ and monotonically declines as the Galaxy evolves. It is the exact models for this monotonic decline that I will now discuss.

The discovery of exact analytic families of solutions (Clayton 1984*b*, 1985*a*) of equation (13) allows those solutions to conveniently serve as standard models for problems in the chemical evolution of galaxies. In this work I will use only the simplest family, those with continuous infall described by (Clayton 1985*a*)

$$f(t) = \frac{kM_{G0}}{\Delta}\left(\frac{t+\Delta}{\Delta}\right)^{k-1} e^{-\omega t}, \tag{17}$$

where k is any positive integer, ω is defined as in equation (4), and Δ is a time parameter whose value is the product of the integer k and the initial time scale for infall to double the total disk mass. This infall rate is pure exponential decline for $k = 1$, but for $k > 1$ can first increase before declining monotonically toward zero. Many other infall rates can also generate exact parameterized families of solutions, as Clayton (1984*b*) described, but this simple family will suffice for this initial study. In fact, I will for numerical evaluation use models having $k = 1$.

The mass of the interstellar medium from equation (5) becomes

$$M_G(t) = M_{G0}\left(\frac{t+\Delta}{\Delta}\right)^k e^{-\omega t}, \tag{18}$$

where M_{G0} is the initial disk mass. The solution of equations (14) and (13) are, respectively,

$$(S - S_f) = (y_S\omega - \lambda S_f)e^{-\lambda t}\left(\frac{\Delta}{t+\Delta}\right)^k I_k(t,\lambda), \tag{19}$$

where the function

$$I_k(t,\lambda) \equiv \int_0^t \left(\frac{t'+\Delta}{\Delta}\right)^k e^{\lambda t'}\,dt', \tag{20}$$

and for the stable nucleus

$$(Z - Z_f) = \frac{y_Z\omega\Delta}{k+1}\left[\frac{t+\Delta}{\Delta} - \left(\frac{t+\Delta}{\Delta}\right)^{-k}\right]. \tag{21}$$

The ratio μ of the interstellar mass to the total mass (gas fraction) is

$$\mu = \frac{M_G}{M_{\rm tot}} = \left(\frac{t+\Delta}{\Delta}\right)^k e^{-\omega t}\left[1 + \frac{k}{\Delta}I_{k-1}(t,-\omega)\right]^{-1}. \tag{22}$$

5.2.1 The Case $k = 1$

These general results take a simpler form for $k = 1$, for which the infall declines exponentially

$$f(t) = \frac{M_{G0}}{\Delta}e^{-\omega t}, \tag{23}$$

$$\mu = (t+\Delta)\omega e^{-\omega t}[\Delta\omega + 1 - e^{-\omega t}]^{-1}, \tag{24}$$

$$Z - Z_f = \frac{y_Z\omega\Delta}{2}\left[\frac{t+\Delta}{\Delta} - \frac{\Delta}{t+\Delta}\right], \tag{25}$$

and

$$S - S_f = \frac{y_S\omega - \lambda S_f}{t+\Delta}\left\{\frac{t+\Delta}{\lambda} - \frac{1}{\lambda^2} - e^{-\lambda t}\left[\frac{\Delta}{\lambda} - \frac{1}{\lambda^2}\right]\right\}. \tag{26}$$

It is furthermore plausible that the gas falling onto the disk is so metal-poor and SUNOCON-poor that we may take $S_f = 0$ and $Z_f = 0$. In this case the SUNOCON remainder for $k = 1$ is simply

$$r_S = \frac{S(t)}{Z(t)} = 2\beta_Z^0\,\frac{\lambda(t+\Delta) - 1 - e^{-\lambda t}(\Delta\lambda - 1)}{\lambda^2 t(t+2\Delta)}, \tag{27}$$

which has the required limits $r_S \to \beta_Z^0$, the initial SUNOCON fraction at early time, and $r_S \to 0$ at very long times.

To ascertain likely values of r_S for our Galaxy we need to determine parameters for $k = 1$ models that are appropriate for it. Table 5.1 lists values of ω, Δ, and galactic age T that give two often used values for today's gas fraction, $\mu = 0.091$ or $\mu = 0.19$, at the solar galactocentric radius. These models are also not in substantial disagreement with other basic observational tests discussed by Clayton (1984*b*, 1985*a*). We will therefore utilize them as reasonable ones for the evaluation of SUNOCON remainders and the associated isotopic anomalies

Table 5.1
Standard Linear Models ($k = 1$, $R = 0.5$)*

	$T = 15$		$T = 10$	
$\mu = 0.091$	$\Delta = 0.1$	$\omega = 0.248$	$\Delta = 0.1$	$\omega = 0.370$
	$\Delta = 1$	$\omega = 0.236$	$\Delta = 1$	$\omega = 0.346$
$\mu = 0.19$	$\Delta = 0.1$	$\omega = 0.181$	$\Delta = 0.1$	$\omega = 0.271$
	$\Delta = 1$	$\omega = 0.172$	$\Delta = 1$	$\omega = 0.252$

*Time T and Δ in 10^9 yr; rates ω in $(10^9\,\text{yr})^{-1}$.

that may be carried along with them. The model listed in Table 5.1 used the value $R = 0.5$ for the return fraction, as argued by Tinsley (1981). However, the SUNOCON remainders depend somewhat on its value for those types of dust that are destroyed primarily by astration.

There is a very interesting feature of equation (27) for the SUNOCON remainder. The astrophysical uncertainties in the sputtering rate λ_s, in the gas decay rate ω, and in the return fraction R are all contained in the effective SUNOCON decay rate λ (equation [12]). The other parameters, the galactic age t and the time parameter Δ, are fixed once one chooses one of the family of standard models for the chemical evolution of the Galaxy. Thus, if we take $T = 15$ for the galactic age and $\Delta = 1$, and if the exponential term can be neglected because $\lambda T > 1$, then equation (27) reduces numerically to

$$r_S = 0.125\beta_Z^0\left(\frac{\lambda - 0.0625}{\lambda^2}\right), \qquad \begin{array}{c}\Delta = 1\\ t = 15\,,\\ \lambda t > 1\end{array} \tag{28}$$

where λ is expressed in units $(10^9\ \text{yr})^{-1}$. Equation (28) is true for any combination of λ_s, ω, R, all of which enter only in the determination of λ. The other unknown, β_Z^0, is the fraction of the newly synthesized supernova nucleosynthesis that condenses into SUNOCON upon ejection. Figure 5.2 displays the ratio r_S/β_Z^0 calculated from equation (28). It is clear that for this quite reasonable model of galactic evolution, the fraction of SUNOCONs surviving is about half for small $\lambda \sim 0.1$ and declines to less than 10% for $\lambda > 1$.

5.3. The Destruction Rate λ

The most effective shocks at destroying grains move with velocities 30–400 km s^{-1} through diffuse clouds ($n_{\text{H}} \sim 10$ cm^{-3} or through intercloud matter ($n_{\text{H}} \sim 0.2$ cm^{-3}). Magnetic fields of 1–3 μG enhance grain-grain collisions via betatron acceleration of charged grains. Shocks slower than 30 km s^{-1} are too infrequent (as viewed by an arbitrary dust grain in the ISM). These conditions are in the range of those frequently encountered in the ISM, so the partial destruction of dust mass is expected to be a significant occurrence there. Indeed, interstellar sputtering is, along with astration in stars, the major dust destruction mechanism. The physics of grain destruction in shocks has been addressed by many workers, including especially Barlow (1978*a*, *b*, *c*), Shull (1977, 1978), Cowie (1978), Draine and Salpeter (1979), and more recently Seab and Shull (1983). The source of the shocks does not much matter insofar as the sputtering

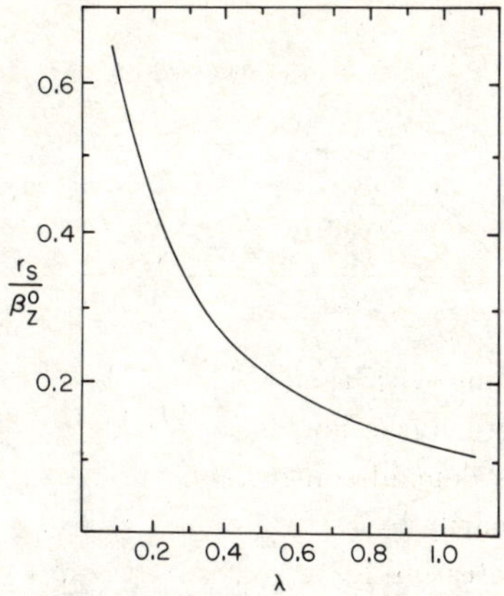

Figure 5.2. The remainder of a superrefractory element contained within SUNOCONs is shown as a function of their destruction rate λ $(10^9\ \mathrm{yr})^{-1}$. This effective SUNOCON destruction rate λ depends according to equation (12) on the sputtering rate λ_s, on the gas consumption rate ω, and on the return fraction R.

physics is concerned, but the most important source is believed to be supernova remnant blast waves. Cloud-cloud collisions, stellar wind shocks, Herbig-Haro objects, Wolf-Rayet stars, and pre-main-sequence outflows in molecular clouds may all play some role, but the evidence is that supernova shocks are the major objects to be considered.

It is important for the topic of this paper to understand why this situation exists, what paradoxes presently face it, and what new considerations may change it, because the evaluation of the sputter rate λ of the SUNOCON's is the major uncertainty. If the mean interval between galactic supernovae if 50 years and if they occur randomly, the average time before an arbitrary interstellar grain experiences a strong shock is approximately the supernova interval divided by the fraction of the galactic volume occupied by the supernova remnant. Evaluating gives about $t_{\mathrm{shock}} \simeq 10^8$ yr. This average intershock interval is not much different in a standard three-phase model of the ISM (McKee and Ostriker 1977). If each shock totally destroyed each grain, the answer would then be 10^8 yr. But the real problem is much more formidable, requiring detailed modeling of the occurrence and evolution of blast waves, of the structural nature of ISM particles, of the rate of erosion of each type of particle in that environment, of the rate of transmutation of each ISM phase to another, of the fragmentation and vaporization attendant to grain-grain collisions, and of even other complications. Several groups have attempted this to some degree. Dwek and Scalo (1980) find average grain lifetimes of 10^9 yr, Draine and Salpeter (1979) find 10^8 yr, near the shock occurrence interval; and Seab, Hollenbach, McKee, and Tielens (in preparation) have more recently found 5×10^8 yr, nearly independent of grain size. In what follows, this last number will be adopted, leading to $\lambda_s \simeq 2(10^9\ \mathrm{yr})^{-1}$.

Severe paradoxes result from the indiscriminate adoption of this number,

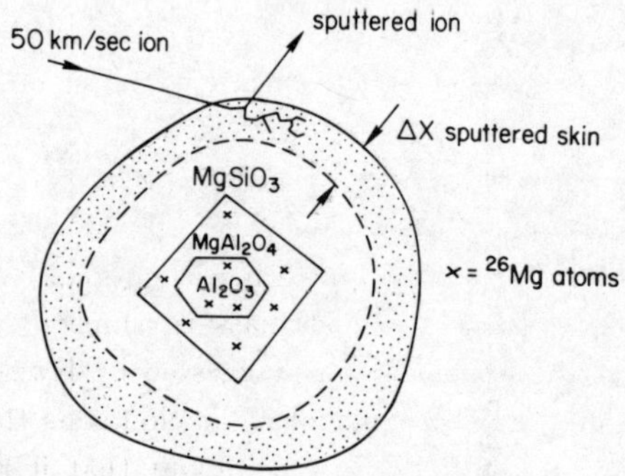

Figure 5.3. The first Al-bearing SUNOCONs are subsequently surrounded by Mg- and Si-bearing SUNOCON mantles, shown here as $MgSiO_3$. More volatile mantles are later added in dense clouds. These mantles protect Al from sputtering more than they protect Mg, explaining the low Al/Mg ratio in interstellar gas. Taken from Clayton (1982*a*).

however. It would require that sputtering be much more important than astration, because equation (12) evaluated with return fraction $R \simeq 0.5$ and an appropriate value of $\omega \simeq 0.2$ (see Table 5.1) would then give $\lambda \simeq \lambda_s + \omega \simeq 2 + 0.2 \simeq \lambda_s$. And yet that does not seem to be the case, as can be seen from two points of view. First, far less Al resides in the gas phase than does Mg or Si, percentagewise. Depletion factors for Mg and Si are variably between 3 and 30, for example, whereas Al depletions range between 10^2 and 10^3, fully ~30 times greater. The problem is that sputtering is not chemically specific; i.e., Al is sputtered away from a grain surface as efficiently as is Mg. Thus Al should be as gaseous as Mg if a simple sputtering picture is adopted. In my view, the problem lies not in the sputtering, but in the structure of the grains. The Al is concentrated within SUNOCON (and, after astration, STARDUST) cores of grains and cannot be sputtered away before the magnesian mantles have been. The situation is illustrated in Figure 5.3, from Clayton (1982*a*), who reviews (pp. 185–87) this argument and its prior history in my work. This grain structure is a direct consequence of the condensation sequence illustrated in Figure 5.1. The reconciliation would thus seem to be that $\lambda_s(\mathrm{Al}) \lesssim 0.1\ \lambda_s(\mathrm{Mg})$, which can only be attributed to this structural effect within the grains. As a conservative estimate, I therefore take $\lambda_s(\mathrm{Al}) = 0.2$ instead of the value 2 that applies to most dust.

There exists a second paradox associated with $\lambda_s = 2$ even for the bulk of the dust, namely, that sputtering destruction rate is so great that the mass of dust observed in the ISM could not easily persist. Dwek and Scalo (1980) seem to have been the first to see this problem clearly. They calculated that about 3×10^9 yr are required to inject into the ISM from stars a mass of grains comparable to the mass residing there. Thus a destruction time of 5×10^8 yr would not permit a steady-state concentration as great as the one observed. The plausibility of their conclusion can be strengthened by the following argument based on the standard models of chemical evolution described in §5.2. The rate of dust astration $\omega \simeq (5 \times 10^9 \text{ yr})^{-1}$ could be argued to be comparable to the rate at which new dust is ejected from stars, at least if stellar ejection is the major source of dust, because the return fraction is about half, and the nonvolatile elements are primarily condensed going into stars and coming back out again. Thus a dust replacement time of 3×10^9 yr from stars is plausible. Clearly a dust destruction time by sputtering as short as 5×10^8 yr therefore affords no steady state unless the dust mass is primarily grown in the ISM itself. That solution to the problem was suggested by Dwek and Scalo (1980), who identified mantle growth within molecular clouds as the relevant process. I regard that solution as plausible, but point out that it requires dust to be primarily composed of abundant volatile elements (H, C, N, O), because the refractories are already so depleted that substantial grain growth in molecular clouds is not possible. The growth of grain mass from volatile elements is in accord with the arguments of Hoyle and Wickramasinghe (1977, 1979) that the interstellar dust is overwhelmingly organic, perhaps even biogenic.

I here suggest some solutions to this paradox. The first is that it may be only an apparent paradox, one constructed by a theoretical blind spot. Suppose nonvolatile elements stick to grains when they hit them, wherever that occurs. Then the entire sputtering rate of dust may be exactly balanced by the sticking of atoms to dust—a washout. That this is not implausible can be seen by noting that the average time required for an arbitrary atom to hit a grain is $\tau_{\text{atom}} \simeq 10^9$ yr$/n_{\text{H}}$, being less than 10^8 yr in most clouds, where the interstellar mass largely is believed to be. Only in the intercloud medium ($n_{\text{H}} \simeq 0.2$) is this time longer than the sputtering time, but the mass of the low density intercloud and hot medium is believed to be small. Thus in this approximation one may be able to ignore sputtering totally insofar as the total mass of dust is concerned, simply balancing it by reaccretion on a mean time scale shorter than the astration time scale. What would remain relevant, of course, is the rate at which specific structures (SUNOCONs) are unidirectionally destroyed by the sputtering. It is for this that I take $\tau_s = 5 \times 10^9$ yr for Al SUNOCONs, as described previously.

There is, it seems to me, at least one other resolution of problems associated with sputtering lifetimes as short as 5×10^8 yr. That is that the supernovae apparently occur in clusters rather than randomly. The estimate that a given interstellar grain will be passed by a supernova shock wave in 10^8 yr rests on the assumption that supernovae occur randomly, so that no portion of the ISM is safe for more than 10^8 yr. However, I see evidence that that simple picture may not be correct. It leads to the three-phase ISM model in a form like that described by McKee and Ostriker (1977), in which the filling factor of the hot phase is large (~50%) owing to the overlap of supernova random remnants into hot tunnels (Cox and Smith 1974). However, Heiles's (1979) data on galactic "superbubbles" shows their filling factor to be small, more like 1%. A new study by Brinks (1984) at Westerborck shows H I "holes" in the ISM of M31, also with filling factors of only 1%. This evidence of a small filling factor for supernova remnants suggests the possibility that the rate of shocking interstellar dust may also be overestimated. The apparent plausibility of a large, hot filling factor may have come about because we have been deceived by our own residence in a local superbubble of hot gas!

Many workers (e.g., Seab and Shull 1983) have identified grain-grain collisions behind the shocks as a major destruction process. This important problem requires exceedingly careful evaluation. Because of the huge abundance of small grains, it seems to me likely that large ones ($a > 1$ μm) will collide overwhelmingly with small grains and that these collisions will only fragment the large grain. Total vaporization will probably not occur, especially of small superrefractory SUNOCONs contained within a larger inhomogeneous particle. What one expects is a cascade of larger particles toward smaller ones (e.g., Dohnanyi 1970), and the question quickly focuses on the source of the largest particles. In the present solar system these are derived fragments of asteroids (Grün et al. 1985). The largest interstellar particles are probably cometary-like fragments spun off from aborted Oort-cloud material generated during star formation; but these contain SUNOCONs only as subunits. The SUNOCONs themselves are by definition grown within supernova interiors during their expansions, but much larger superrefractory particles may be aggregated in the ISM as a result of the shock waves there (Clayton 1981*a;* Elmegreen 1981). And each of these particles repeatedly accrete mantles during their cold cloud residences.

This discussion has ranged over several topics in order to clarify the effort needed to ascertain the rate of destruction of a specific SUNOCON. Taking for better or for worse the preliminary estimate $\lambda_s = 0.2$ for Al SUNOCONs, taking stellar return fraction $R = 0.5$ despite its uncertainty, and taking the relevant gas fraction in the ISM near the Sun to be 0.19 for that part of the disk that has participated in Population I chemical evolution (Tinsley 1981), the destruction rate for SUNOCONs is

$$\lambda = \lambda_s + \frac{\omega R}{1-R} = 0.2 + 0.172 = 0.37 \text{ Gyr}^{-1}, \tag{29}$$

where the particular choice $\omega = 0.172$ is taken from Table 5.1 for $\Delta = 1$ and age $T = 15$. Equation (29) numerically implies a mean SUNOCON core lifetime $\tau \simeq 2.7 \times 10^9$ yr, with roughly equal rates of destruction by astration and by sputtering. I argue that this result is, for SUNOCONs, much more plausible than recent estimates of shorter sputtering lifetimes for dust generally.

Because the product $\lambda T = 5.5$ then greatly exceeds unity, equation (28) is the applicable form of the more general equation (27) for the $k = 1$ models having $\Delta = 1$, $T = 15$. The value $\lambda = 0.37$ Gyr^{-1} then gives today

$$r_S = 0.28\ \beta_Z^0 \tag{30}$$

for the remainder of element Z within SUNOCONs. For the superrefractory elements Al, Ca, and Ti the initial fraction condensing within SUNOCONs is

$\beta^0 = 1$, so that 28% of those elements reside today within interstellar SUNOCONs. It will also be clear from equation (27) that the dependence on both T and Δ is rather weak. The value at $t = 10$, when the solar system formed, is, in the same ($k = 1$, $\Delta = 1$, $\omega = 0.172$, $\lambda = 0.37$) model, $r_S(t = 10) = 0.41\ \beta^0$, about 47% greater than the value today in the ISM. This result differs little from my estimate of a few years ago (Clayton 1982*a*, p. 185), but the present calculation is better formulated. The major uncertainty remains the value of the sputtering rate.

5.4. SUNOCON Remainders and Isotopic Anomalies

With the resultant 41% of superrefractory elements remaining in SUNOCONs when the solar system formed, it is possible to understand how it is possible that very large isotopic anomalies can have been present in samples formed in the early solar system. In this concluding section I will discuss three of these, excess ^{16}O, excess ^{26}Mg, and excess ^{50}Ti. I have discussed these previously (Clayton 1981*b*, 1982*a*), but the present study provides a firmer quantitative and physical foundation. It is important to realize that no simple picture will suffice and that continuing refinements will be necessary in almost every aspect of this problem. What I do regard as likely is that the general viewpoint for interpreting these isotopic anomalies is correctly expressed. It is a quite different viewpoint than the one based on inhomogeneous injection from a supernova trigger of the collapse of the solar cloud (Cameron and Truran 1977), one of Al Cameron's many bold views and one that dominated the thinking of most researchers for about five years. The use of "cosmic chemical memory" as a guiding picture (Clayton 1978, 1982*a*) will be adopted in what follows.

5.4.1 Excess ^{16}O

It is probably no accident that the most ^{16}O-rich meteoritic samples are also Al-rich. That fundamental fact demands explanation. The Al SUNOCONs that condense during the almost adiabatic outflows from the stellar shells of burning carbon and neon where the overwhelming bulk of aluminum was synthesized contain isotopically pure ^{16}O (see Figure 5.1). Therefore any dust collection will be ^{16}O-rich in comparison with the gas, and an Al-rich dust collection will be even more so. The immediate objective, therefore, is to calculate the ^{16}O richness of a dust collection with the simplifying assumption that nonaluminous dust is isotopically normal. Although this suppresses many interesting details of the dust and although it does not of itself provide a realistic chemical picture of the physical origin of the CaAl-rich inclusions, it does set the order of magnitude to be expected for ^{16}O-rich dust aggregates. That is surely a useful first step.

The aluminum solids (Al_2O_3 alone or bonded subunits within mineral hosts) carry a ^{16}O excess set by the value of r_S(Al); i.e., the three oxygen atoms bonded with a pair of Al atoms are monoisotopic ^{16}O in the fraction r_S and are (by oversimplifying assumption) isotopically normal in the fraction $1 - r_S$, so that as a group they carry an excess, here designated $^{16}O^*$, equal to

$$^{16}O^*/^{16}O_\odot = r_S/1 - r_S) \text{ in } Al_2O_3, \tag{31}$$

where $^{16}O_\odot$ is a notation for the number of ^{16}O atoms accompanied by the normal solar isotopic complement of $^{17,18}O$. Now suppose that the refractory portion of interstellar dust consists of mineral-like structures (though perhaps amorphous rather than crystalline) containing Na_2O, MgO, Al_2O_3, and SiO_2, omitting CaO because of the small abundance of calcium and omitting FeO because much of the metal probably resides in metallic droplets or sulfides in the ISM (Clayton and Ramadurai 1976). The bulk ^{16}O richness of such a refractory aggregate is then

$$\frac{^{16}O^*}{^{16}O_\odot} = \frac{1.5\, r_S N(\mathrm{Al})}{0.5N(\mathrm{Na}) + N(\mathrm{Mg}) + 1.5(1 - r_S)N(\mathrm{Al}) + 2N(\mathrm{Si})}\,, \tag{32}$$

where $N(Z)$ is the abundance of that element in the aggregate. If the abundances are solar, the denominator is dominated by Mg and Si, with the approximate result

$$[^{16}O^*/^{16}O_\odot]_\odot = 0.04\, r_S \simeq 0.016, \text{ solar mix.} \tag{33}$$

This is the correct order of magnitude, though a factor of 3 too small, to account for the 5% $^{16}O^*$ found in the parent aggregates of CaAl-rich inclusions—that is, before the oxygen was partially exchanged with ambient oxygen (R. N. Clayton et al. 1977). That this difficult theoretical number is even approximately right suggests to me that the physical concepts are correct.

If the abundances of the dust aggregate in equation (32) are enriched in Al relative to Mg and Si, as the inclusions undeniably are, the percentage of $^{16}O^*$ rises abruptly toward its limit expressed in equation (31), about 70%. It is evident from equation (33) that only a factor 3 enrichment in the Al concentration above solar would suffice to reach the observed level of 5%. But it would not be appropriate to invoke the measured Al richness of the inclusions, because that richness may have been established by evaporative loss in a hot (~2000 K) environment. (Oxygen exchange also occurs during such a distillation, further complicating the argument.) It is the more modest Al enrichments of the roughly homogeneous aggregate before distillation that one should fairly use in this comparison. That enrichment is unknown, as are the sorting processes during aggregation that cause an aggregate to have nonsolar bulk abundances. The importance of such a sorting process will again be evident in the discussion of excess ^{26}Mg and excess ^{50}Ti.

5.4.2 Excess ^{50}Ti

A 10% excess of ^{50}Ti has recently been discovered (Fahey et al. 1985) in a different type of CaAl-rich inclusion, namely a hibonite crystal from the Murray carbonaceous meteorite. Although much smaller (typically 0.3%) excesses in ^{50}Ti were already known to be quite common (Niemeyer and Lugmair 1981), the huge size of this new discovery poses an interesting challenge to the chemical memory theory (described for ^{50}Ti by Clayton 1981*b*). The first question, whether there exists a logically distinct SUNOCON for bearing ^{50}Ti, was almost immediately answered in the affirmative (Clayton 1981*b*). The second question, whether a significantly high fraction of those SUNOCONs can have survived galactic evolution, has been answered in the affirmative in this paper: 41% can have survived. The two remaining questions are much the hardest. They cannot be answered here, but there is some merit in posing them. Because the other Ti isotopes also reside with comparable probability in SUNOCONs, a fractionation mechanism for augmenting the ^{50}Ti-bearing SUNOCON concentration with respect to the others is needed. But even a 1 mm particle requires 10^9 dust particles of 1 μm to assemble its mass. Statistical fluctuations are therefore hopelessly small. A systematic enhancement owing to some distinguishing property of the SUNOCON is needed. Clayton (1981*b*) pointed out the promising likelihood that the ^{50}Ti-bearing SUNOCON is metallic, rather than oxidized or sulfidized. Their mass density was therefore significantly greater, enabling any of several sorting mechanisms responsive to the ratio of inertia to drag. A separate possibility would be the derivation of this hibonite on a parent body by utilizing a metallic-phase carrier for ^{50}Ti. The proper answer depends upon the last and hardest question, namely how the superrefractory minerals were physically created. At the time of this writing there exists some favoritism for regarding these as evaporative residues, or aggregates of same, in the hottest portions of a solar disk (Wark and Lovering 1982); but parent-body aggregates of these superrefractory residues, which can then accommodate parent-body processes to form the minerals found today, are also not excluded. Be that as it may, a 10% excess of ^{50}Ti appears to require some process for enhancing the remaining ^{50}Ti-carrying SUNOCONs by about 25% above their average concentration.

4.3 Excess ^{26}Mg

Gray and Compston (1974) and Lee, Papanastassiou, and Wasserburg (1977) first showed that CaAl-rich inclusions have excess ^{26}Mg in certain Al-rich minerals and that its concentration (^{26}Mg*) within minerals of a single inclusion stands in constant proportion to Al, usually $^{26}\mathrm{Mg}^*/\mathrm{Al} = 5 \times 10^{-5}$ regardless of the Al/Mg elemental ratio. That finding can be rationalized by assuming that the inclusions formed with $^{26}\mathrm{Al}/^{27}\mathrm{Al} = 5 \times 10^{-5}$ and that *in situ* decay of that ^{26}Al ($\tau_{1/2} = 7.3 \times 10^5$ yr) has produced the observed correlation. That interpretation has been almost universally accepted, despite continuing challenges by this writer. Considerations of stellar condensation (Clayton 1975*b*) predicted that Al-rich interstellar dust should also be ^{26}Mg-rich even before the meteoritic correlation was regarded as well established. Later Clayton (1977*b*, 1982*b*) described chemical scenarios for generating the observed correlation with the aid of a "corundum-ice-cube model" for trapping the ^{26}Mg* in Al_2O_3 structures. This theory gained few adherents owing to the nonequilibrium chemistry that had to be postulated to explain these nonequilibrium objects, an unconfirmed ad hoc special chemical assumption. Nonetheless that theory might be correct. What support it garnered derived from the apparent frivolity of the ^{26}Mg excesses. Not all inclusions carried the same ratio ^{26}Mg*/Al, some seemed characterized by more than one value for that ratio, and, most troublesome of all, some even seemed to have no excess ^{26}Mg. (For example, the inclusion HAL [Lee, Russell, and Wasserburg 1979] showed ^{26}Mg*/Al $\simeq 0$ even though Al/Mg $> 10^3$.) This variability poses less problems for a fossil theory than for a theory based on *in situ* decay. The fossil theory is not obliged to regard apparent time differences as real.

The situation remained locked in this impasse until the recent detection of live ^{26}Al in today's interstellar medium (Mahoney et al. 1984). The unexpectedly large flux of 1809 keV photons (4.8×10^{-4} cm^{-2} s^{-1}) has now received a welcome confirmation from the γ-ray spectrometer aboard the *Solar Maximum Mission* (Share et al. 1985). We showed in our analysis (Clayton 1984*a;* Leising and Clayton 1985) of that flux that it cannot easily be envisioned as a product of supernova nucleosynthesis throughout the Galaxy, because supernovae occur too infrequently and the supernova production ratio $P(^{26}\mathrm{Al}/P(^{27}\mathrm{Al}) \simeq 10^{-3}$ is two orders of magnitude too small. Clayton and Hoyle (1976) had argued that nova explosions produce ^{26}Al, and the yields seem about right (Clayton 1984*a*). Cameron (1984) was quick to discover that ^{26}Al-laden winds are expected from low mass asymptotic giant-branch (AGB) stars, and he devised a wind-driven variation of his earlier "supernova-trigger" theory (Cameron and Truran 1977). And more recently Dearborn and Blake (1985) have argued that Wolf-Rayet winds are also ^{26}Al-rich, perhaps enough so to account for most of the 3.4 $M_\odot$ of ^{26}Al required per million years. So very quickly it has become apparent that at least three sources of ^{26}Al have production rates far in excess of that expected from supernovae. Leising and Clayton (1985) studied the angular distribution of γ-rays to be expected from different sources with an eye toward identifying the correct source by observations with the Gamma-Ray Observatory. They also showed that the isotopic ratio expected today at the solar galactocentric radius probably lies in the range $^{26}\mathrm{Al}/^{27}\mathrm{Al} = 0.5$–$1.0 \times 10^{-5}$. This ratio is 5 to 10 times smaller than the correlation most frequently found in CAI, so it raises doubts that the average ISM concentration was as high as 5×10^{-5} when the solar system formed. Either that value was a fluctuation associated with the collapsing solar cloud (e.g., Cameron 1984), or the interstellar average ratio was implausibly larger 5×10^9 yr ago than it is today. Although the γ-ray discovery makes the live-^{26}Al interpretation of the meteoritic data much more believable, the average magnitude is nonetheless slightly off.

Curiously enough, the high interstellar abundance of ^{26}Al also increases the plausibility of the fossil theory by eliminating the requirement of a special nonequilibrium chemistry! It is to this rejuvenation of the fossil theory that I now turn as another application of SUNOCON survival. The increased plausibility of a revised fossil theory for excess ^{26}Mg, based on the γ-ray observations, was discovered by Clayton (1985*c, d*). What I provide here is a more careful discussion of the astrophysics, concentrating on nucleosynthesis within the solar annulus within the context of analytic models of the chemical evolution of the solar neighborhood. For purposes of illustration I will utilize the models having continuing infall (Clayton 1985*a*), the same models introduced in §5.2.

I choose the $k = 1$ model having $\Delta = 1$ and $\omega = 0.172$, which yields a gas mass fraction $\mu = 0.19$ today ($t = 15$) in the solar neighborhood, as seen in Table 5.1. The choice of a smaller gas fraction, which would be appropriate if the entire dynamical mass has participated in the chemical evolution of the Population I system, would not much alter the quantitative conclusions. The gas mass today is taken as $M_G = 8.1\ M_\odot\ \mathrm{pc}^{-2}$, and the local star formation rate today $\psi(15) = 2.8\ M_\odot\ \mathrm{pc}^{-2}\ \mathrm{Gyr}^{-1}$, consistent with equation (4) for $\omega = 0.172$ and a stellar return fraction $R = 0.5$. This model has the property that the total mass density ($43\ M_\odot\ \mathrm{pc}^{-2}$) of the local Population I system (not the total dynamic mass) has increased by a factor 6.37 times the initial mass density of the local disk owing to the continuous infall over the past 15 Gyr. Primary nucleosynthesis products grow according to equation (25) in $k = 1$ models. Taking infall $Z_f = 0$, $t = 15$, and $\Delta = 1$ then gives today $Z = 1.37\ y_Z$. I take the abundances today to be solar, believing that the Sun was slightly metal-rich for its formation epoch and believing that a mean of observations today reveals solar abundances in H II regions and in young stars. This fixes the required yield $y_Z = Z_\odot/1.37$. With these numbers it is elementary to calculate, consistent with the instantaneous recycling approximation, the rates at which both new and old aluminum are injected into the local interstellar medium. The rate of new aluminum creation, to be specific, is

$$\begin{aligned}\dot{\mathrm{Al}}_{\mathrm{new}} \equiv y_Z(1-R)\psi = y_Z\omega M_G &= 1.4\ y_Z M_\odot\ \mathrm{pc}^{-2}\ \mathrm{Gyr}^{-1} \\ &= 1.0\ Z_\odot M_\odot\ \mathrm{pc}^{-2}\ \mathrm{Gyr}^{-1}\end{aligned} \tag{34}$$

whereas the return of old aluminum from stars is

$$\dot{\mathrm{Al}}_{\mathrm{old}} \equiv R\psi Z_\odot = \omega M_G Z_\odot R/(1-R) = 1.4\ Z_\odot M_\odot\ \mathrm{pc}^{-2}\ \mathrm{Gyr}^{-1}, \tag{35}$$

where $Z_\odot(\mathrm{Al}) = 6\times10^{-5}$ is the star mass fraction of aluminum. This new Al condenses into SUNOCONs with almost 100% efficiency, and the old Al condenses into STARDUST with equally high efficiency. Without these assumptions, it becomes difficult to explain the huge depletion of Al from the gas phase, as discussed earlier.

We wish to compare these input rates for stable Al with the observed input rate of radioactive ^{26}Al. Averaged over the entire Galaxy we know the latter rate to be about $3.4\ M_\odot/10^6$ yr in order to account for the γ-ray emissivity; however, the input rate into the solar annulus depends upon the spatial distribution of ^{26}Al sources. Leising and Clayton (1985) have studied that problem in some detail, calculating the $^{26}\mathrm{Al}/^{27}\mathrm{Al}$ ratio at the solar radius that would be consistent with spatially different sources of ^{26}Al production (novae, Wolf-Rayet, etc.). They also stress, by the way, the importance of measuring the angular distribution of the ^{26}Al γ radiation with the Gamma-Ray Observatory in order to better define the physical source of the ^{26}Al. Their finding is that $^{26}\text{Al}/^{27}\text{Al}$ locally is expected to lie between a lower limit of 2×10^{-6} if novae are its source to an upper limit of 9×10^{-6} if the source is proportional to either the total light of the disk or the total mass of gas. Giving extra weight to the latter possibility because Wolf-Rayet or AGB production is probably better known than nova production at present, I take $^{26}\text{Al}/^{27}\text{Al} = 8 \times 10^{-6}$ at the solar position. This is easily seen to require a local input rate

$$^{26}\dot{\text{A}}\text{l} = 8 \times 10^{-6}\ Z_\odot(\text{Al})M_G(10^6\ \text{yr})^{-1} = 0.065\ Z_\odot\ \text{pc}^{-2}\ \text{Gyr}^{-1}. \tag{36}$$

The production ratio for aluminum isotopes is given by the ratio of equation (36) to equation (34):

$$P(^{26}\text{Al})/P(^{27}\text{Al}) = 0.065. \tag{37}$$

This ratio is 65 times greater than the value expected from supernovae alone, reaffirming Clayton's (1984*a*) argument that the observed ^{26}Al is not a supernova product, at least not if its occurrence is galactic-wide. It is also of interest that equation (37) and the solar abundance ratio $(^{26}\text{Mg}/^{27}\text{Al})_\odot = 1.39$ together imply that 4.7% of all ^{26}Mg was initially synthesized as ^{26}Al parent if the ratio of ^{26}Al production to general nucleosynthesis has been a constant over galactic time. It is the unexpectedly large size of these numbers, thrust upon us directly by the γ-ray observations, that so increases the plausibility of any fossil theory of excess ^{26}Mg in aluminous meteoritic matter. But to address this quantitatively requires some assumption about the *chemical form* of the ^{26}Al ejecta, for that is the basis of the cosmic chemical memory theory (Clayton 1982*a*). Clearly, if this ^{26}Al is ejected in gaseous form, it will also decay in gaseous form and the correlation of $^{26}\text{Mg}^*$ with Al will not exist. But that is decidedly not to be expected. It was the thesis of Clayton and Hoyle's (1976) work that novae ejecta condense as dust, that observations confirm that conclusion, that aluminum will especially do so, and that $^{26}\text{Mg}^*$-rich aluminum NOVACONs will result. But the fate of the ^{26}Al should be little different if it is instead produced primarily in Wolf-Rayet stars or in the hydrogen-burning shells of AGB stars. Its highly refractory nature guarantees its efficient condensation, just as efficiently as in other STARDUST. I will therefore bring this section to conclusion by assuming that all types of aluminum ejecta condense with unit efficiency into similar chemical structures, which I have called SUNOCONs in the preceding sections of this paper.

The magnitude of the interstellar excess of $^{26}\text{Mg}^*$ in Al dust can be seen in two different ways. Taking the production ratio equation (37) to represent average SUNOCON production rates in the ISM, and taking the SUNOCON remainder to be $r_S = 0.41$ at the time the solar system formed, one obtains

$$^{26}\text{Mg}^*/^{27}\text{Al} = 0.027 \tag{38}$$

within a representative sample of Al-bearing dust in the ISM.

This same number is also obtained if we evaluate the ratio today of the ejection rate of ^{26}Al to that of the ejection rate of all aluminum, new plus old. That comparison yields

$$\frac{^{26}\dot{\mathrm{Al}}}{\dot{\mathrm{Al}}_{\mathrm{old}}+\dot{\mathrm{Al}}_{\mathrm{new}}}=\frac{0.065}{1.40+1.02}=0.027, \tag{39}$$

which again yields equation (38) if we assume that the ^{26}Al decays overwhelmingly in the Al-bearing dust components (all of which are treated identically). This number is slightly smaller than the value 0.040 obtained by Clayton (1985*c*) because of my present concentration on the more secure characteristics of chemical evolution in the solar neighborhood.

Assume next that bulk interstellar Mg has normal isotopic composition, being a composite of ^{26}Mg* in aluminum dust according to equation (38) and a homogeneous pool of Mg with a slight ^{26}Mg deficit (named a "ghost" by Clayton 1977*b*). The ^{26}Mg/^{24}Mg isotopic ratio in a mechanical aggregate of dust can then on average be characterized by the aluminum richness of that dust collection. Since the fraction of interstellar ^{26}Mg residing as ^{26}Mg* in aluminous dust is $^{26}\mathrm{Mg}^*/^{26}\mathrm{Mg}=0.019$, the isotopic composition of a dust aggregate is easily seen to be

$$[^{26}\mathrm{Mg}/^{24}\mathrm{Mg}]=0.019\,[\mathrm{Al/Mg}]+0.981, \tag{40}$$

where I use the common astronomical convention $[a/b]\equiv a/b\div(a/b)_{\odot}$. Equation (40) states (*a*) that an average solar mix of dust, characterized by [Al/Mg] = 1, has a normal bulk isotopic composition [26/24] = 1; (*b*) that the nonaluminous interstellar Mg has a negative ^{26}Mg ghost equal to −19 ppm (parts per mil = parts per thousand, the usual convention of isotopic chemists); and (*c*) that an Al-rich aggregate has a bulk ^{26}Mg excess:

$$^{26}\delta_a=19\,([\mathrm{Al/Mg}]-1)_a\ \mathrm{ppm}. \tag{41}$$

The subscript $a\equiv$ aggregate designates this isotopic composition as the bulk composition of a dust aggregate that may be an ancestor of a CaAl-rich inclusion. I prefer to leave open the questions of where this Al-rich dust aggregate formed and what were the mechanical processes that sorted the dust. But the aggregate may have been formed in the interstellar medium itself (Clayton 1981*a;* Elmegreen 1981), in the turbulent collapse of the protosolar cloud (Cameron 1973), or in the solar planetary disk (Clayton 1982*a;* Morfill 1983). Suffice it to imagine that mechanical dust aggregates enriched in Al/Mg can be assembled. The next question is how a dust aggregate becomes very Al-rich, as it must if it is to become a CaAl-rich inclusion (CAI), which typically have Al/Mg ≃ 1–2 (that is, [Al/Mg] ≃ 10–25). To this question also no answer can be given at the present time. Meteoriticists have waged a lengthy debate during the past decade over the origin of the CAI, based on their morphological and mineralogical properties. But a strong argument has been made (e.g., Wark and Lovering

1982; Kornacki and Wood 1984) that the extreme enrichments of Al (and Ca and other superrefractory elements) were established by evaporation of less volatile elements such as Mg. That is, the bulk chemistry of the inclusions is what it is because they or their immediate precursors were superrefractory-rich residues of evaporation loss, probably in the hottest portions of a turbulent preplanetary disk (Morfill 1983). If we take that idea as a guide, we can construct a scenario for generating the correlation between $^{26}Mg^*$ and Al that is found in CAIs.

Suppose a dust aggregate described by equations (40) and (41) is heated to the point (perhaps 1800° C) where most of the Mg is evaporated. The Al/Mg ratio in bulk will thereby be greatly increased, but the Mg isotopic composition will remain the same if all three Mg isotopes evaporate at the same rate. Then equation (41) still applies. Values of $[Al/Mg] \simeq 5$, not incredibly large, are needed to provide a roughly 8% excess of ^{26}Mg, as observed in many inclusions (e.g., Wasserburg and Papanastassiou 1982). The linear correlation between $^{26}Mg/^{24}Mg$ and Al/Mg can then be generated by dilution of the large ^{26}Mg excess in the initial evaporative residue with ambient Mg taken up from the environment during cooling. For such an explanation to work, it is necessary that the Mg-rich minerals having low $^{26}Mg/^{24}Mg$ excess be formed by the uptake of ambient magnesium and that the extremely Al-rich anorthite be the true residue. It is at the moment not clear whether the suite of minerals can be petrologically described in this manner. But if it can, Clayton (1985*c, d*) has shown that the resulting correlation can be described by

$$\frac{^{26}Mg}{^{24}Mg} = \left(\frac{^{26}Mg}{^{24}Mg}\right)_a \frac{Al/Mg}{(Al/Mg)_0} + g'\left[1 - \frac{Al/Mg}{(Al/Mg)_0}\right], \qquad (42)$$

where $(Al/Mg)_0$ is the initial ratio in the evaporative residue, before uptake of ambient Mg lowers its value to Al/Mg, and where g' is the isotopic ratio in the ambient magnesium (presumably almost solar). The celebrated linear relationship observed in Allende inclusions is seen in this equation as well.

If it is not chemically possible for the origins of CAIs (of certain types) to be described in terms of the uptake of ambient Mg by a cooling Al-rich residue, the fossil theories must revert to their original form (Clayton 1977*b*, 1982*b*). In this form, it is more difficult to extract ^{26}Mg from aluminum than it is for the other isotopes because of the $^{26}Mg^*$ locked up in Al-rich subunits (corundum, Al_2O_3). This conjecture is yet to be tested. In either case the known large interstellar abundance of ^{26}Al has rendered the fossil theory more plausible than it was initially.

The calculations made here may prove to be in quantitative or physical error, perhaps irrelevant to cosmogonical processes. But A. G. W. Cameron has repeatedly helped us understand that all of physics and astrophysics must be carefully evaluated if we are to understand even the simplest stones that fall from the sky.

Acknowledgments

This research has been supported in part by NASA grant NAG 9-100 BA-

SIC, by the Robert A. Welch Foundation, and by the Max Planck Institut für Kernphysik, Heidelberg, where I was in residence between 5 May and 5 August 1985.

References

Barlow, M. J. 1978*a*, *M.N.R.A.S.*, **183**, 367.
——. 1978*b*, *M.N.R.A.S.*, **183**, 397.
——. 1978*c*, *M.N.R.A.S.*, **183**, 417.
Brinks, E. 1984, Ph.D. thesis, University of Leiden.
Cameron, A. G. W. 1973, *Icarus*, **18**, 407.
——. 1984, *Icarus*, **60**, 416.
Cameron, A. G. W., and Truran, J. W. 1977, *Icarus*, **30**, 447.
Clayton, D. D. 1975*a*, *Ap. J.*, **199**, 765.
——. 1975*b*, *Nature*, **257**, 36.
——. 1977*a*, *Icarus*, **32**, 255.
——. 1977*b*, *Earth Planet. Sci. Lett.*, **35**, 398.
——. 1978, *Moon and Planets*, **19**, 109.
——. 1981*a*, *Ap. J.*, **251**, 374.
——. 1981*b*, *Proc. Lunar Planet. Sci. Conf.*, **12**, 1781.
——. 1982*a*, *Quart. J.R.A.S.*, **23**, 1978.
——. 1982*b*, *Lunar Planet. Sci.*, **13**, 115.
——. 1984*a*, *Ap. J.*, **280**, 144.
——. 1984*b*, *Ap. J.*, **285**, 411.
——. 1985*a*, in *Nucleosynthesis: Challenges and New Developments*, ed. W. D. Arnett and J. W. Truran, p. 65 (Chicago: University of Chicago Press).
——. 1985*b*, *Ap. J.*, **290**, 428.
——. 1985*c*, *Ap. J.*, submitted.
——. 1985*d*, *Meteoritics*, **20**, 626.
Clayton, D. D., and Hoyle, F. 1976, *Ap. J.*, **203**, 490.
Clayton, D. D., and Ramadurai, S. 1976, *Nature*, **265**, 427.
Clayton, R. N.; Onuma, N.; Grossman, L.; and Mayeda, T. K. 1977, *Earth Planet. Sci. Lett.*, **34**, 209.
Consolmagno, G. J., and Cameron, A. G. W. 1980, *Moon and Planets*, **23**, 3.
Cowie, L. L. 1978, *Ap. J.*, **225**, 887.
Cox, D. P., and Smith, B. W. 1974, *Ap. J. (Letters)*, **189**, L105.
Dearborn, D. S. P., and Blake, J. B. 1985, *Ap. J. (Letters)*, **288**, L21.
Dohnanyi, J. S. 1970, *J. Geophys. Res.*, **75**, 3468.
Draine, B. T., and Salpeter, E. E. 1979, *Ap. J.*, **231**, 438.
Dwek, E., and Scalo, J. M. 1980, *Ap. J.*, **239**, 193.
Elmegreen, B. G. 1981, *Ap. J.*, **251**, 820.
Fahey, A. J.; Goswani, J. N.; McKeegan, K. D.; and Zinner, E. 1985, *Ap. J. (Letters)*, **296**, L17.
Gray, C. M., and Compston, W. 1974, *Nature*, **251**, 495.
Grün, E.; Zook, H. A.; Fechtig, H.; and Giese, R. H. 1985, *Icarus*, in press.

Heiles, C. 1979, *Ap. J.*, **229**, 533.

Hoyle, F., and Wickramasinghe, N. C. 1970, *Nature*, **226**, 62.

——. 1977, *Nature*, **268**, 610.

——. 1979, *Ap. Space Sci.*, **66**, 77.

Kornacki, A. S., and Wood, J. F. 1984, *Geochim. Cosmochim. Acta*, **48**, 166.

Lee, T., Papanastassiou, D. A., and Wasserburg, G. J. 1977, *Ap. J. (Letters)*, **211**, L107.

Lee, T., Russell, W. A., and Wasserburg, G. J. 1979, *Ap. J. (Letters)*, **228**, L93.

Leising, M. D., and Clayton, D. D. 1985, *Ap. J.*, **294**, 591.

Mahoney, W. A.; Ling, J. C.; Wheaton, W. A.; and Jacobson, A. S. 1984, *Ap. J.*, **286**, 578.

McKee, C. F., and Ostriker, J. P. 1977, *Ap. J.*, **218**, 148.

Morfill, G. E. 1983, *Icarus*, **53**, 41.

Niemeyer, S., and Lugmair, G. 1981, *Geochim. Cosmochim. Acta*, **48**, 1401.

Seab, C. G., and Shull, J. M. 1983, *Ap. J.*, **275**, 652.

Share, G. H.; Kinzer, R. L.; Kurfess, J. D.; Forrest, D. J.; and Chupp, E. L. 1985, *Ap. J. (Letters)*, **292**, L61.

Shull, J. M. 1977, *Ap. J.*, **215**, 805.

——. 1978, *Ap. J.*, **226**, 858.

Tinsley, B. 1981, *Ap. J.*, **250**, 758.

Wark, D. A., and Lovering, J. F. 1982, *Geochim. Cosmochim. Acta*, **46**, 2581.

Wasserburg, G. J., and Papanastassiou, D. A. 1982, in *Essays in Nuclear Astrophysics*, ed. C. A. Barnes, D. D. Clayton, and D. N. Schramm, p. 77 (New York: Cambridge University Press).

6

Supernova Nucleosynthesis

Wolfgang Hillebrandt

6.1. Introduction

When in 1957 E. M. Burbidge, G. R. Burbidge, W. A. Fowler, and F. Hoyle and independently A. G. W. Cameron taught us that the chemical elements and their isotopes are formed in stars and stellar explosions, they probably did not envision that now, almost 30 years later, this field of research would still be and likely remain one of the most exciting areas in astrophysics. Moreover, although based on the very limited number of astronomical and nuclear physics data available at that time, their theory did withstand all challenges that were imposed over the years, and more than a generation of scientists were needed to work out some of the details of their fundamental papers.

In my contribution to this book, which is intended to honor Al Cameron on the occasion of his 60th birthday, I will try to review the status of one particular aspect of stellar nucleosynthesis, namely the transmutations of nuclear abundances in supernova explosions.

Because over the last couple of years there has been an increasing flux of information on abundances in supernovae and young supernova remnants ranging from infrared to X-ray observations, I will first discuss the evidence for explosive nucleosynthesis based on the modeling of spectra. It will become obvious, however, that the interpretation of the observational data is still controversial, mostly because all abundance determinations are strongly model-dependent. Nevertheless it has become clear that several young supernova remnants are indeed enriched in the products of explosive nucleosynthesis, thus supporting our ideas of how the chemical elements are formed.

I then will review the present status of supernova models. It will be shown that the most promising candidates for type I supernovae are exploding white dwarfs in binary systems although alternative models have certainly not yet been ruled out. The main evidence comes from the fact that excellent fits to observed early time spectra of type I supernovae can be obtained from a particular class of these models only by assuming a certain amount of mixing of the processed matter. Models of type II supernovae, on the other hand, are much more sensitive to the input physics, and therefore it is not surprising that so far we have not found an explosion mechanism accepted by all people working in the field. It will be demonstrated, however, that very likely at least two distinct classes of type II supernovae exist. The first class contains stars of around $10 \pm 2\ M_\odot$ on the main sequence. In this case a rather low mass iron core of low entropy forms at the end of hydrostatic nuclear evolution. It collapses to neutron star densities, and a hydrodynamical shock wave generated from the rebounding proto-neutron star is thought to eject the stellar envelope with an energy of about 10^{51} ergs. While this type of model may well account for most of the observed type II supernovae, in particular Crab-like objects, the bulk of the heavy elements cannot come from those rather low mass stars. It is more likely that they stem from stars more massive than 15 $M_\odot$, for which the simple core-bounce mechanism fails to work. This has led to a revival of the idea that neutrino transport out of the proto-neutron star may be the cause of an explosion. The advantages and disadvantages of both pictures will be considered in some detail.

In §6.4 I will concentrate on predictions made by the various supernova models for nuclear abundance yields. Type I supernovae will prove to be a good source for the synthesis of iron and nickel. This finding may explain the apparent delay between the formation of oxygen and iron inferred from observed abundances of metal-poor stars. Type II supernova models, on the other hand, predict that approximately the relative proportions of elements from oxygen to iron will be ejected that are observed in the solar system. It is not clear at all, however, whether they can also account for the abundances of neutron-rich isotopes beyond the iron group, the so-called r-process nuclei. The site of the r-process remains one of the mysteries of the theory of nucleosynthesis.

6.2. Observed Abundances in Supernovae and Supernova Remnants

6.2.1 Type I and Type I Remnants

The very young type I supernova remnants Tycho (SN 1572) and Kepler (SN 1604) have been studied extensively both by *HEAO 2* (the *Einstein Observatory*) and recently by the Japanese X-ray satellite *Tenma* (Becker et al. 1980*a*, *b*; Tsunemi 1985). The spectra of the two supernova remnants look very similar, showing strong emission lines from helium-like ionization states of Si, S, Ar, and Ca. The interpretation of these excellent and simple spectra, however, is not unique as far as elemental abundances are concerned, because in order to model the observed spectra one has to make several assumptions which may not always be correct. Two major sources of uncertainties have to be considered. First, in order to subtract the X-ray continuum from the spectra, one has to assume abundances of the light elements H, He, C, N, and O, which cannot be measured directly. It is generally assumed that the abundances of the light elements are solar, which is not a bad assumption if the exploding star was indeed an approximately 1 $M_\odot$ white dwarf and the supernova blast wave has swept up a large amount of interstellar matter. The other source of uncertainty is probably more severe. To compute a synthetic spectrum one has to model both the hydrodynamical and the thermodynamic properties of the supernova remnant. It is quite questionable, for example, whether a simple blast

Table 6.1
Range of Abundances in Tycho's and Kepler's Supernova Remnant as Inferred from X-Ray Data (brackets denote abundances relative to their solar value)

[Ne]	[Mg]	[Si]	[S]	[Ar]	[Ca]	[Fe]
$\lesssim 1$	$\lesssim 2$	4–20	6–20	2–4	2–3	2–5

wave model is applicable or whether the assumption of a collisional ionization equilibrium correctly describes the thermodynamic state of matter. Following the original abundance determinations of Becker et al. (1980*a, b*) several people have reanalyzed the Einstein data by relaxing one or the other of the assumptions mentioned above (Shull 1982, 1983; Pravdo and Nugent 1983; Holt 1983; Szymkowiak 1984; Hughes and Helfand 1984) with the general tendency that the estimated abundances of several of the heavier elements (Ar, Ca, Fe) decreased if non-ionization equilibrium effects were taken into account. A typical range of abundances as inferred for both Tycho and Kepler from the references given above is presented in Table 6.1. Taking into consideration that the uncertainties in the numbers are at least of the order of a factor of 2, we can conclude that the two youngest galactic type I remnants do now show any dramatic overabundances of heavy nuclei, with the possible exception of Si and S. But there is nevertheless a clear indication that nuclear processing has changed the composition of the exploded star.

The somewhat older type I remnant SN 1006 has also been observed by both the *Einstein Observatory* and *EXOSAT* (Becker et al. 1980*c;* Jones, Pye, and Culhane 1984), but no line emission has been detected. Only some excess in the X-ray flux below 2 keV may be attributed to O VII and O VIII emission (Vartanian, Lum, and Ku 1984). This result can be caused either by an underabundance by at least a factor of 2 in the heavy elements if collisional ionization equilibrium is assumed (Vartanian et al.) or by abundances that are about solar with non-ionization equilibrium effects playing nevertheless an important role despite the fact that the remnant is fairly old (Aschenbach 1985).

An alternative way of estimating the abundances of type I supernovae is to fit early and late time spectra. This approach has been used by Axelrod (1980*a, b*), Weaver, Axelrod, and Woosley (1980), and Woosley, Axelrod, and Weaver (1984) for SN 1972e, and by Branch and his collaborators (Branch et al. 1981, 1983, 1985; Branch 1985) and Harkness (1984) for SN 1981b. The general conclusion from this procedure is that the early time spectra show strong absorption lines of O I, Mg II, Si II, S II, Ca II, and Co II, and also possibly Na I, which can be fitted by assuming solar relative abundances of the elements from O to Ca and by increasing the ratio of Ca to Fe by about a factor of 10 relative to its solar value, and two Co III emission lines seem also to be present in the late time spectra of SN 1972e (Axelrod 1980*a, b*). The presence of Co in the spectra of type I supernova explosions is the first direct indication that the radioactive decay of ^{56}Ni to ^{56}Co and to ^{56}Fe indeed powers the light curves of type I supernovae, as originally suggested by Colgate and McKee (1969) and later discussed by Arnett (1979), Colgate, Petschek, and Kriese (1980), and Weaver et al. (1980). But again one must keep in mind that abundance determinations from synthetic spectra are strongly model-dependent. In particular they depend on the assumed velocity, temperature, and density profile of the exploding star (see, e.g., Harkness 1984). Moreover, one must also explain why there seems to be so much Co in the spectra of type I supernovae but the overabundance of Fe in their remnants is apparently small.

6.2.2 Type II Supernovae and Their Remnants

It is commonly believed that type II supernovae have a larger impact on galactic nucleosynthesis than type I explosions, because their light curves indicate that their progenitor stars are evolved massive red supergiants. It should then follow that young type II remnants will show heavy elements in roughly solar proportions but enhanced relative to H and He. But unfortunately this is only partially the case.

The Crab Nebula, which is probably the result of a very luminous type II explosion, has been studied in the UV, optical, and near-IR emission (Davidson et al. 1982; Henry and McAlpine 1982; Henry, McAlpine, and Kirshner 1984) and in X-rays (Becker et al. 1979). While no lines have been detected in X-rays, features in the other energy bands show that He is largely overabundant, He/H $\gtrsim 1$, O is underabundant relative to its solar value by at least a factor of 2, the C to O ratio lies in the range from 0.4 to 1.5, Ne, S, and Fe may be slightly underabundant $[\mathrm{Ne}] \simeq 1$, $[\mathrm{S}] \simeq 1$, $[\mathrm{Fe/S}] \simeq 0.5$ (where the brackets denote ratios relative to solar values), and Ni is strongly overabundant relative to S ($[\mathrm{Ni/S}] \simeq 18$). So at least the Crab Nebula does not look like the prototype of an object responsible for galactic nucleosynthesis. Moreover, it has still not been confirmed that the Crab Nebula is surrounded by a hydrogen-rich halo (Murdin and Clark 1981), which should be the case if the progenitor star had indeed a mass of more than about 4 $M_\odot$.

A better case for explosive nucleosynthesis is Cas A. Optical data obtained from fast moving knots (Chevalier and Kirshner 1979) show large abundance inhomogeneities but virtually no H, He, C, and N. These knots are depleted in Ne and Mg, but some are significantly enriched in S, Ar, and Ca relative to O ($[\mathrm{S/O}] \simeq 1$–26, $[\mathrm{Ar/O}] \simeq 2$–20, $[\mathrm{Ca/O}] \simeq 0.3$–$24$). The results obtained from optical studies have meanwhile been confirmed by interpretations of X-ray observations from *HEAO 1* (Pravdo and Smith 1979), the *Einstein Observatory* (Becker et al. 1979; Holt 1983; Pravdo and Nugent 1983), and *EXOSAT* (Peacock 1984; Jansen et al. 1984). The abundances for the whole remnant given in Table 6.2 summarize these results. In particular the overabundances of S, Ar, and Ca indicate that the exploding star has at least burned oxygen to heavier elements during the explosion, which in turn means that it was an evolved massive object. Although again the abundance determinations for Cas A are model-dependent and in particular interpretations of the X-ray data are sensitive to the assumed amounts of C, N, O, and Ne, the fact that both types of studies lead to a

Table 6.2
Elemental Abundances of Cas A Inferred from X-Ray Data Relative to Solar Values

[Mg]	[Al]	[Si]	[S]	[Ar]	[Ca]	[Fe]
0.5	1.2	1.7	3.3	6.9	3	$\lesssim 0.5$

consistent set of abundances is very encouraging.

Puppis A, another young supernova remnant that presumably is the result of a type II explosion, has also been studied extensively by both the *Einstein Observatory* and *EXOSAT* (Winkler et al. 1981; Winkler, Canizares, and Bromley 1983; Pfeffermann et al. 1981). The interpretation of the data, however, is somewhat controversial. While Szymkowiak (1984) finds a reasonable fit to the X-ray spectra by assuming solar abundances, Canizares and Winkler (1981) and Canizares et al. (1983) obtain from their fits an overabundance of O relative to Fe of about a factor of 3 to 5 and $[O/Ne] \simeq 0.5$–1, consistent with the picture that the progenitor star was rather massive ($M \simeq 25\ M_\odot$).

6.3. Supernova Models

6.3.1 Models of Type I Supernovae

Theories and numerical experiments that model type I supernovae have the great advantage that from the observational data discussed in §6.2.1 it seems to be well established that the cause of the explosion is the thermonuclear burning of some fuel to iron group nuclei and in particular to ^{56}Ni. Moreover, from the similarities of type I light curves one can conclude that in all events about the same amount of ^{56}Ni is produced. The best fits of the light curves are obtained if it is assumed that about 0.5 to 1.0 $M_\odot$ of ^{56}Ni is ejected (see, e.g., Arnett 1982).

Finally, from the missing hydrogen lines in the spectra, from the presence of heavy nuclei near maximum light, and from the shape of the light curves, one can infer that the exploding stars are rather compact, presumably white dwarf–like objects.

All observational constraints are best met if it is assumed that the progenitors of type I supernovae are accreting C-O white dwarfs in binary systems, and this scenario is now favored by most people working in the field. Stellar evolution scenarios that will lead to the right initial conditions for a type I supernova explosion have been discussed in great detail by Iben and Tutukov (1984) and also by Iben in this volume, so I will only describe the final fate of these stars.

There are three possible ways in which C-O white dwarfs can ignite their nuclear fuel and explode, depending on the mass of the white dwarf, the accretion rate, and the composition of the accreted material. Nomoto (1984*a*) has nicely reviewed all the details, so I will discuss only the general ideas here.

6.3.1.1 *Double Detonation Models*

Double detonation models have been computed and discussed by Woosley and Weaver (1981), Woosley, Weaver, and Taam (1980), Weaver, Axelrod, and Woosley (1980), and Nomoto (1980, 1982, 1984*a*). In these models it is assumed that either He is accreted onto a white dwarf at a rate of about 10^{-8} $M_\odot$ yr^{-1}, or H-rich matter is accreted but the H-shell flashes are weak. The latter condition is more easily met for the accretion rates under consideration if the mass of the white dwarf is small, about 0.5 $M_\odot$. Due to the moderate accretion rates, He will be ignited first in a shell, driving a He detonation wave outward and a C detonation wave inward. In both detonation waves the nuclear fuel is burned all the way up to iron group nuclei (mostly ^{56}Ni) and the white dwarf is disrupted. By the time the explosion goes off, about 0.1–0.6 $M_\odot$ of the He has accumulated on top of the C-O white dwarf, and therefore about 1.0–1.2 $M_\odot$ of ^{56}Ni is ejected in this case. It has been demonstrated by Weaver, Axelrod, and Woosley (1980) that typical light curves of type I supernovae can be obtained from such a model (e.g., SN 1972e), but from the fact that the detonation waves burn the nuclear fuel at very high temperatures ($T \gtrsim 5 \times 10^9$ K) it is obvious that the early time spectra cannot be fitted, because no intermediate mass nuclei are synthesized.

6.3.1.2 *Single Detonation Models*

If the mass of the white dwarf is somewhat larger (0.8 to 1.1 $M_\odot$) and the accretion rate lower (10^{-9} $M_\odot$ yr^{-1}), again the accreted He ignites first, but this time the burning front propagating inward does not incinerate the entire carbon-oxygen core (Nomoto 1982; Woosley, Axelrod, and Weaver 1984). If the inward moving front extinguishes before it reaches the center of the C-O white dwarf, a white dwarf remnant of about 0.5 to 1 $M_\odot$ is left over, but the accreted He is essentially burned to ^{56}Ni (Woosley, Axelrod, and Weaver 1984). In this type of explosion, again a sufficient amount of ^{56}Ni is produced to explain the observed type I light curves, but apart from Ni only some C and O is ejected, raising the same problems with early time spectra that the double detonation models are facing.

6.3.1.3 *Carbon Deflagration Models*

At present the most popular models for type I supernovae utilize the fact that, if the accretion rate is very high (more than 10^{-8} $M_\odot$ yr^{-1} of H-rich matter or He) or if the accretion rate is somewhat lower but the C-O dwarf is rather massive ($M \gtrsim 1.1$ $M_\odot$), quiet nuclear burning will raise the mass of the degenerate core to the Chandrasekhar limit (1.4 $M_\odot$). The star then contracts and ignites C at the center. The further evolution is characterized by the fact that the C-burning front propagates outward with subsonic velocity, which has been demonstrated in a two-dimensional computer simulation, avoiding the uncertainties of time-dependent mixing-length theory, by Müller and Arnett (1982) (see also Müller and Arnett 1985). Such a subsonic burning front is usually called a deflagration wave, indicating that the energy transport is due to convection rather than shock heating as in the case of a detonation wave.

Carbon deflagration models of type I supernovae have been reviewed by Wheeler (1982), and many of the details have been worked out by Nomoto (1980, 1982, 1984*a*, *c*), Nomoto and Sugimoto (1977), Weaver, Axelrod, and Woosley (1980), Woosley, Weaver, and Taam (1980), Woosley and Weaver

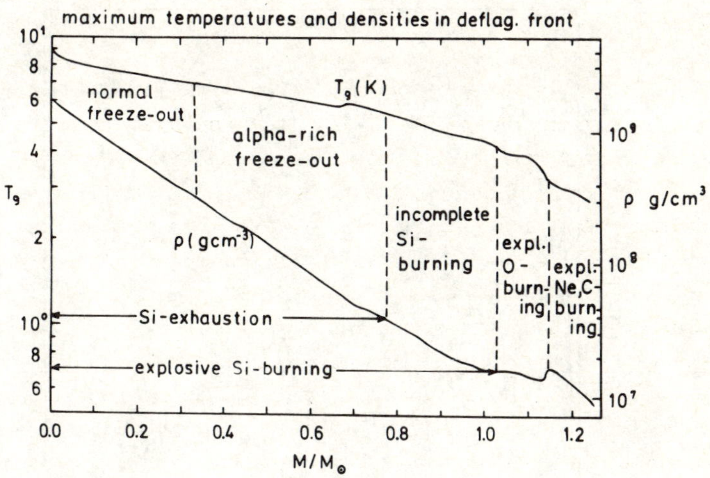

Figure 6.1. Properties of a carbon deflagration type I supernova model (from Thielemann, Nomoto, and Yokoi 1984). The temperature scale is given in units of 10^9 K.

(1981), Woosley, Axelrod, and Weaver (1984), Iben and Tutukov (1984), and Sutherland and Wheeler (1984). The general outcome is that the energy release in the explosive burning of carbon is sufficiently large to overcome the gravitational binding of the white dwarf, which is completely disrupted. Due to the somewhat lower temperatures in the deflagration front (as compared to a detonation wave), only the inner 0.7 to 0.8 $M_\odot$ of the white dwarf is incinerated into iron group nuclei; the rest undergoes incomplete Si burning, explosive O burning, Ne and C burning, respectively (see Figure 6.1). It therefore does not come as a surprise that these models can explain both the light curves and the early time spectra of type I supernovae (see, e.g., Branch et al. 1984). There may, however, be some problems with certain predictions of nucleosynthetic yields, which will be discussed later.

The overall success of the carbon deflagration models makes it tempting to believe that they indeed are the correct explanation of type I supernovae. One has to keep in mind, however, that the models are based on the assumption that both mixing and convective energy transport can be treated by a mixing-length type of theory. This in turn means that the typical scale for convection is assumed to be related to the pressure scale height. As has been shown by Müller and Arnett (1982, 1985), the convective flow in the carbon deflagration wave is likely to be highly turbulent and may even be nonspherical on large scales. Any relation to the pressure scale height is therefore at least questionable. Although the general picture of a carbon deflagration wave giving rise to a type I supernova explosion may not be affected by these considerations, detailed predictions such as isotopic abundances and synthetic spectra should be made with some care. Branch et al. (1984), for example, had to assume considerably more mixing than was predicted from their hydrodynamical model in order to obtain a good fit to the spectrum of SN 1981b (see Figure 6.2).

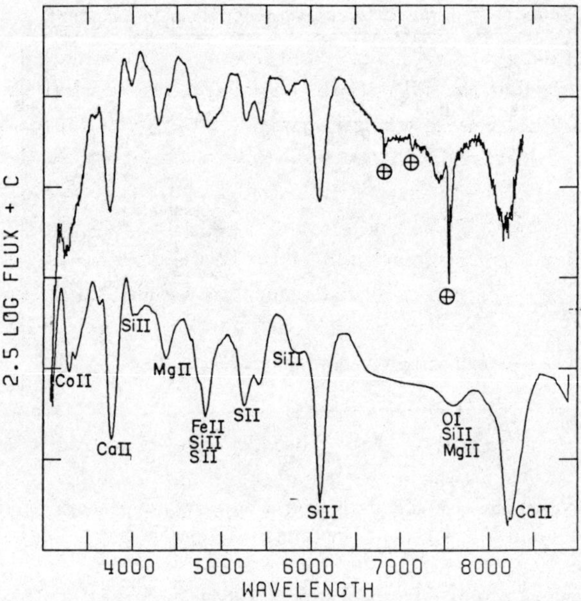

Figure 6.2. A comparison of the maximum-light spectrum of SN 1981b with the computed synthetic spectrum of a carbon deflagration model (from Branch et al. 1985). The vertical displacement between the two spectra is arbitrary. Crosses denote terrestrial absorption features in the observed spectrum.

6.3.1.4 *Alternative Models*

Cores of intermediate mass stars with main-sequence masses between about 6 and 9 $M_{\odot}$ may evolve to the stage of degenerate C burning in very much the same way as was described for C deflagration models based upon accreting C-O white dwarfs (Arnett 1969; Buchler and Mazurek 1975; Nomoto, Sugimoto, and Neo 1976). It seems more likely, however, that they lose their envelopes before the core has grown to the Chandrasekhar mass. If this turns out to be true, they should end their lives as C-O white dwarfs rather than type I supernovae, as is indicated by recent investigations of white dwarfs in young clusters (Weidemann and Koester 1983) and by the absence of very luminous stars on the asymptotic giant branch (Iben, this volume). On the other hand, it may be that the mass loss mechanism depends on the metallicity of the star, in which case in the early history of the Galaxy intermediate mass stars may also have contributed to the type I supernova rate (see, e.g., Tornambé 1984).

Next one can think of He white dwarfs as the progenitors of type I supernovae (Mazurek 1973; Nomoto and Sugimoto 1977), but these models face the same problems I have discussed in the context of single and double detonation models, because degenerate ignition of He always leads to a detonation wave and therefore the ashes of nuclear burning will be essentially pure ^{56}Ni. Moreover, the velocities of the ejecta predicted from He detonation models are much higher than those observed in type I supernovae.

Finally, O-Ne-Mg white dwarfs with a mass near the Chandrasekhar limit may also be thought of as progenitors of type I events. In this case electron captures on Mg and Ne will initiate the collapse of the white dwarf and Ne or O burning will not release enough energy to stop the collapse. A neutron star will be formed in very much the same way as discussed in the next section. Parts of the outer layers of the white dwarf will be ejected with rather high energies ($\gtrsim 10^{51}$ ergs), but the light curve and the spectrum will not resemble those of a typical type I explosion, because very little ^{56}Ni will be produced and the abundances of intermediate mass nuclei are likely to be small. Moreover, this scenario is in contradiction with the fact that in none of the young type I supernova remnants has a neutron star been found.

6.3.2 Models of Type II Supernovae

Supernovae of type II are now generally believed to originate from massive stars ($M \gtrsim 8\ M_\odot$ on the main sequence). Some arguments in favor of this hypothesis have been given in §6.2.1. A conclusion then is that they cannot be powered directly by thermonuclear burning (if the total mass is lower than about 100 $M_\odot$) and that they have to get their energy, at least partially, from another source. It is therefore intriguing to relate the type II phenomenon to the formation of neutron stars, an idea that goes back to Baade and Zwicky (1934) and is supported by the fact that several supernova remnants, including the Crab Nebula, do indeed contain neutron stars.

Detailed numerical models of core-collapse supernovae were first computed by Colgate and White (1966) and Arnett (1967). At that time it was believed that neutrinos leaking out of the proto-neutron star may either transfer momentum to the stellar envelope or heat the infalling matter sufficiently to cause an explosion that leaves behind it a neutron star. But already at that time the issue was quite controversial, mainly because only about a percent of the gravitational binding energy of a neutron star is needed in order to explain the observed energies of the outburst. Thus the mechanism has to be very inefficient and consequently is difficult to model. After the discovery of weak neutral currents it became clear, however, that the time scale for the diffusion of neutrinos out of the proto-neutron star is much longer than the hydrodynamical time scale, and therefore it was difficult to imagine a situation in which neutrinos could have any significant impact on the stellar explosion (Mazurek 1975; Sato 1975; Arnett 1977).

Because several reviews are available which describe the situation as it appeared to be until very recently and also give a comprehensive list of references (Trimble 1982, 1983; Hillebrandt 1982, 1984), I will only discuss here the general features of the core-collapse models.

Stars more massive than about 8 $M_\odot$ on the main sequence burn C non-degenerately forming central cores of O, Ne, and Mg. The further evolution is strongly dependent on the stellar entropy, which in turn depends on the main-sequence mass and the core mass. For He core masses below 3 $M_\odot$ (corresponding to main-sequence masses below 10 to 12 $M_\odot$) electron captures on Ne and Mg cause a rapid contraction of the central O-Ne-Mg core. If the core mass exceeds the critical mass for Ne ignition (1.37 $M_\odot$), Ne and O are both burned in a shell incinerating the matter from about 0.6 to 1.4 $M_\odot$ into a nuclear statistical equilibrium composition (Wilson et al. 1985). Slightly smaller cores contract until the central density exceeds 2×10^{10} g cm^{-3}, oxygen burning is ignited, and a deflagration front again transforms the matter into nuclear statistical equilibrium (Nomoto 1984*a;* Hillebrandt, Nomoto, and Wolff 1984). In both cases, however, the core mass exceeds the Chandrasekhar mass, which drops with decreasing electron concentration. Both types of models therefore predict that the core will collapse to neutron star densities. They also have in common that the iron core mass at the onset of collapse is rather small (around 0.7 to 1.4 $M_\odot$) and that their entropy is low ($\lesssim 1$ $k_{\rm B}$ nucleon^{-1}).

More massive stars, $M \gtrsim 12$ $M_\odot$, on the other hand, burn all their nuclear fuel nondegenerately. The mass inside the Si-burning shell depends on the main-sequence mass but is also sensitive to the $^{12}{\rm C}(\alpha,\gamma)^{16}{\rm O}$ reaction rate. Recent evolutionary calculations which use a rate for this reaction consistent with measurements of the Münster group (Kettner et al. 1982) predict that only stars with main-sequence masses around 15 $M_\odot$ will still have small core masses, whereas more massive stars ($M \gtrsim 20$ $M_\odot$) evolve iron cores of at least 1.9 $M_\odot$ (Wilson et al. 1985), significantly larger than was obtained before (Weaver, Woosley, and Fuller 1985). For these large iron cores the collapse is triggered by nuclear photodissociations rather than electron captures, but again the entropy is sufficiently low ($S \lesssim 2$ $k_{\rm B}$ nucleon^{-1}) that the collapse will not be stopped before nuclear matter density.

6.3.2.1 *General Features of the Core-Collapse Phase*

During core collapse the entropy per nucleon stays low, and therefore nucleons remain bound in heavy nuclei. Consequently the electron gas dominates the pressure and the adiabatic index is only slightly below the critical value of 4/3, as was first discussed by Bethe et al. (1979). This means that about a Chandrasekhar mass will collapse homologously (velocity proportional to distance from the star's center) (Goldreich and Weber 1980). The Chandrasekhar mass, however, is proportional to the square of the electron concentration Y_e. At the onset of collapse, Y_e is around 0.40 to 0.46. During the early stages of collapse neutrinos can still freely escape from the star, and therefore electron captures reduce Y_e to about 0.35 to 0.39 by the time neutrino diffusion times become comparable to the collapse time at a central density of a few 10^{11} g cm^{-3}. From then on neutrinos are essentially trapped in the collapsing cores, the neutrino Fermi sea is filled, and electron and neutrino captures approach β-equilibrium with only some 20% of the leptons being neutrinos. During this phase very little entropy is generated because most of the electrons are captured by free protons.

At the time the central density reaches nuclear matter density, the homologous inner core of the collapsing star contains about 0.7 to 0.8 $M_\odot$ and the outer part of the original iron core moves inward with supersonic velocity. Since beyond nuclear matter density the nuclei dissolve into a homogeneous fluid of nucleons, the equation of state is now dominated by the nucleons and becomes very stiff. The central zones of the star are stopped, and a sound wave carries this information to the sonic point at the edge of the homologous core. A shock wave forms while the matter inside the sonic point comes to rest in a few tenths of a millisecond. The total energy given to the shock wave can be estimated from the kinetic energy of the inner part of the core at the time the innermost mass zones begin to decelerate (Brown, Bethe, and Baym 1982). Typical values found in numerical computations are in the range from 4 to 8×10^{51} ergs.

6.3.2.2 *Prompt Explosions*

The matter that passes through the outward propagating shock front is heated to entropies of at least 6 k_B nucleon^{-1} and is therefore dissociated into free neutrons and protons. The energy taken from the shock is therefore 8×10^{18} ergs g^{-1} of dissociated matter. This in turn means that even the most energetic shocks found in numerical calculations can dissociate only about 0.5 $M_\odot$. The crucial quantity for successful shock propagation is thus the difference between the mass of the original iron core and its part that remains unshocked (approximately the homologous core at bounce). As has been discussed before, the mass of the iron core depends not only on the main-sequence mass of the star and certain nuclear reaction rates but also on its entropy, convective energy transport, neutrino cooling, and Y_e at the onset of collapse. Values that are found from stellar evolution codes range from 1 to 2 $M_\odot$. The mass of the homologous core, on the other hand, depends on details of the equation of state, on e^--capture rates, on the ν-transport model, and again on the entropy at the onset of collapse. Although different numerical simulations predict slightly different masses of the homologous core, they agree that it cannot be much larger than about 0.7 $M_\odot$ (Bowers and Wilson 1982; Hillebrandt 1984; Hillebrandt and Wolff 1985; Bruenn 1985). One therefore comes to the conclusion that at best for models with iron core masses of about 1.2 $M_\odot$ the shock wave contains enough energy for a prompt explosion. This then means that the core-bounce explosion mechanism is limited to a very narrow range of main-sequence masses, namely 9 ± 1 $M_\odot$. In one computation for such a low mass stellar model indeed a quite energetic explosion was found (Hillebrandt, Nomoto, and Wolff 1984), but this result has not been confirmed yet by an independent computation. There is general agreement, however, that more massive stars will not explode by this mechanism.

6.3.2.3 *Delayed Explosions*

Recently Wilson (1985) found that even if the hydrodynamical shock wave does not have enough energy to eject the envelope of a massive star, this may not exclude a supernova explosion, but that the neutrino heating may revive the shock on time scales of 0.5 to 1 s. His mechanism can be understood as

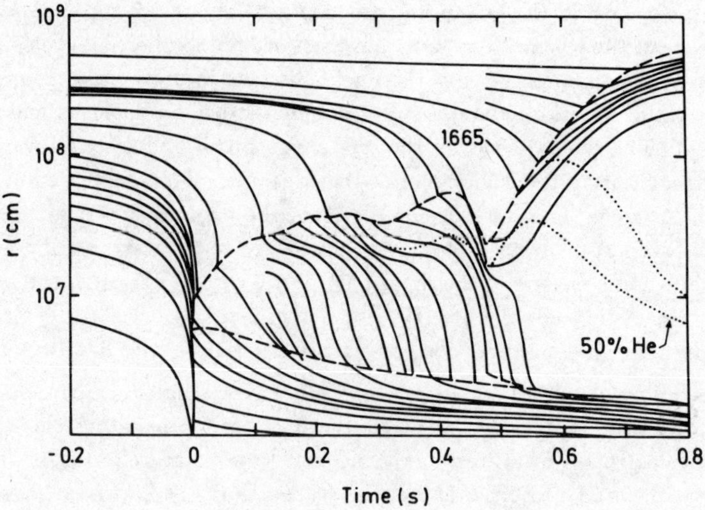

Figure 6.3. Radius versus time for selected mass zones of a 25 $M_\odot$ supernova model (from Wilson 1985). Time is measured from core bounce. The position of the shock front (*upper dashed curve*) as well as of the ν-sphere (*lower dashed curve*) is also shown.

follows (see Figure 6.3). A few hundred milliseconds after core bounce the shock has become a nearly standing accretion shock at a radius of several hundred kilometers. The mass inside the shock front is about 1.4 $M_\odot$, the density just behind it is roughly 10^7 g cm^{-3}, and the temperature has dropped to less than 1 MeV. The shocked matter being composed of free neutrons and protons is irradiated by a neutrino flux originating from the neutrino sphere at a radius of about 30 to 70 km, a density of about 10^{11} g cm^{-3}, and a temperature of about 3 to 5 MeV. Most of these neutrinos are created by e^- and e^+ captures on protons and neutrons in the shocked matter behind the ν-sphere but outside the unshocked inner core. One can estimate the energy density of the neutrinos and antineutrinos from this process from the following expression:

$$Q_{e^-,e^+} \simeq 5 \times 10^{10} n_{n,p} \left(\frac{kT}{m_e c^2} \right)^6 F_5 \left(\frac{\mu_{e^-,e^+}}{kT} \right) \text{ [ergs cm}^{-3}\text{ s}^{-1}],$$

where $n_{n,p}$ are the neutron and proton number densities, respectively, m_e is the electron rest mass, μ_{e^-} and μ_{e^+} are the electron and positron chemical potentials, and F_5 is the fifth relativistic Fermi integral. All other quantities have their usual meaning. Since in the shock gas electron and positron densities are approximately equal, their chemical potential is roughly zero and $F_5(0) \simeq 120$. Assuming also that $n_n \simeq n_p$, one finds

$$Q_{e^-} \simeq Q_{e^+} \simeq 2.4 \times 10^{28} x_n T_{\text{MeV}}^6 \rho_{10} \text{ [ergs cm}^{-3}\text{ s}^{-1}],$$

where x_n is now the neutron concentration. By integrating over the shocked gas behind the ν-sphere one finds a neutrino (and antineutrino) luminosity of about (2–4) $\times 10^{52}$ ergs s^{-1}, in good agreement with numerical results (see

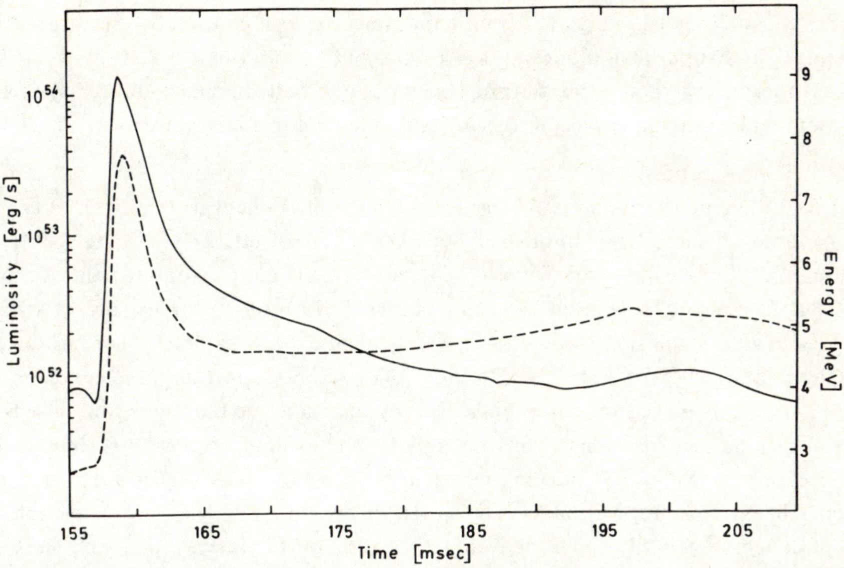

Figure 6.4. Neutrino luminosity (*solid curve*) and average neutrino energy (*dashed curve*) versus time for a 20 $M_\odot$ supernova model (from Hillebrandt and Müller 1984). Time is measured from the start of the computations. The peak luminosity corresponds to the time when the shock front crosses the ν-sphere.

Figure 6.4). It should be noted that on time scales less than 1 s the neutrinos leaking out of the unshocked inner core can contribute little to the neutrino luminosity, since the diffusion time scales are too long. Also, the luminosity from plasma neutrinos is at least a factor of 4 smaller. Some of these neutrinos and antineutrinos will be absorbed by neutrons and protons far outside the ν-sphere and will be reemitted with lower energy in the form of a blackbody gas. The maximum temperature the matter can reach by this heating process is a certain fraction of the neutrino temperature (see, e.g., Lattimer and Burrows 1984; Bethe and Wilson 1985), and it is therefore obvious that the temperature at the ν-sphere is crucial for this heating mechanism. In Wilson's computations the radius of the ν-sphere shrinks to about 30 km after a few hundred milliseconds, and the neutrino temperatures he finds are about 5 MeV. These properties of his models seem to be just sufficient for weak explosions, which, however, can hardly explain typical type II light curves unless energy is added from nuclear burning in the oxygen shell (Wilson et al. 1985). I recently computed the evolution of the 20 $M_\odot$ model of Weaver, Woosley, and Fuller (1985) for about 0.6 s after core bounce and did not find an explosion, probably because in my computation the radius of the ν-sphere stayed at around 70 km and the neutrino temperature there was only 3.6 MeV. Further investigations therefore seem to be necessary to confirm Wilson's results.

6.4. Implications for Models of Nucleosynthesis

From what has been discussed in the preceding section it is clear that predictions of nucleosynthesis yields from supernova explosions are still quite uncertain. Type I supernova models, for example, suffer from the fact that convective transport in thermonuclear burning fronts is not well understood. For type II supernovae even the explosion mechanism is not known beyond doubt.

6.4.1 Type I Supernovae

Detailed nucleosynthesis studies based on carbon deflagration models of type I supernovae have been published by Woosley, Axelrod, and Weaver (1984), Nomoto, Thielemann, and Yokoi (1984), and Thielemann, Nomoto, and Yokoi (1984) (see also Thielemann 1984 for a review). In order to be consistent with light curve information, only those models should be considered that produce at least about 0.5 $M_\odot$ of ^{56}Ni. It then follows that type I supernovae should predominantly eject iron group nuclei if they indeed are exploding white dwarfs. In addition, contributions from incomplete Si burning, explosive O burning, and explosive Ne and C burning can also be expected (see Figure 6.1). If one normalizes the composition of the ejecta in such a way that ^{56}Fe has solar abundance, several isotopes of iron (^{57}Fe and, particularly, ^{54}Fe) and nickel (^{58}Ni and ^{62}Ni) are overproduced by about a factor of 2 to 5. Proton-rich isotopes of Si, S, Ar, and Ca are underproduced by about a factor of 2 to 3, whereas the rest of the isotopes of the elements from carbon to germanium are unproduced by at least a factor of 5 (see Figure 6.5). If, therefore, type I supernovae were the main producers of ^{56}Fe in the Galaxy, they could also contribute to the abundances of ^{28}Si, ^{32,33}S, ^{36}Ar, and ^{40}Ca. The predicted overabundances of ^{54}Fe and ^{58,62}Ni would then be a problem that cannot easily be removed. They result from the fact that the neutron excess in the inner nuclear statistical equilibrium zones of the exploding white dwarf is somewhat large, which is a consequence of both e^- captures at high densities and the initial metallicity. The latter has been assumed to be solar in all model computations, an assumption that may not be true if the exploding white dwarfs stem from an old population of stars. But even if one reduces the initial metallicity (^{22}Ne abundance) significantly, the problem is only partially removed because most of the ^{54}Fe and ^{58,62}Ni is formed in the inner zones of the star where the neutron excess is mostly a result of e^- captures (Thielemann 1984).

Two more arguments can be given in favor of the idea that C deflagration supernovae are indeed the main producers of Fe and Ni in the Galaxy. First, the models that predict that per event more than 0.5 $M_\odot$ of iron group elements is ejected not only fit the light curves very well but also give very good fits to the spectra (see Figure 6.2) (Branch et al. 1984). Second, the observed overabundance of O relative to Fe in metal-poor stars indicates that at least some of the iron has been produced with some time delay during the history of the Galaxy (see Pagel, this volume, for a review). If type I supernovae are exploding white dwarfs, this fact would have a natural explanation, because their progenitors are intermediate mass stars and they may be quite old.

The nucleosynthesis products of off-center detonation models, finally, are

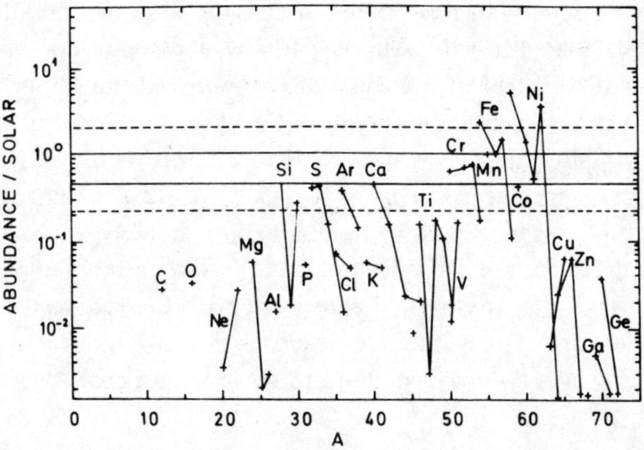

Figure 6.5. Predicted abundances from a carbon deflagration type I supernova model relative to solar abundances (from Thielemann, Nomoto, and Yokoi 1984). Isotopes of the same element are connected.

not very interesting because they produce essentially only ^{56}Fe via ^{56}Ni. They are facing the problem that this prediction is inconsistent with the spectral information one has from type I supernovae, and therefore they cannot explain a typical event.

6.4.2 Type II Supernovae

At present it seems that there may be at least two distinct classes of type II supernovae: those which originate from a narrow mass range around 9 $M_\odot$ and explode by the core-bounce mechanism and a second group of less explosive energy from more massive progenitor stars, for which the explosion is caused by neutrino heating. Although a large fraction of the observed type II supernovae may belong to the first class, they will not contribute much to galactic nucleosynthesis.

Detailed computations of the yields from the model of Hillebrandt, Nomoto, and Wolff (1984) have not been performed yet, but one can obtain rough estimates from both the composition of the pre-supernova star and the properties of the model when the shock has reached the edge of the O-Ne-Mg core. From the fact that in those stars the mass inside the C-burning shell is less than 1.4 $M_\odot$ and the neutron star that is forming will have at least 1.25 $M_\odot$, one can conclude that at most about 0.1 $M_\odot$ of heavy elements can be ejected in such models. Because, on the other hand, the shock is quite strong and the temperatures will certainly exceed 5×10^9 K for the matter inside the core, these ejecta will be burned into a nuclear statistical equilibrium and therefore will be composed of iron group nuclei. Some percent of this matter will be very neutron-rich, as was found by Hillebrandt, Nomoto, and Wolff (1984), and may even be a site of the r-process (see also the next section).

The main ejecta, however, will consist of the unprocessed H-He rich envelope of the pre-supernova star, which has partially undergone CNO burning and maybe also some He burning. Nomoto (1984*b*) finds that for He core masses below 2.3 $M_\odot$ very little He is burned in a shell. The ejecta will therefore be enhanced in N (a factor of ~4 when mixed with the rest of the star), slightly enhanced in C (a factor of ~1.5), and depleted in O (a factor of ~2). If, on the other hand, the He core is more massive than 2.3 $M_\odot$, He burns in a shell. In this case the enhancement of N is reduced to a factor of ~2, C is enhanced by a factor of 2.2, and the mass fraction of O becomes approximately solar. Although these enhancement factors are certainly too small to have a large impact on galactic nucleosynthesis, they are in good agreement with the composition observed in the Crab Nebula. It may therefore well be that a large fraction of the type II supernova events do not contribute to the bulk nucleosynthesis yields in the Galaxy.

The composition of the ejecta of more massive stars exploding by the neutrino-heating mechanism has recently been computed by the California group (Wilson et al. 1985). They again find two distinct cases. A model of 15 $M_\odot$ with an initial iron core mass of 1.33 $M_\odot$ gave rise to a very weak explosion and the formation of a neutron star of 1.42 $M_\odot$ with the possibility that an additional 0.7 $M_\odot$ may reimplode onto the neutron star. Due to the low explosion energy the ejecta had essentially the composition of the pre-supernova stellar model, with the exception of most of the iron group nuclei which were buried in the neutron star. The net result of this computation was a large enhancement (by a factor of 20 to 50) of the elements from Si to Ca, but much smaller enhancement factors ($\lesssim 5$) for all other elements. The large enhancement of Si through Ca would disappear, however, if the inner zones did reimplode.

A 25 $M_\odot$ model computed by the same group, on the other hand, showed remarkable agreement with solar system abundances, the abundances being enhanced relative to their solar system values on the average by a factor of about 10 with a scatter of about a factor of 2. This was caused by the fact that, although the initial iron core mass was larger (2.2 $M_\odot$ due to the revised $^{12}C(\alpha,\gamma)^{16}O$ reaction rate) and so was the neutron star (or black hole?) mass (2.44 $M_\odot$), the shock energy was sufficiently high (10^{51} ergs) for explosive Si and O burning. It is interesting to note that if O, Si, S, Ca, and Fe, which have the largest enhancement factors, are produced in stars of this mass range, C cannot come from the same objects, because relatively little C is produced, again due to the increased $^{12}C(\alpha,\gamma)$ rate. Moreover, if O is predominantly produced in stars around 25 $M_\odot$, Si, S, Ca, and Fe group elements must come from these sources as well unless a significant fraction of the products of explosive Si burning are reimploding onto the neutron star (or black hole). This latter possibility cannot be excluded at present because of the uncertainties in the supernova models.

6.4.3 *r*-Process Nucleosynthesis in Supernovae

So far I have only considered supernovae as the potential sources of the elements from carbon to the iron group. But since publication of the classic papers of Al Cameron (1957) and B^2FH (1957), it has also been believed that the neutron-rich isotopes of the heavier elements are of supernova origin. The process thought to be responsible is the rapid neutron capture (r-) process. The observed solar system r-process abundance distribution shows very pronounced maxima at atomic masses $A \simeq 80$, 130, and 195. The straightforward interpretation is that these maxima are a consequence of closed neutron shells at $N = 50$, 82, and 126, which means that the process that has led to their formation must have involved nuclei some 20 mass numbers away from the valley of β-stability on the neutron-rich side. This, on the other hand, is only possible if a certain seed composition is exposed to an extremely strong neutron flux such that neutron capture times become at least comparable to β^--decay half-lives.

If we assume for the moment that, under r-process conditions, the neutron density and the temperatures are sufficiently high to establish an (n,γ)-(γ,n) equilibrium in each isotopic chain on the time scales of typical β^--decays, the r-process path can be obtained from the condition $(\partial n(Z,A,T)/\partial A)_{Z,T=0}$, where $n(Z,A,T)$ denotes the number density of isotopes in a chain Z at given temperature T. By using the Boltzmann gas expressions for the number densities and by assuming constant partition functions, one can obtain a rough relation between the neutron number density, the temperature, and the neutron separation energy S_n, namely,

$$n_n \simeq 1.23 \times 10^{34} T_9^{-3/2} \exp\{-11.6\ S_n/T_9\},$$

where T_9 is the temperature in units of 10^9 K. In order to fit the positions of the observed abundance peaks, the neutron separation energy along the r-process path has to be around 3 MeV for any reasonable mass formula, which means that in a ($\log T$)-($\log \rho$) diagram one can draw lines of constant neutron concentration indicating roughly the required r-process conditions (see Figure 6.6). As can be seen from Figure 6.6, the neutron densities needed to drive the r-process increase very rapidly with increasing temperature but decrease with increasing density. It is therefore easier to find the right conditions at relatively low temperatures. These considerations, however, are only very qualitative because they neglect the fact that a certain number of neutrons per seed nucleus have to be available for a long enough time that β-decays can transform a given seed composition into an r-process distribution.

Neutron sources that have been discussed in some detail in the literature are either neutrons from electron captures on free protons at high densities and temperatures or neutrons from α captures on ^{22}Ne, ^{18}O, or ^{13}C in explosive He burning.

The first alternative can only work if the neutrino chemical potential in the matter under consideration is essentially zero so that the inverse reaction $\nu + n \rightarrow e^- + p$ can be neglected. In the case of type II supernova models this means that the density has to be smaller than about 10^{10} g cm^{-3}. In addition, electron capture times have to be smaller than the hydrodynamic expansion time scale, which is only the case if the density is at least of the order of 10^{10} g cm^{-3}. It therefore follows that this type of r-process can only happen in a very narrow

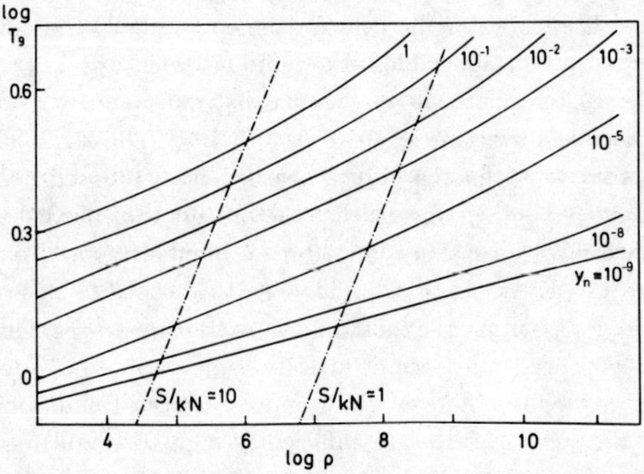

Figure 6.6. Neutron concentrations (Y_n) required in order to fit the solar r-process abundance peaks for a given density ρ (in g cm^{-3}) and temperature T (in 10^9 K). It can be seen that low densities and high temperatures imply high neutron concentrations whereas for lower temperatures ($T_9 \lesssim 1$) only moderate neutron concentrations are needed. Lines of constant entropy are also given.

density range near the neutron star that is forming, and in most cases this matter will not be ejected. Hillebrandt, Nomoto, and Wolff (1984) found that only in extremely strong shocks may some fraction of this neutronized matter be ejected, and they found indeed three mass zones with together $2 \times 10^{-2}\ M_\odot$ which obtained escape velocity with neutron concentrations between 0.2 and 0.3. There are, however, large numerical uncertainties, because these results are strongly dependent on the zoning, the neutrino transport scheme, etc., and it is therefore difficult to make any reliable predictions. Moreover, if 2×10^{-2} $M_\odot$ of r-process nuclei were ejected in some very energetic type II events, those supernovae would have to be extremely rare (less than a percent of all events) in order to avoid large overproductions. It is therefore hard to believe that the highly neutronized matter is the site of the r-process. One may speculate, however, that the parameter that selects the few r-process producing supernovae from the overwhelming majority is not the stellar mass but that only in jets from rapidly rotating, highly magnetized cores some neutron-rich matter is ejected (Meier et al. 1976; Symbalisty, Schramm, and Wilson 1984). Those events then might indeed be rare.

Motivated by these difficulties of the standard r-process scenario, investigators have searched for alternative sites. Hillebrandt and Thielemann (1977) and Truran, Cowan, and Cameron (1978) have suggested that the supernova shock front passing through the He-burning shell of a massive star might liberate enough neutrons by the $^{22}\mathrm{Ne}(\alpha,n)^{25}\mathrm{Mg}$ reaction to drive an r-process. In contrast to the standard scenario the seed composition on which the r-process operates will not be neutron-rich iron group isotopes but a solar system–like abundance distribution, modified by slow neutron captures and β^--decays during hydrostatic shell He burning. This means that the strong neutron flux expected during the passage of the shock front only has to transform an s-process–like abundance distribution into an r-process one.

From Figure 6.6 one can see that, for typical conditions of explosive He burning in massive stars ($\rho_{\max} \simeq (10^4–10^5)$ g cm^{-3}, $T_{\max} \simeq (5–8) \times 10^8$ K), a neutron concentration of at least 10^{-10} to 10^{-9} is required for about 0.5 to 1 s to reproduce the r-process abundance peaks. Numerical models show, however, that the ^{22}Ne neutron source cannot supply enough neutrons for more than a tenth of a second (Blake et al. 1981; Cameron, Cowan, and Truran 1983, 1985; Klapdor et al. 1981), in particular since at the above conditions neutron capture times cannot be neglected and the assumption of a $(n,\gamma) \rightleftharpoons (\gamma,n)$ equilibrium breaks down. The good fit obtained by Klapdor et al. (1981) from their model was merely caused by the assumption of an (n,γ)-(γ,n) equilibrium.

Therefore Cowan, Cameron, and Truran (1983) have suggested that ^{13}C may also act as a neutron source in explosive He burning. ^{13}C can be formed in a He shell if some protons are mixed in via the reactions $^{12}\mathrm{C}(p,\gamma)^{13}\mathrm{N}(\beta^+)^{13}\mathrm{C}$. They found that an admixture of a few percent by mass of ^{13}C is necessary to obtain a good fit to the solar system r-process abundances from explosive He burning (see Figure 6.7). It is also obvious from their figure, however, that although the positions and the relative height of the r-process abundance peaks are well reproduced by their model the abundances in between the peaks are certainly too low, which may result from β-decay half-lives that are too short. In addition, it remains to be shown that such a large amount of ^{13}C can actually be present in the He shells of massive stars. So in conclusion I think it is fair to say that the site of the r-process is not known at present, but one has to keep in mind that among the various possibilities only a very limited range of conditions has been investigated in some detail.

6.4.4 Some Remarks on ^{48}Ca

One of the puzzles in nucleosynthesis is still the large abundance ratio of ^{48}Ca to ^{46}Ca observed in the solar system material ($^{48}\mathrm{Ca}/^{46}\mathrm{Ca} \simeq 56$ [Cameron 1982]). In addition, isotopic anomalies (Niederer, Papanastassiou, and Wasserburg 1981) are also found for these Ca isotopes. Cameron (1969) and recently Hartmann, Woosley, and El Eid (1984) have noted that the fact that ^{48}Ca has such a large abundance is caused by binding energy effects, ^{48}Ca is a doubly magic nucleus, and therefore it should have been formed in a neutron-rich nuclear statistical equilibrium environment or in neutron-rich Si burning. In particular, Hartmann et al. could show that both the solar system ^{48}Ca and the isotopic anomalies in the Allende FUN inclusions can be reproduced if some supernova ejecta are sufficiently neutron-rich (neutron excess ~0.1–0.2) and have passed through a nuclear statistical equilibrium phase. They also concluded that for an average supernova this process should only involve less than 10^{-3} $M_\odot$ in order to avoid overproductions, which, similar to the r-process discussed earlier, makes theoretical predictions from a particular supernova model very difficult.

Sandler, Koonin, and Fowler (1982), on the other hand, have suggested that

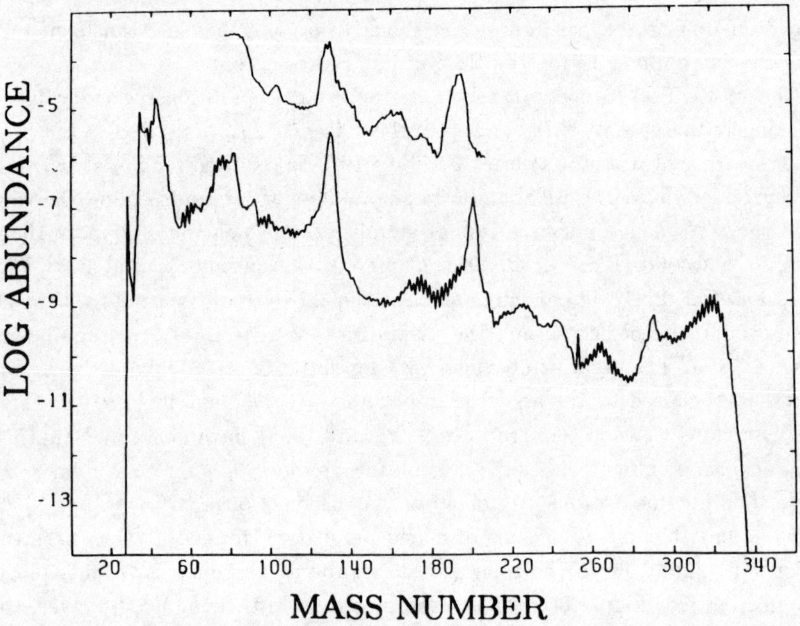

Figure 6.7. Computed r-process abundances (*lower curve*) compared to solar system values (from Cameron, Cowan, and Truran 1985). The yield curve was obtained from an r-process in the helium shell of an exploding massive star and a ^{13}C initial abundance of 5.2% by mass.

the isotopic anomalies found in the Ca isotopes as well as those of neutron-rich Ti isotopes should be attributed to an n-process in which a solar system seed composition is subject to a neutron exposure of about 10^{-4} moles cm^{-3}s, which for a neutron density of 10^{-7} moles cm^{-3} would correspond to an exposure time of 10^3 s. Under such conditions both neutron capture and β^--decay rates are of the same order and the nucleosynthesis path bypasses ^{46}Ca if the neutron capture cross section of ^{46}K predicted from the statistical model is increased by a factor of 10. Similarly, agreement for ^{49}Ti is obtained in the same astrophysical scenario if the ^{49}Ca$(n,\gamma)^{50}$Ca reaction rate is increased by the same factor. Sandler et al. argue that a resonance in ^{47}K and ^{50}Ca near the neutron separation energy can lead to such an increase if the γ-widths are a substantial fraction of a Weisskopf unit. In the case of ^{50}Ca such a resonance has indeed been identified at 155 keV (Kratz et al. 1983), but upper limits found for the γ-width of this state are too small to have the desired effect on the neutron capture cross section of ^{49}Ca (Kratz et al. 1985).

These difficulties led us to investigate a slightly different mechanism (Kratz et al. 1985). If a solar composition of elements from S to Ca is irradiated by a strong neutron flux with an exposure of a few 10^{-5} moles cm^{-3}s for about 1 s, similar to what one would like to have for r-process scenarios, the abundances in the sulfur isotopic chain will peak at ^{44}S (closed neutron shell $N = 28$), at Cl the $N = 28$ shell will be crossed, the Ar isotopes will peak at $A = 48$, K will be

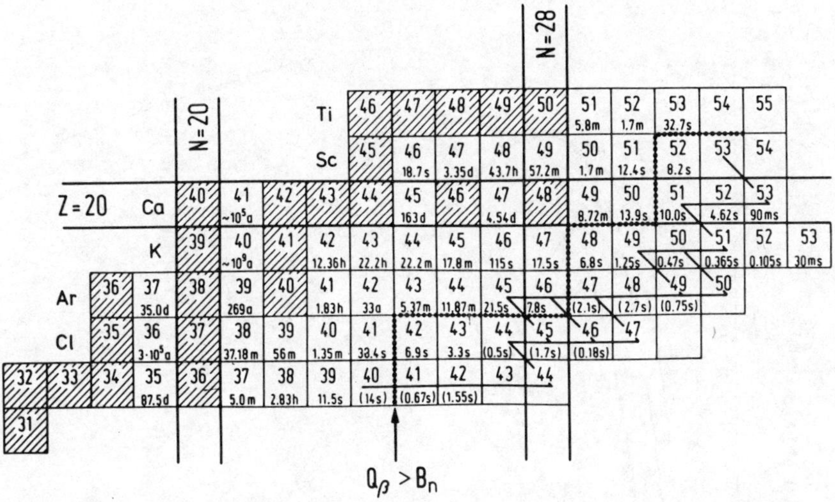

Figure 6.8. Nucleosynthesis path for neutron exposures of 10^{-5} to 10^{-4} moles cm^{-3}s. The dotted line indicates the regime where neutron separation energies become smaller than the β^--decay Q-value. Isotopes on the right have a finite probability to emit neutrons after β^--decay.

Table 6.3
Delayed Neutron Emission Probabilities (P_n) for Several Nuclei of Importance for Ca and Ti Isotopic Abundances (the P_n values are given in percent)

^{41}S	0.10	^{42}Cl	2×10^{-3}	^{47}Ar	0.44	^{48}K	1.14
^{42}S	13	^{43}Cl	3.0	^{48}Ar	27	^{49}K	90
^{43}S	0.24	^{44}Cl	40	^{49}Ar	2.1	^{50}K	30
^{44}S	32	^{45}Cl	7.0	^{50}Ar	20	^{51}K	50
^{45}S	28	^{46}Cl	45	^{51}Ar	7.5		
^{46}S	95	^{47}Cl	43				

transformed into ^{49,50,51}K, and so on. Due to the fact that for those nuclei the β^--decay Q-value is larger than the neutron separation energy, delayed neutron emission after β^--decay should play an important role and should significantly alter the final abundances obtained from such a process. It is obvious, however, that ^{46}Ca will predominantly be formed as ^{46}Cl and ^{47}Cl $(\beta^- n)$ and ^{48}Ca as ^{48}Ar (see Figure 6.8). So already from the solar ratio of Ar to Cl one can expect that the ^{48}Ca to ^{46}Ca ratio will also be large.

In order to analyze this process in more detail we have computed the delayed neutron emission probabilities for all nuclei of interest in the present context by using the RPA-shell-model code of Krumlinde and Möller (1984) (see Table 6.3 for a selection of important P_n values). We then assumed that β^--decays can be neglected on the short time scale of the neutron pulse as well as (γ,n) reactions. The composition in each isotopic chain was computed for a given neutron exposure, and the final abundances were obtained by summing over all decay channels (including delayed neutrons).

The results are shown in Figures 6.9 to 6.11. As could be expected, for a neutron exposure of about 5×10^{-5} moles cm^{-3}s, the mass 48 is predominantly

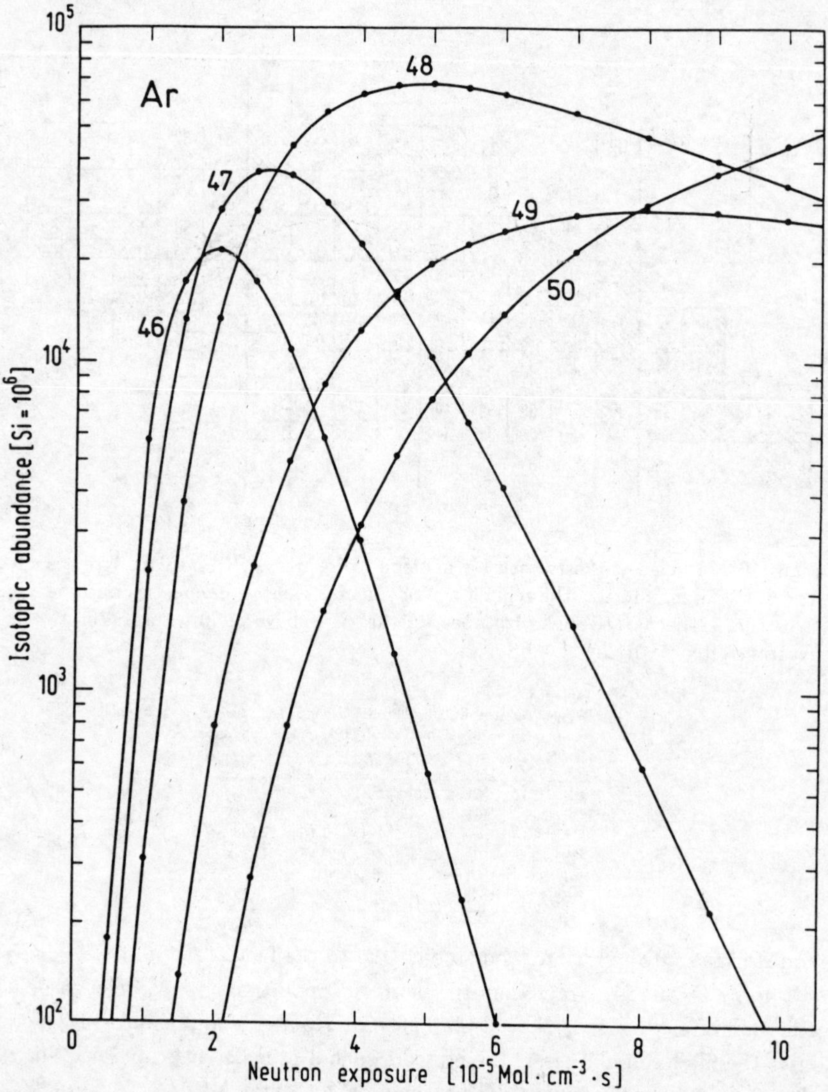

Figure 6.9. Abundances of Ar isotopes as a function of neutron exposure. For a rather broad range of exposures the most abundant isotope is ^{48}Ar which has a low P_n value as well as the decay product ^{48}K (see Table 6.2). Therefore ^{48}Ar will end predominantly as ^{48}Ca.

formed as ^{48}Ar and ends as ^{48}Ca. For the same neutron exposure the mass 46 comes mostly from ^{47}Cl($\beta^- n$) with a small contribution from ^{46}Cl. The ratio ^{48}Ca/^{46}Ca is found to be 40 for 5×10^{-5} moles cm^{-3}s and 64 for 6×10^{-5} moles cm^{-3}s, in very good agreement with the solar system ratio of 56. Our numbers will certainly change somewhat if a more extended reaction network is used in the computations, but it is clear that for a certain range of neutron exposures abundance ratios close to the solar system value will be obtained. It is also interesting to note that from our simple model we can predict isotopic

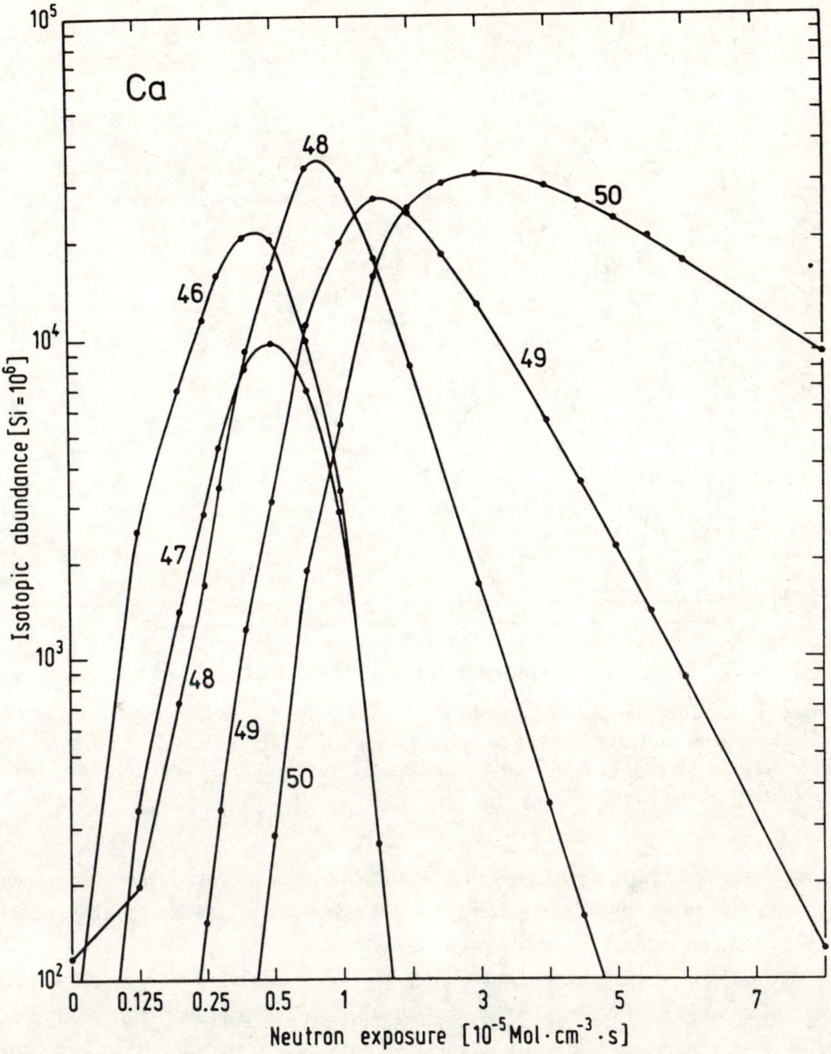

Figure 6.10. Abundances of Ca isotopes as a function of neutron exposure. For neutron exposures in excess of 2×10^{-5} moles $cm^{-3}s$ the most abundant isotopes are ^{49}Ca and ^{50}Ca, whereas the abundance of ^{48}Ca decreases very rapidly with increasing exposure.

anomalies in ^{46,48}Ca and ^{47,49,50}Ti *not* very different from those found in the FUN inclusions if a slightly higher neutron exposure of about 7×10^{-5} moles $cm^{-3}s$ is assumed (see Figure 6.12).

6.5. Summary and Open Questions

In this article I have reviewed the present status of our ideas about nucleosynthesis in supernovae, the main conclusion being that, although there are no indications that the basic theory developed in the late 1950s is not correct, many of the details have not been worked out yet. One reason is that a lot of the

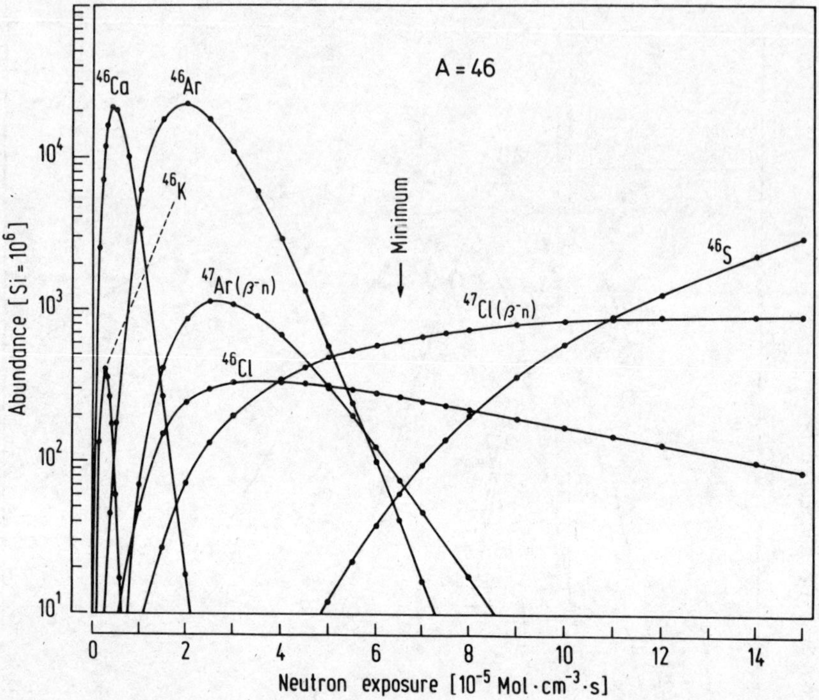

Figure 6.11. Contributors to mass 46 (^{46}Ca) for different neutron exposures. Since ^{47}Cl has a delayed neutron emission probability of 43% (see Table 6.2), the final ^{46}Ca abundance for an exposure of 7×10^{-5} moles cm^{-3}s will be roughly a factor of 2 lower than shown in the figure.

nuclear physics input data are still not known with sufficiently high accuracy, but also the supernova models have not yet reached the level of quality which would allow us to make detailed predictions.

Although it is very likely that C deflagration models are the correct explanation for most of the observed type I events, nucleosynthesis yields computed in this framework are still quite uncertain. Part of this uncertainty arises from the fact that a range of white dwarf models can explain the light curves and the spectra equally well but their ignition densities are quite different. Therefore the degree of neutronization also varies and thus the isotopic composition of the ejecta. These uncertainties in particular influence the abundance of ^{54}Fe and ^{58}Ni, which are overproduced in recent computations. A more severe problem is the numerical description of the deflagration front which so far has only been treated with parametrized models in one dimension or by two-dimensional codes with insufficient spatial resolution. Nevertheless, if the problem of the overproduction of ^{54}Fe and ^{58}Ni could be resolved, type I supernovae could contribute significantly to the cosmic iron abundance but very little to the rest of the heavy elements. Models of type II supernovae, on the other hand, predict that stars in the mass range from 8 $M_\odot$ to maybe a little over 10 $M_\odot$ on the main sequence

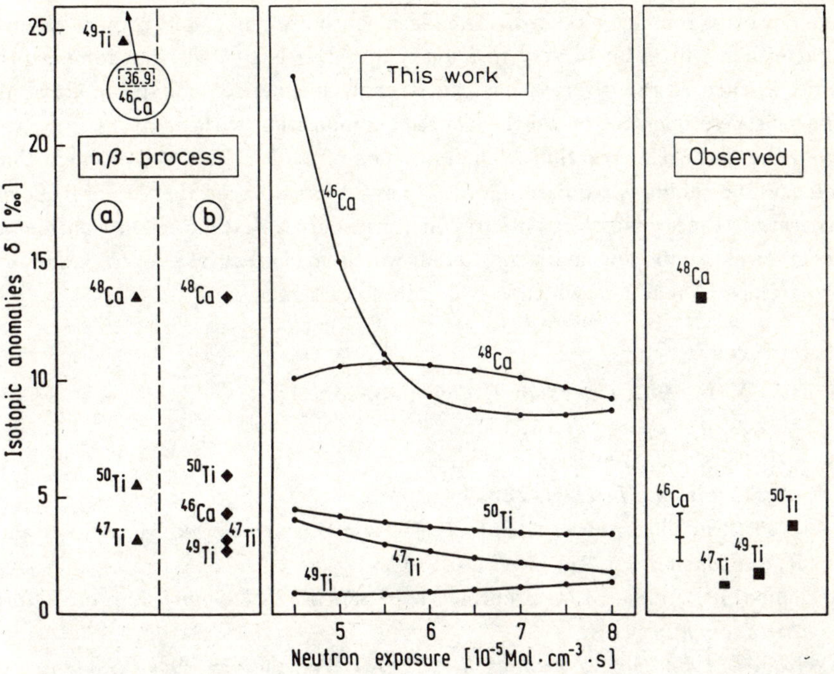

Figure 6.12. Isotopic anomalies predicted for Ca and Ti isotopes for different neutron exposures compared with the observed anomalies in EK-1-4-1 (Niederer, Papanastassiou, and Wasserburg 1981) and the $n\beta$-process of Sandler, Koonin, and Fowler (1982). (a) Results for standard Hauser-Feshbach neutron capture rates. (b) Capture rates for ^{46}K and ^{49}Ca, increased by a factor of 10.

explode by the core-bounce mechanism leaving behind neutron stars. The ejecta of those stars will have Crab-like composition and therefore will not contribute significantly to the heavy element production, with the possible exception of neutron-rich isotopes of certain elements. The main problem with these models is that their predictions are very sensitive to the pre-supernova stellar models, the input physics, and the numerical methods.

The pre-supernova composition of more massive stars, finally, is very similar to that of the solar system, and they will eject the elements from O to the iron group with enhancement factors sufficiently large to explain the abundances in the galactic material, provided the explosion energy is about 10^{51} ergs. In particular, stars more massive than about 20 $M_\odot$ have this property. It is interesting to note that these types of stars possess enough nuclear fuel to power the supernova outburst by nuclear burning (in particular by O burning). Therefore even low energy shocks as predicted by the neutrino heating mechanism will finally give rise to quite energetic explosions and to ejecta that have the right chemical composition.

Therefore, while it seems to be likely that supernovae from massive stars are the main producers of the elements from oxygen to iron and nickel, the site of the r-process remains unknown. The ideas which are pursued at present assume that either a little bit of very neutron-rich matter is ejected from some exotic objects, such as rapidly rotating and strongly magnetized collapsing cores, or the r-process operates on a solarlike seed composition with neutrons liberated from thermonuclear reactions. The first class of models faces the problem that a lot of fine-tuning is required in order to avoid large overproduction of r nuclei, whereas in the second class potential neutron sources that have been considered seem to release too few neutrons. Here it would be of great interest to remeasure the $^{22}Ne(\alpha,n)$ reaction, which is still quite uncertain.

References

Arnett, W. D. 1967, *Canadian J. Phys.,* **45,** 1621.

——. 1969, *Ap. Space Sci.,* **5,** 180.

——. 1977, *Ap. J.,* **218,** 815.

——. 1979, *Ap. J. (Letters),* **230,** L37.

——. 1982, in *Supernovae: A Survey of Current Research,* ed. M. J. Rees and R. J. Stoneham, p. 221 (NATO-ASI C90).

Aschenbach, B. 1985, MPE Preprint 31, in *18th ESLAB Symposium on X-Ray Astronomy,* in press.

Axelrod, T. 1980*a,* in *Proceedings of the Texas Workshop on Type I Supernovae,* ed. J. C. Wheeler, p. 80 (Austin: University of Texas Press).

——. 1980*b,* Ph.D. thesis, University of California at Santa Cruz.

Baade, W., and Zwicky, F. 1934, *Phys. Rev.,* **45,** 138.

Becker, R. H.; Holt, S. S.; Smith, B. W.; White, N. E.; Bolt, E. A.; Mushotzky, R. F.; and Serlemitsos, P. J. 1979, *Ap. J. (Letters),* **234,** L73.

——. 1980*a, Ap. J. (Letters),* **235,** L5.

Becker, R. H.; Bolt, E. A.; Holt, S. S.; Serlemitsos, P. J.; and White, N. E. 1980*b, Ap. J. (Letters),* **237,** L77.

Becker, R. H.; Szymkowiak, A. E.; Bolt, E. A.; Holt, S. S.; and Serlemitsos, P. J. 1980*c, Ap. J. (Letters),* **246,** L33.

Bethe, H. A.; Brown, G. E.; Applegate, J.; and Lattimer, J. M. 1979, *Nucl. Phys.* **A324,** 487.

Bethe, H. A., and Wilson, J. R. 1985, *Ap. J.,* **295,** 14.

Blake, J. B.; Woosley, S. E.; Weaver, T. A.; and Schramm, D. N. 1981, *Ap. J.,* **248,** 315.

Bowers, R., and Wilson, J. R. 1982, *Ap. J.,* **263,** 366.

Branch, D. 1985, in *Nucleosynthesis: Challenges and New Developments,* ed. W. D. Arnett and J. W. Truran, p. 261 (Chicago: University of Chicago Press).

Branch, D.; Dogget, J. B.; Nomoto, K.; and Thielemann, F.-K. 1985, *Ap. J.,* **294,** 619.

Branch, D.; Falk, S. W.; McCall, M. L.; Rybski, P.; Nomoto, K.; and Wills, B. J. 1981, *Ap. J.,* **244,** 780.

Branch, D.; Lacy, C. H.; McCall, M. L.; Sutherland, P. G.; Nomoto, K.; Wheeler, J. C.; and Wills, B. J. 1983, *Ap. J.,* **270,** 123.

Brown, G. E., Bethe, H. A., and Baym, G. 1982, *Nucl. Phys.*, **A375**, 481.

Bruenn, S. W. 1985, *Ap. J. Suppl.*, **58**, 771.

Buchler, J. R., and Mazurek, T. J. 1975, *Mem. Soc. R. Sci. Liège*, **8**, 435.

Burbidge, E. M.; Burbidge, G. R.; Fowler, W. A.; and Hoyle, F. 1975, *Rev. Mod. Phys.*, **29**, 547 (B^2FH).

Cameron, A. G. W. 1957, *Pub. A.S.P.*, **69**, 201.

——. 1969, *Ap. J. (Letters)*, **230**, L53.

——. 1982, in *Essays in Nuclear Astrophysics*, ed. C. A. Barnes, D. D. Clayton, and D. N. Schramm (Cambridge University Press).

Cameron, A. G. W., Cowan, J. J., and Truran, J. W. 1983, *Ap. Space Sci.*, **91**, 235.

——. 1985, in *Nucleosynthesis: Challenges and New Developments*, ed. W. D. Arnett and J. W. Truran, p. 190 (Chicago: University of Chicago Press).

Canizares, C. R., and Winkler, P. F. 1981, *Ap. J. (Letters)*, **246**, L33.

Canizares, C. R.; Winkler, P. F.; Markert, T. H.; and Berg, C. 1983, in *IAU Symposium 101, Supernova Remnants and Their X-Ray Emission*, ed. J. Danziger and P. Gorenstein, p. 205 (Dordrecht: Reidel).

Chevalier, R. A., and Kirshner, R. P. 1979, *Ap. J.*, **219**, 931.

Colgate, S. A., and McKee, C. 1969, *Ap. J.*, **157**, 623.

Colgate, S. A., Petschek, A. G., and Kriese, J. T. 1980, *Ap. J. (Letters)*, **237**, L81.

Colgate, S. A., and White, R. H. 1966, *Ap. J.*, **228**, 682.

Cowan, J. J., Cameron, A. G. W., and Truran, J. W. 1983, *Ap. J.*, **265**, 429.

Davidson, K.; Gull, T. R.; Maran, S. P.; Stecher, T. P.; Fesen, R. A.; Parise, R. A.; Harvel, C. A.; Kafatos, M.; and Trimble, V. L. 1982, *Ap. J.*, **253**, 696.

Goldreich, P., and Weber, S. 1980, *Ap. J.*, **238**, 991.

Harkness, R. 1984, *Proceedings of the Harvard-Smithsonian Workshop on Supernovae as Distance Indicators*, in press.

Hartmann, D., Woosley, S. E., and El Eid, M. F. 1984, *Ap. J.*, in press.

Henry, R. B. C., and McAlpine, G. M. 1982, *Ap. J.*, **258**, 11.

Henry, R. B. C., McAlpine, G. M., and Kirshner, R. P. 1984, *Ap. J.*, **278**, 619.

Hillebrandt, W. 1982, in *Supernovae: A Survey of Current Research*, ed. M. J. Rees and R. J. Stoneham, p. 123 (NATO-ASI C90).

——. 1984, in *Proceedings XI Texas Symposium on Relativistic Astrophysics* (*Ann. N.Y. Acad. Sci.*, **422**, 197).

Hillebrandt, W., and Müller, E. 1984, in *Neutrino Physics and Astrophysics*, ed. K. Kleinknecht and E. A. Paschos, p. 229 (World Science Publishing).

Hillebrandt, W., Nomoto, K., and Wolff, R. G. 1984, *Astr. Ap.*, **133**, 175.

Hillebrandt, W., and Thielemann, F.-K. 1977, *Mitt. Astr. Gesellschaft*, **43**, 234.

Hillebrandt, W., and Wolff, R. G. 1985, in *Nucleosynthesis: Challenges and New Developments*, ed. W. D. Arnett and J. W. Truran, p. 131 (Chicago: University of Chicago Press).

Holt, S. S. 1983, in *IAU Symposium 101, Supernova Remnants and Their X-Ray*

Emission, ed. J. Danziger and P. Gorenstein, p. 17 (Dordrecht: Reidel).

Hughes, J. P., and Helfand, D. J. 1984, *Ap. J.,* **291,** 544.

Iben, I., Jr., and Renzini, A. 1983, *Ann. Rev. Astr. Ap.,* **21,** 271.

Iben, I., Jr., and Tutukov, A. V. 1984, *Ap. J. Suppl.,* **55,** 335.

Jansen, F. A.; McKechnie, S. P.; de Korte, P. A. J.; Bleeker, J. A. M.; Gronenschild, E.; Peacock, A.; Manzo, G.; Branduardi-Reymont, G.; and Kellett, B. 1984, COSPAR, Xray Astronomy, Graz.

Jones, L. R., Pye, J. P., and Culhane, J. L. 1984, in *Proceedings Symposium "X-Ray Astronomy '84,"* ed. Y. Tanaka (Bologna).

Kettner, K. V.; Becker, H. W.; Buchmann, L.; Görres, J.; Krähwinkel, H.; Rolfs, C.; Schmalbrock, P.; Trautvetter, H. P.; and Vliks, A. 1982, *Zs. Phys.,* **A308,** 73.

Klapdor, V.; Oda, T.; Metzinger, J.; Hillebrandt, W.; and Thielemann, F.-K. 1981, *Zs. Phys.,* **A299,** 213.

Kratz, K. L.; Hillebrandt, W.; Krumlinde, J.; Möller, P.; Thielemann, F.-K.; Wiescher, M.; and Ziegert, W. 1985, in *Nucleosynthesis and Its Implications on Nuclear and Particle Physics* (5th Moriond Astrophysics Meeting, Les Arcs).

Kratz, K. L.; Ziegert, W.; Hillebrandt, W.; and Thielemann, F.-K. 1983, *Astr. Ap.,* **125,** 381.

Krumlinde, J., and Möller, P. 1984, *Nucl. Phys.,* **A417,** 419.

Lattimer, J. M., and Burrows, A. 1984, in *Problems of Collapse and Numerical Relativity,* ed. D. Bancel and M. Signore, p. 147 (NATO-ASI C134).

Mazurek, T. J. 1973, *Ap. Space Sci.,* **23,** 365.

——. 1975, *Ap. Space Sci.,* **35,** 117.

Meier, D. L.; Epstein, R. I.; Arnett, W. D.; and Schramm, D. N. 1976, *Ap. J.,* **204,** 869.

Müller, E., and Arnett, W. D. 1982, *Ap. J. (Letters),* **261,** L109.

——. 1985, in *Nucleosynthesis: Challenges and New Developments,* ed. W. D. Arnett and J. W. Truran, p. 235 (Chicago: University of Chicago Press).

Murdin, P., and Clark, D. H. 1981, *Nature,* **294,** 543.

Niederer, F. R., Papanastassiou, D. A., and Wasserburg, G. J. 1981, *Geochim. Cosmochim. Acta,* **45,** 1017.

Nomoto, K. 1980, in *Proceedings of the Texas Workshop on Type I Supernovae,* ed. J. C. Wheeler, p. 164 (Austin: University of Texas Press).

——. 1982, *Ap. J.,* **257,** 780.

——. 1984*a,* in *Stellar Nucleosynthesis,* ed. C. Chiosi and A. Renzini, pp. 205 and 238 (Dordrecht: Reidel).

——. 1984*b, Ap. J.,* **277,** 791.

——. 1984*c,* in *Problems of Collapse and Numerical Relativity,* ed. D. Bancel and M. Signore, p. 89 (NATO-ASI C134).

Nomoto, K., and Sugimoto, D. 1977, *Pub. Astr. Soc. Japan,* **29,** 765.

Nomoto, K., Sugimoto, D., and Neo, S. 1976, *Ap. Space Sci.,* **39,** L37.

Nomoto, K., Thielemann, F.-K., and Yokoi, K. 1984, *Ap. J.,* **286,** 644.

Peacock, A. 1984, cited by Aschenbach (1985).

Pfeffermann, E.; Aschenbach, B.; Bräuninger, H.; Heinecke, N.; Ondrusch, A.; and Trümper, J. 1980, *Bull. AAS,* **11,** 789.

Pfeffermann, E.; Aschenbach, B.; Bräuninger, H.; and Trümper, J. 1981, *Space Sci. Rev.,* **30,** 251.

Pravdo, S. H., and Nugent, J. J. 1983, in *IAU Symposium 101, Supernova Remnants and Their X-Ray Emission,* ed. J. Danziger and P. Gorenstein, p. 29 (Dordrecht: Reidel).

Pravdo, S. H., and Smith, B. W. 1979, *Ap. J. (Letters),* **234,** L195.

Sandler, D. G., Koonin, S. E., and Fowler, W. A. 1982, *Ap. J.,* **259,** 908.

Sato, K. 1975, *Prog. Theoret. Phys.,* **54,** 1352.

Shull, J. M. 1982, *Ap. J.,* **262,** 308.

——. 1983, in *IAU Symposium 101, Supernova Remnants and Their X-Ray Emission,* ed. J. Danziger and P. Gorenstein, p. 99 (Dordrecht: Reidel).

Sutherland, P., and Wheeler, J. C. 1984, *Ap. J.,* **280,** 282.

Symbalisty, E. M. D., Schramm, D. N., and Wilson, J. R. 1984, *Ap. J.,* in press.

Szymkowiak, A. 1984, Ph.D. thesis, University of Maryland.

Thielemann, F.-K. 1984, *Adv. Space Res.,* **4,** no. 2-3, 67.

Thielemann, F.-K., Nomoto, K., and Yokoi, K. 1984, *Astr. Ap.,* in press.

Tornambé, A. 1984, *M.N.R.A.S.,* **206,** 867.

Trimble, V. 1982, *Rev. Mod. Phys.,* **54,** 1183.

——. 1983, *Rev. Mod. Phys.,* **55,** 522.

Truran, J. W., Cowan, J. J., and Cameron, A. G. W. 1978, *Ap. J. (Letters),* **22,** L63.

Tsunemi, H. 1985, talk given at MPE Garching.

Vartanian, M. H., Lum, K. S. K., and Ku, W. H.-M. 1984, *Ap. J. (Letters),* in press.

Weaver, T. A., Axelrod, T. S., and Woosley, S. E. 1980, in *Proceedings of the Texas Workshop on Type I Supernovae,* ed. J. C. Wheeler, p. 113 (Austin: University of Texas Press).

Weaver, T. A., Woosley, S. E., and Fuller, G. M. 1985, in *Numerical Astrophysics,* ed. J. Centrella, J.-Le Blanc, and R. Bowers, p. 374 (Boston: Jones & Bartlett).

Weidemann, V., and Koester, D. 1983, *Astr. Ap.,* **121,** 77.

Wheeler, J. C. 1982, in *Supernovae: A Survey of Current Research,* ed. M. J. Rees and R. J. Stoneham, p. 167 (NATO-ASI C90).

Wilson, J. R. 1985, in *Numerical Astrophysics,* ed. J. Centrella, J.-Le Blanc, and R. Bowers, p. 422 (Boston: Jones & Bartlett).

Wilson, J. R.; Mayle, R.; Woosley, S. E.; and Weaver, T. A. 1985, in *Proceedings XII Texas Symposium on Relativistic Astrophysics* (Jerusalem), in press.

Winkler, P. F., Canizares, C. R., and Bromley, B. C. 1983, in *IAU Symposium 101, Supernova Remnants and Their X-Ray Emission,* ed. J. Danziger and

P. Gorenstein, p. 245 (Dordrecht: Reidel).

Winkler, P. F.; Canizares, C. R.; Clark, G. W.; Markert, T. H.; Kalata, K.; and Schnopper, H. W. 1981, *Ap. J. (Letters),* **246**, L27.

Woosley, S. E., Axelrod, T. S., and Weaver, T. A. 1984, in *Stellar Nucleosynthesis,* ed. C. Chiosi and A. Renzini, p. 263 (Dordrecht: Reidel).

Woosley, S. E., and Weaver, T. A. 1981, *Ann. N.Y. Acad. Sci.,* **375**, 357.

Woosley, S. E., Weaver, T. A., and Taam, R. E. 1980, in *Proceedings of the Texas Workshop on Type I Supernovae,* ed. J. C. Wheeler, p. 96 (Austin: University of Texas Press).

Part III

Exotic Stars

7

Binary Star Evolution and Type I Supernovae

Icko Iben, Jr.

Abstract

Many ideas that have influenced the development of the theory of type I supernovae have been contributed by A. G. W. Cameron and his students. These include: the likelihood that binary stars are precursors of type I supernovae, the probable implication of a nuclear fuel which ignites in degenerate matter, and the recognition that the Urca process may be critical in the development of the explosion. This review will sketch a currently popular scenario for producing type I supernovae which makes use of these ideas.

7.1. Why Not Single Star Precursors?

An interpretation of the observations using modern theoretical models of evolving stars tells us that single stars less massive initially than about 9 $M_\odot$ become white dwarfs which are significantly less massive than the Chandrasekhar mass of 1.4 $M_\odot$. Theory tells us that stars initially more massive than 9 $M_\odot$ experience core collapse followed by an explosion which has the characteristics of type II supernovae.

The major observational evidence against the conversion of low and intermediate mass single stars into supernovae is the extreme paucity in the Magellanic Clouds and in our Galaxy of asymptotic giant branch (AGB) stars brighter than $L_{\max} \sim 2 \times 10^4$ $L_\odot$. AGB stars consist of a central, hot white dwarf surrounded by a helium-rich layer and a hydrogen-rich envelope. All single stars less massive than 9 $M_\odot$ pass through this stage of evolution. During most of the lifetime of an AGB star, hydrogen burns quiescently at the base of the envelope, adding mass to the helium-rich layer below it. Periodically, when the mass of the helium-rich

---

*The precise value of the mass which marks where evolutionary behavior diverges following the exhaustion of central helium depends sensitively on the composition (Becker and Iben 1979) as well as on the input physics (Paczyński 1970; Castellani et al. 1985). In particular, the manner in which mixing is assumed to occur near the outer edge of the convective core during core helium burning plays an important role and the outcome depends critically on the choice of opacity. Nevertheless, the existence of a critical (composition-dependent) initial mass $M_{\rm up}$ below which a star develops an electron-degenerate, carbon-oxygen core and above which it exhausts central carbon before developing an electron-degenerate core has been established, and in this paper the choice $M_{\rm up} \sim 9$ $M_\odot$ will be adopted for discussion purposes. Stars initially less massive than $M_{\rm up}$ will be called low and intermediate mass stars.

layer reaches a critical value, a helium-burning runaway occurs. The observations suggest that, when the luminosity of an AGB star exceeds a critical value larger than about $2 \times 10^4\ L_\odot$, either during a thermal runaway or during the preceding quiescent hydrogen-burning phase, it develops a very powerful "superwind" that ejects most of the hydrogen-rich matter from the star. During the quiescent hydrogen-burning phase, the stellar luminosity and the mass of the central white dwarf are closely related by $L \sim 6 \times 10^4\ L_\odot\ (M_{\rm core}/M_\odot - 0.5)$, and the critical luminosity of $2 \times 10^4\ L_\odot$ corresponds to a core mass of about $0.85\ M_\odot$. Thus, single stars that become AGB stars with core masses as large or larger than $0.85\ M_\odot$ are destined to become white dwarfs of mass essentially equal to the mass of the core they first develop on the AGB. This initial core mass varies from $0.85\ M_\odot$ for a star of main-sequence mass about $5\ M_\odot$ to $1.1\ M_\odot$ for a star of main-sequence mass about $9\ M_\odot$.

The behavior of single stars in the H-R diagram (as we currently believe it to be) is sketched in Figure 7.1, and the evolution in the density-temperature plane of matter at the centers of these stars is shown in Figure 7.2. Note particularly (Figure 7.2) that matter at the stellar center does not reach the carbon ignition line.

The realization that mass loss plays an important role in the life of low and intermediate mass stars during the AGB phase has grown steadily over nearly two decades, but the possibility that mass loss *terminates* this phase for *all* AGB stars prior to the development of explosive conditions in the electron-degenerate core has not been entertained seriously until relatively recently, with the establishment of a paucity of bright AGB stars in the Magellanic Clouds providing the critical observational evidence.

An example of the sort of thinking that was popular just a few years ago is shown in Figure 7.3. In this figure the final mass of a hypothetical AGB star and the core mass of this same star are given as a function of its initial mass $M_{\rm MS}$ and as a function of an assumed recipe for mass loss. That stars must lose mass at a substantial rate is clear from the fact that, if no mass loss at all were to occur, every intermediate mass star of mass larger than $1.4\ M_\odot$ would explode as a supernova (see, e.g., Hoyle and Fowler 1960; Arnett 1969; Nomoto, Sugimoto, and Neo 1976; Nomoto, Thielemann, and Yokoi 1984; Woosley, Axelrod, and Weaver 1984) and the frequency of supernovae would be a substantial fraction of the stellar birthrate. For example, with a stellar birthrate of about 1 $\rm yr^{-1}$ in our Galaxy and a Salpeter mass function (translated to mean a stellar birthrate per unit mass proportional to $M^{-2.35}$), no mass loss would imply about 1 supernova every 4 yr, compared to an estimated type I supernova rate of approximately one every 50 to 100 yr.

Using the full Reimers (1975) mass loss rate coupled with a prescription for planetary nebula ejection (Wood and Cahn 1977; Iben and Truran 1978; Renzini and Voli 1981), all stars less massive initially than $6.6\ M_\odot$ would become white dwarfs, whereas those initially of mass between 6.6 and $9\ M_\odot$ would produce

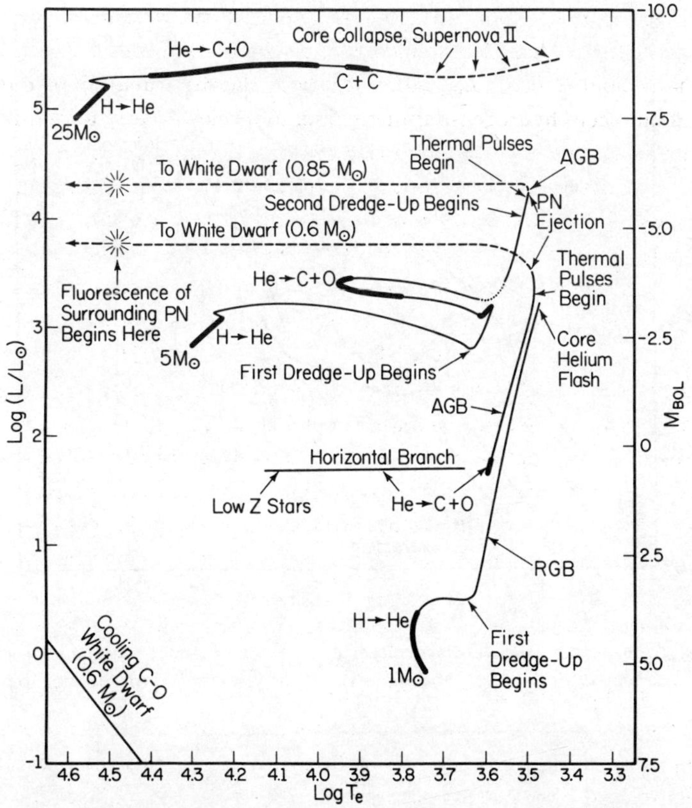

Figure 7.1. Evolution in the H-R diagram of single stars of low (1 $M_\odot$), intermediate (5 $M_\odot$), and high (25 $M_\odot$) mass during major nuclear-burning phases. Following the core helium-burning phase, stars of low and intermediate mass develop an electron-degenerate core and experience thermal pulses as AGB stars. They also lose mass rapidly via a stellar wind, eventually ejecting most of their hydrogen-rich envelope above the degenerate core. The remnant central star contracts, and photons from its surface light up the ejected material which is seen as a planetary nebula. Thereafter the remnants evolve into white dwarfs. After exhausting central helium, stars initially more massive than about 9 $M_\odot$ go on to burn carbon under nondegenerate conditions. Those more massive than about 12 $M_\odot$ continue to burn nuclear fuel until they have formed an electron-degenerate core composed of iron and nickel. This core collapses into a neutron star, and the envelope is ejected in consequence of deposition of energy into the base of the envelope by neutrinos emitted from the collapsing core. Stars in the 9–12 $M_\odot$ range also form neutron stars and eject their envelopes with energy supplied by explosive nuclear burning. Thus all single stars produce compact remnants (Iben 1985).

supernovae at a rate of about one per 50 yr. This prescription, then, would indicate that single AGB stars of initial mass in the range 6.6–9.0 $M_\odot$ might be the precursors of type I supernovae at the observed rate and that those less massive than 6.6 $M_\odot$ would, in consequence of wind mass loss, become white dwarfs of mass between about 0.6 $M_\odot$ and about 1.4 $M_\odot$. The observed

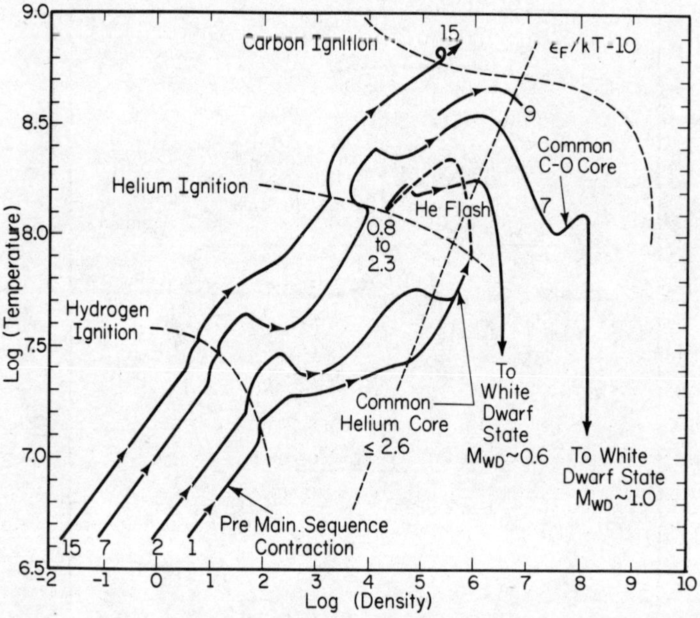

Figure 7.2. The path in the density-temperature plane of matter at the centers of low (1–2 $M_\odot$), intermediate (7 $M_\odot$), and massive (15 $M_\odot$) stars (adapted from Iben 1974).

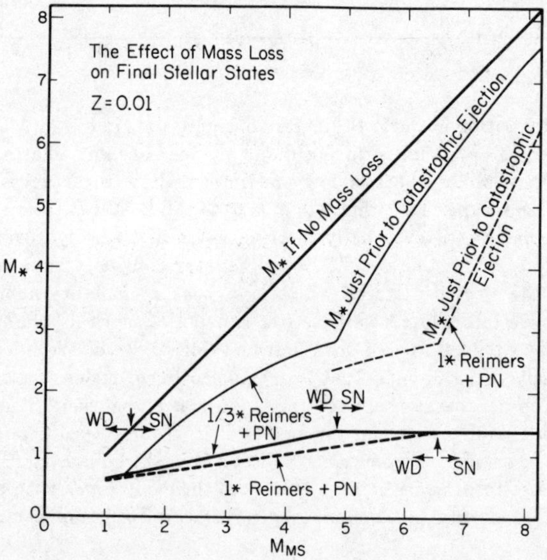

Figure 7.3. Remnant white dwarf mass as a function of initial mass and mass loss algorithm (Iben 1981).

similarity among type I outbursts could be ascribed to the common core mass of 1.4 $M_\odot$ shared by the presumed SNI precursors. The fact that, in this picture, the larger the initial main-sequence mass of the precursor, the larger the mass of the hydrogen-rich envelope at outburst, however, remains a strong argument against commonality. Thus the indication from the observations that single low and intermediate mass stars cannot achieve explosive conditions removes several difficulties (at the same time that it creates new ones).

7.2. The Binary Star Scenario—Overview

Long before it became almost inescapably evident that single low and intermediate mass stars might not achieve explosive conditions, binary systems containing such stars were recognized as potential precursors of type I supernovae. The fact that SNeII do not occur in elliptical galaxies is easily explicable if one believes that progenitors of most SNeII are single stars of initial mass larger than 9 $M_{\odot}$ and if one accepts the spectral characteristics of ellipticals as evidence that all stars in these galaxies which are actively burning nuclear fuel are less massive than about 1 $M_{\odot}$. However, single stars of mass less than about 1.4 $M_{\odot}$ are not expected to evolve into explosive configurations, regardless of their initial structure, and yet SNeI occur in ellipticals. This paradox led Truran and Cameron (1971) to suggest that binary systems might somehow be implicated in the SNI phenomenon. Whelan and Iben (1973) proposed a specific binary model consisting of an intermediate mass star and of a low mass star close enough together that the more massive star would overflow its Roche lobe after having developed an electron-degenerate core of mass near 1 $M_{\odot}$. They assumed that (1) the Roche lobe overflowing star would be stripped of all matter outside the degenerate core, that (2) most of this matter would be lost from the binary system, and that (3), in consequence of the loss of orbital angular momentum associated with system mass loss, the orbital separation of the system would shrink to such an extent that the secondary would fill its Roche lobe before igniting helium at its center. The final act then consisted of (1) a transfer of hydrogen-rich matter onto the massive white dwarf, (2) the growth in mass of the white dwarf due to the conversion of hydrogen into helium and of helium into carbon and oxygen at its surface, and (3) an explosion when the mass of the dwarf reached 1.4 $M_{\odot}$. In this scenario, the "clock mechanism" which ensures that a supernova explosion is delayed until long after active star formation has ceased in the elliptical host galaxy is the nuclear evolution of the low mass secondary.

This specific scenario and its variants (most prominently, the scenario which leads to the formation of cataclysmic variables as an intermediate stage) have encountered many damaging and probably fatal blows to their credibility, ranging from estimates of formation frequency which are considerably below the estimated SNeI rate to the realization that a white dwarf which accretes hydrogen- and/or helium-rich material at anticipated and observed rates probably loses more mass in each nova outburst than it gains between outbursts (see, e.g., Iben and Tutukov 1984 for discussion and references).

With the passage of time, quite a few additional scenarios have been constructed (see, e.g., Wheeler 1982; Trimble 1982; Iben and Tutukov 1984). A better understanding of the manner in which a binary system can lose mass and angular momentum during Roche lobe overflow and a better understanding of the consequences of these losses for orbital shrinkage and change in component masses have been achieved with the introduction and development of the concepts of a "common envelope" and of "frictional" heating between this envelope and the orbiting stars within it (e.g., Sparks and Stecher 1974; Paczyński 1976; Tutukov and Yungelson 1979; Taam, Flannery, and Faulkner 1980; Meyer and Meyer-Hofmeister 1979). Application of these concepts has led to the recognition that systems consisting initially of main-sequence stars of low and intermediate mass can evolve into systems consisting of two white dwarfs which are close enough to each other that angular momentum loss due to gravitational wave radiation can lead to Roche lobe contact and, if the masses of the two dwarfs are large enough, possibly to a SNI explosion (Tutukov and Yungelson 1979; Webbink 1979). Two observed binary systems, HZ 29 (Faulkner, Flannery, and Warner 1972) and G61−29 (Nather, Robinson, and Stover 1981)—whose orbital characteristics may best be understood if both components are helium white dwarfs of very low mass, the lighter of which fills its Roche lobe—demonstrate that this theoretical inference is not entirely fanciful. Very strong, though indirect, evidence for the role of gravitational wave radiation in leading to mergers of white dwarf pairs is the absence of a peak at low mass ($\sim 0.25\ M_{\odot}$) in the distribution of white dwarfs in number versus mass (Iben and Tutukov 1986). Several recent papers (Iben and Tutukov 1984; Webbink 1984; Paczyński 1985) have enthusiastically embraced the double degenerate scenario as the most likely mode of SNI production.

In the following sections we examine the characteristics of candidate precursor systems in some detail and present some of the theoretical problems yet to be solved and observational implications yet to be explored before the scenario can be said to be established as a viable mode of SNI production.

7.3. Details of Degenerate Dwarf Formation

The outcome for the mass donor of a mass loss episode that occurs because of Roche lobe overflow depends primarily on the evolutionary state of the interior of the donor at the beginning of the mass loss episode. The outcome for the companion and for the orbital characteristics of the binary system depend upon both the evolutionary state of the donor, which strongly influences the attempted mass transfer rate, and the evolutionary state of the companion (the potential accretor), which helps determine whether this companion accretes efficiently or whether it rejects most of the proffered matter. If proffered matter is rejected, a common envelope is formed and orbital shrinkage occurs, regardless of the mass ratio of the two components. If the donor possesses a deep convective envelope and if its mass exceeds about six-tenths of the mass of its companion, then a common envelope is likely to be formed, regardless of the evolutionary state of its companion. In the absence of a common envelope, orbital shrinkage is likely to occur as a consequence of mass transfer if the donor is more massive than the

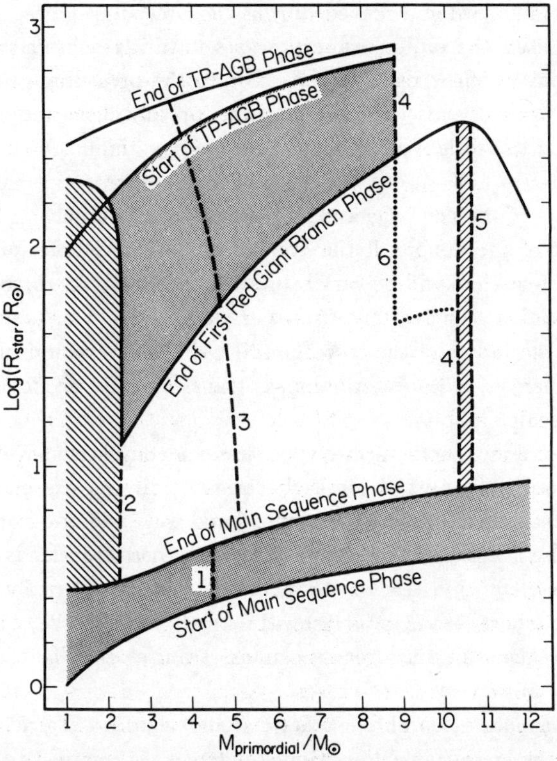

Figure 7.4. The radius of a single star or component of a binary (before it fills its Roche lobe) as a function of its mass and evolutionary state (Iben and Tutukov 1985).

accretor or if the donor is less massive than the accretor and angular momentum loss due to a magnetic stellar wind or gravitational wave radiation occurs.

As shown in Figure 7.4 the radius of a single star or of a binary component that has not yet filled its Roche lobe is related to the evolutionary state of the interior. In the absence of mass loss, stellar radius increases monotonically during most of the core hydrogen-burning phase, decreases briefly after central hydrogen is exhausted, and then increases again monotonically until helium is ignited at the center. The radius decreases during most of the subsequent core helium-burning phase, increases again toward the end of core helium burning, and eventually surpasses its value at helium ignition. The radius continues to increase as helium burns in a shell until hydrogen is reignited and the star enters the thermally pulsing (TP) AGB phase. During most of the TP-AGB phase, hydrogen burns quiescently and the radius increases monotonically with the growth in mass of the helium zone above the carbon-oxygen core. At intervals, which decrease in length roughly as the eighth power of the core mass, a helium-burning runaway occurs and the radius first drops abruptly and then increases abruptly to reach a value larger than that obtaining prior to the runaway. During the subsequent quiescent helium-burning phase, the radius decreases considerably below its maximum value achieved during the preceding pulse. Finally, when hydrogen reignites, the radius again increases until it reaches a value modestly larger than that which is reached just prior to the preceding pulse. With the occurrence of another pulse, the radius drops precipitously and then rises just as precipitously to a briefly maintained relative maximum.

The manner in and rate at which the donor can transfer mass are related to the structure of the envelope and to the rate at which nuclear fuel is burned in the interior of the donor. If the donor first fills its Roche lobe as a main-sequence star, mass loss will be very rapid until the donor becomes less massive than its companion. Thereafter, if mass transfer were to be controlled by the rate at which the radius of the corresponding single star would increase due to composition changes in the interior, mass transfer would occur on the nuclear burning time scale.

If either the donor or the accretor possesses a convective envelope and supports a magnetic stellar wind, or if the binary pair is close enough that orbital angular momentum is lost by gravitational wave radiation on a time scale shorter than the nuclear burning time scale, then mass transfer is controlled by the magnetic stellar wind (e.g., short period WUMa stars) or by gravitational wave radiation (short period cataclysmic binaries). In any case, when the donor is a main-sequence star, the process of mass transfer is called "case A" mass transfer.

Without the mediation of a magnetic stellar wind or of gravitational wave radiation, orbital separation will continue to increase in response to the evolution of the donor, but the process may ultimately be interrupted by the evolution of the companion if the companion is also a main-sequence star. In the presence of a magnetic stellar wind or gravitational wave radiation, the donor may eventually be entirely "eaten."

If the donor has already developed a compact, hydrogen-exhausted core before it fills its Roche lobe, and if Roche lobe filling occurs before helium is ignited and radial shrinkage sets in, the ensuing mass loss episode is called historically a "case B" mass transfer event.

In general, the donor engaged in a case B mass transfer episode at first has a deep convective envelope, and it loses mass on the thermal time scale of this envelope until its mass has been reduced below about 0.6 times the mass of its companion. It continues to lose mass until matter that has been processed through hydrogen burning begins to be exposed at the surface. It then shrinks within its Roche lobe and remains there as a compact object until helium is exhausted in its core.

An example of the evolution of a star which experiences case B mass transfer is shown in Figure 7.5. Note that the model star expands again after exhausting central helium and again fills its Roche lobe briefly. The accompanying mass loss episode is called a case BB episode. In the particular instance shown, the star retains some hydrogen at its surface when it detaches from its Roche lobe for

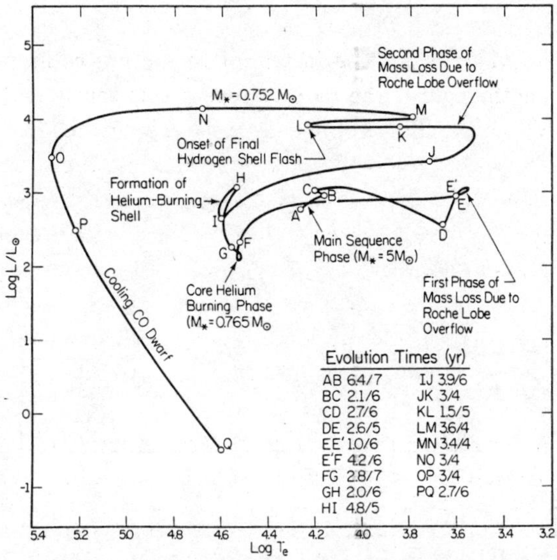

Figure 7.5. The evolution of a star of initial mass 5 $M_\odot$ which undergoes case B mass loss after exhausting central hydrogen and then undergoes case BB mass loss after exhausting central helium. Lines of constant radius may be imagined; they are nearly parallel to the last portion of the cooling white dwarf track shown (adapted from Iben and Tutukov 1985).

the last time. In the case of stars somewhat more massive, all surface hydrogen is lost in the second, case BB, mass transfer episode. Examples of the motion in the density-temperature plane of matter at the centers of such stars and at the center of the 5 $M_\odot$ star are shown in Figure 7.6.

Stars which are slightly less massive than 5 $M_\odot$ when they experience a case B mass loss episode do not expand again after exhausting central helium and thus do not lose further mass in a case BB episode. An example of such evolution is shown in Figure 7.7. Note that the lifetime during core helium burning exceeds the main-sequence lifetime of the star. This fact makes it likely that a majority of subdwarf B and O stars may be binary components in this phase of evolution. In the particular example shown, helium shell flashes occur even though the surface hydrogen layer is very thin and hydrogen-burning processes matter at an average rate that is over 100 times less rapid than the rate at which helium-burning processes matter.

When all nuclear burning in the 0.378 $M_\odot$ remnant of the 3 $M_\odot$ star is finished, the abundances of the dominant elements in the star are as shown in Figure 7.8. Note that considerable helium remains in the remnant. This is in contrast to the case of stars initially more massive than about 4.5 $M_\odot$. In remnants of such stars, the overwhelming constituents are ^{12}C and ^{16}O, with ^{4}He remaining only in a very thin layer near the surface. As initial mass is reduced below 3 $M_\odot$, the proportion of ^{4}He remaining in the remnant increases

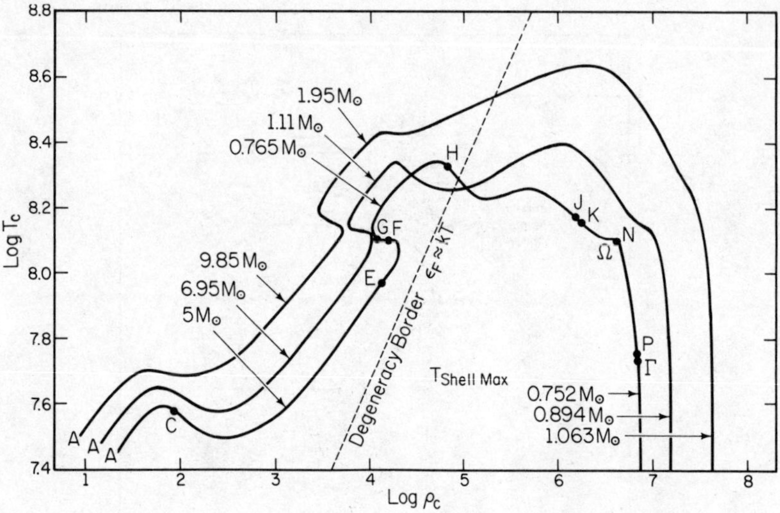

Figure 7.6. Tracks in the central density–central temperature plane of stars which undergo both case B and case BB mass transfer (Iben and Tutukov 1985).

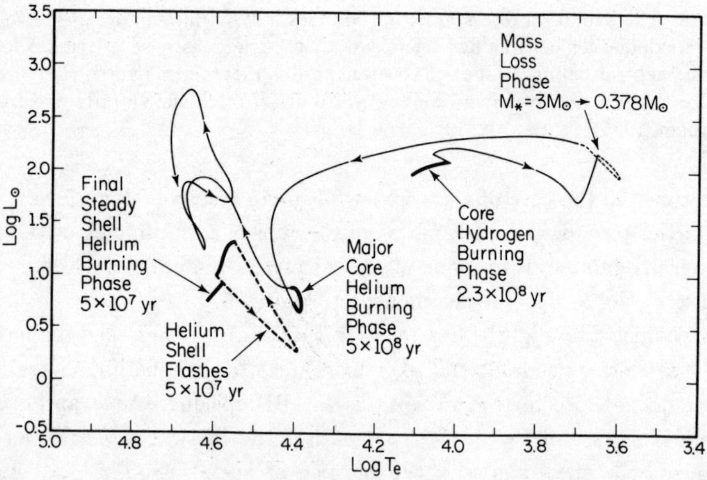

Figure 7.7. Evolution in the H-R diagram of a star which experiences case B mass transfer when stellar mass is reduced from 3 $M_\odot$ to 0.378 $M_\odot$ but remains compact following core helium exhaustion (adapted from Iben and Tutukov 1985).

rapidly, with no ^{12}C and ^{16}O at all remaining in stars initially less massive than about 2 $M_\odot$. From the point of view of what happens when two binary white dwarfs merge, these abundance distributions are crucial.

The motion in the density-temperature plane of matter at the centers of two model components (initially of masses 3 $M_\odot$ and 4 $M_\odot$) which experience a case B event but not a subsequent case BB event is shown in Figure 7.9. Also shown are paths of two models (initially of masses 11 $M_\odot$ and 12 $M_\odot$) which ignite carbon at their centers during a case BB event and go on to become neutron

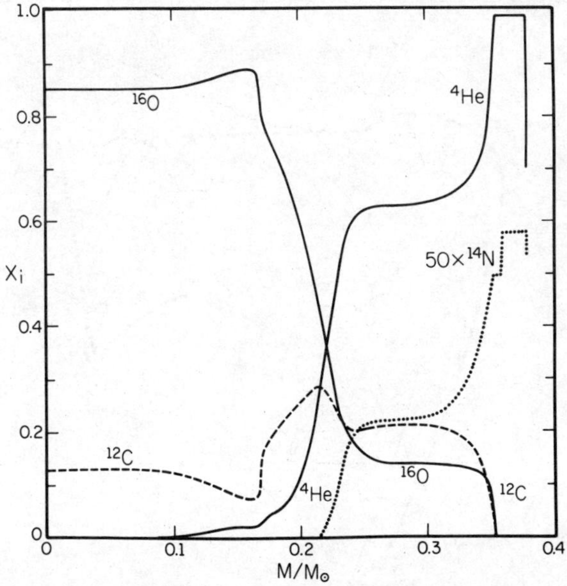

Figure 7.8. Element abundances versus mass in the remnant of a case B mass transfer experienced by a star of initial mass 3 $M_\odot$ (Iben and Tutukov 1985).

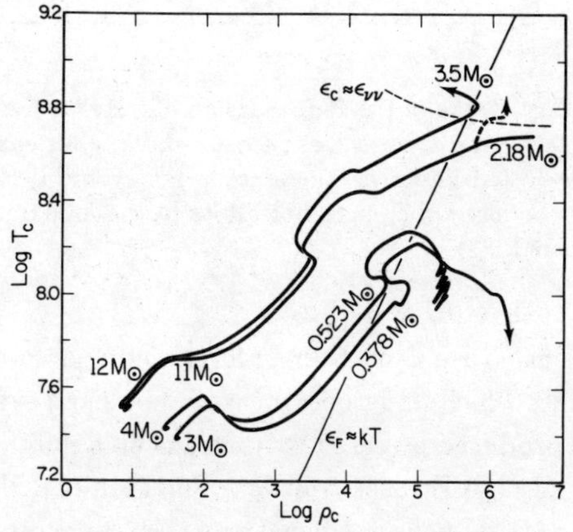

Figure 7.9. Tracks in the central density–central temperature plane of stars which become white dwarfs after experiencing just one (case B) mass transfer episode (3 $M_\odot$ and 4 $M_\odot$) and of stars which ignite and burn carbon at their centers before detaching from their Roche lobes during a case BB mass transfer episode (11 $M_\odot$ and 12 $M_\odot$). These stellar models will go on to explode as "electron-capture" supernovae, leaving neutron star remnants (Iben and Tutukov 1985).

stars.

If a component does not fill its Roche lobe for the first time until after it has exhausted helium at its center and before it has reignited hydrogen in a shell, it is said to experience an "early case C" mass loss episode when it fills its Roche lobe. In Figure 7.10 the evolutionary behavior of a star of initial mass 4 $M_\odot$

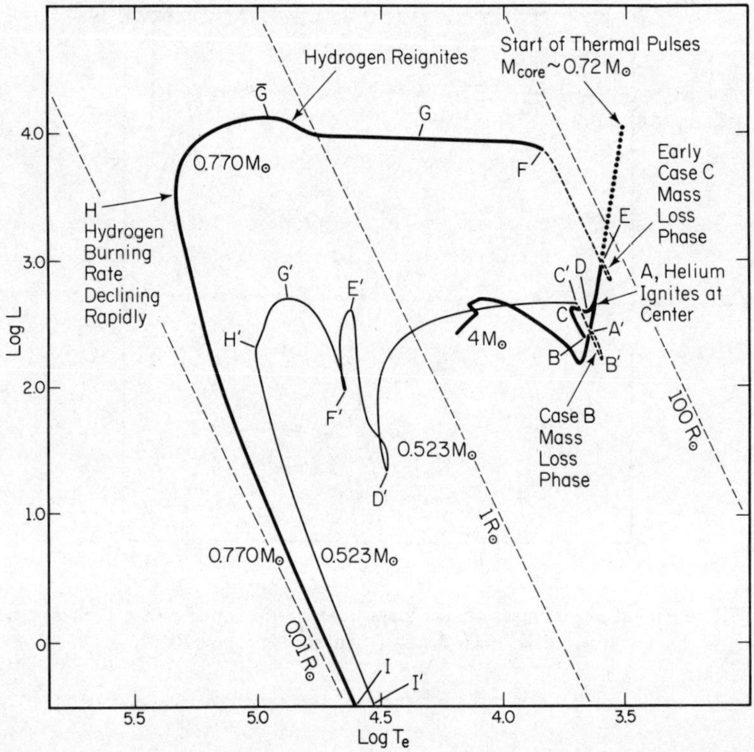

Figure 7.10. Evolutionary tracks in the H-R diagram of a stellar model of initial mass 4 $M_\odot$ which undergoes early case C mass loss (*heavy solid line*) or case B mass loss (*light solid line*). The heavy dotted line gives the track which would be followed by a single star or by a binary component which does not fill its Roche lobe until after hydrogen is reignited in a shell (Iben 1986).

which undergoes an early case C mass loss phase is compared with that of a star of the same initial mass which undergoes a case B mass loss event. Note that the mass of the remnant produced by case C mass transfer is significantly larger than that of the remnant of case B mass transfer. Furthermore, the larger remnant has very little hydrogen and helium left near its surface, being composed almost exclusively of ^{12}C and ^{16}O. These features distinguish the results of early case C evolution from the results of case B (and of case BB, if relevant) evolution for other initial masses as well.

If Roche lobe filling is delayed until after the star has reached the TP-AGB phase, then, once the stellar radius first exceeds the Roche lobe radius during a thermal pulse, mass will be lost very rapidly during this pulse and during the next few pulses until, after a very few pulses, only the remnant carbon-oxygen core (with a thin topping of helium-rich matter) will be left. The star is said to have experienced a "late case C" mass loss episode.

In Figure 7.11 are shown the masses of the white dwarf remnants of Roche lobe-induced mass loss from stars of different initial masses.

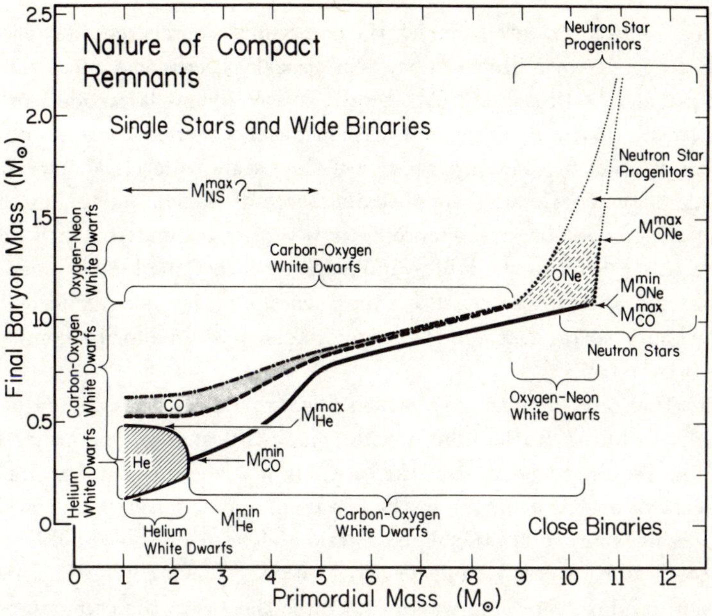

Figure 7.11. The mass of a white dwarf remnant as a function of the initial mass of the parent star for various classes of mass loss brought about by Roche lobe overflow (Iben and Tutukov 1985).

7.4. Progenitor Systems

Having mapped out the relationship between the initial mass and the final mass and composition of the remnant that is produced by Roche lobe overflow, we are now in a position to search for those combinations of initial component masses and separations which may lead to a final pair of white dwarfs which can merge and explode in a finite time.

In order that the products be white dwarfs, the initial mass of each component must be less than about 10 $M_\odot$. Since an explosion will require that the total mass of the merging white dwarf pair be close to or larger than about 1.4 $M_\odot$, it is further clear that the initial mass of each component must exceed something on the order of 4–5 $M_\odot$.

Two important questions which must be addressed are (1) how far apart may the initial components be if the final system is to consist of two white dwarfs which are close enough together that gravitational wave radiation can bring them to within merging distance within the lifetime of the Galaxy? and (2) how close may the initial components be if Roche lobe overflow by the initially more massive component results in a white dwarf rather than in dissolution? These questions are at present impossible to answer in a quantitatively precise way because they require a solution of the very complicated physics that is introduced by the formation of a common envelope. The best one can do at the moment is to proceed intuitively, using energy and time scale arguments.

A common envelope is formed when the rate at which mass is lost from the donor exceeds the rate at which the companion can accrete matter without swelling to fill its own Roche lobe. Numerical experiments must, of course, be performed to estimate critical accretion rates for stellar models in various stages of evolution, but, in general, expansion and therefore Roche lobe filling will occur if mass transfer proceeds on a time scale which is shorter than the time scale of the envelope of the potential accretor. This implies that the more compact the companion is, the more likely it is to reject matter which the donor transfers through its Roche lobe. Similarly, the larger the linear dimensions of the donor are when it first overflows its Roche lobe, the more rapid will be the rate at which it loses mass and the more difficult it will be for its companion to accrete this matter.

When the companion to the star which is losing mass through its Roche lobe also swells to fill its Roche lobe, matter must be lost from the entire system. The energy required to drive out the escaping matter must come ultimately at the expense of orbital energy, and the two components must spiral toward each other. One envisions a "frictional" interaction between the more rapidly orbiting stellar cores and the less rapidly orbiting outward-flowing gas. This frictional interaction produces pressure waves that force the gas in the common envelope outward.

A "zeroth order" estimate of the extent of orbital shrinkage by common envelope action follows from simple energy considerations. In principle, one wishes to evaluate the energy necessary to remove from the system the matter donated by the mass-losing star and to equate this to the difference between the binding energy of the initial binary pair and the binding energy of the remnant pair. In practice, the exercise is not trivial, and it is convenient (Tutukov and Yungelson 1979) to parameterize the process by setting $GM_{1R}M_2/A_{\rm final}$ equal to a free parameter, "α," times $GM_1^2/A_{\rm initial}$, where M_1 is the initial mass of the mass donor, M_{1R} is the remnant mass following the common envelope phase, M_2 is the mass of the companion, $A_{\rm initial}$ is the initial orbital separation, and $A_{\rm final}$ is the final orbital separation. This rough relationship expresses the intuitive expectation that (1) the greater the extent of mass loss, the more pronounced the degree of orbital shrinkage produced by common envelope action and (2) the larger the mass ratio between components prior to mass transfer, the more vigorous the degree of common envelope action, due to the difference in thermal time scales of the underlying stars.

As an example, suppose that we begin with an 8 $M_\odot$ star and a 5 $M_\odot$ star separated by about 500 $R_\odot$. After about 3×10^7 yr of evolution, the more massive star will exhaust helium at its center, fill its Roche lobe, and lose mass until the underlying carbon-oxygen core of mass $\sim M_\odot$ is revealed. Applying our simple recipe, the separation between this remnant and its 5 $M_\odot$ companion will be of the order of 40 $R_\odot$. In another 4×10^7 yr, when its radius has expanded to about 15 $R_\odot$, the 5 $M_\odot$ star will fill its Roche lobe. Since it will have developed a deep convective envelope and since core contraction and shell hydrogen burning are forcing a rapid expansion of this envelope, the 5 $M_\odot$ star will begin to transfer mass at a very high rate, $\dot{M} \gtrsim 10^{-5}$ $M_\odot$ yr^{-1}. In response to accretion at a rate larger than about 10^{-7} $M_\odot$ yr^{-1}, a white dwarf of 1 $M_\odot$ will develop an envelope which swells to giant dimensions. Hence a common envelope must be formed and the mass transferred by the donor through its Roche lobe will be lost from the system. This will continue until the donor decreases in mass to about 0.75 $M_\odot$, at which point matter which has been extensively processed by hydrogen burning appears at its surface and the donor shrinks within its Roche lobe. Again applying our simple recipe for common envelope action, we find that the orbital separation between the two newly formed CO white dwarfs is of the order of 1.2 $R_\odot$.

The rate at which the orbital separation A will decrease due to the loss of angular momentum carried away by gravitational waves is given by $dA/dt = -(64G^3/5c^5)M_1M_2(M_1+M_2)/A^3$, where M_1 and M_2 are component masses. Applying this expression, we have that, in about 10^8 yr, our two CO dwarfs will have spiraled inward toward each other to such an extent that the 0.75 $M_\odot$ dwarf fills its Roche lobe and the first stage of the merging process is launched. Adding in the nuclear-burning lifetime of the initial secondary, we have that the total elapsed time between the formation of the initial system of two main-sequence stars of intermediate mass and the beginning of the merging process between two fairly massive white dwarfs is about 2×10^8 yr. Since this is short compared to the age of a typical elliptical galaxy in which SNeI occur, it is clear that this particular system will not explain the occurrence of SNeI in elliptical galaxies. But, simply by enlarging the separation between the two main-sequence components in the initial system, we can increase the total elapsed time between inception and merging to as long as we wish. So, for example, by choosing the initial separation to be about 1000 $R_\odot$, we obtain an elapsed time of about 3×10^9 yr, and by choosing $A \sim 1500$ $R_\odot$, we obtain an elapsed time of about 2×10^{10} yr.

In Figures 7.12 and 7.13 are shown additional examples of combinations of initial masses and orbital separations that may lead to the merging of two white dwarfs of sufficient total mass to produce an explosion. It must be emphasized that all of these examples are to be viewed only as suggestive rather than quantitatively believable. The examples have been constructed by deliberately oversimplifying the physics of evolution and ignoring even the diversity of phenomena that appear when one examines individual cases in detail. For example, in the case of an initial binary in which the components are of comparable mass and the primary overflows its Roche lobe before developing a deep convective envelope, it is expected that, once the mass of the initial primary decreases below the mass of the secondary, the rate of mass transfer will drop considerably and the secondary will begin to accrete matter rather than reject it into a common envelope. Further mass transfer may therefore lead to the growth in mass of the secondary and an increase in orbital separation that partially offsets the

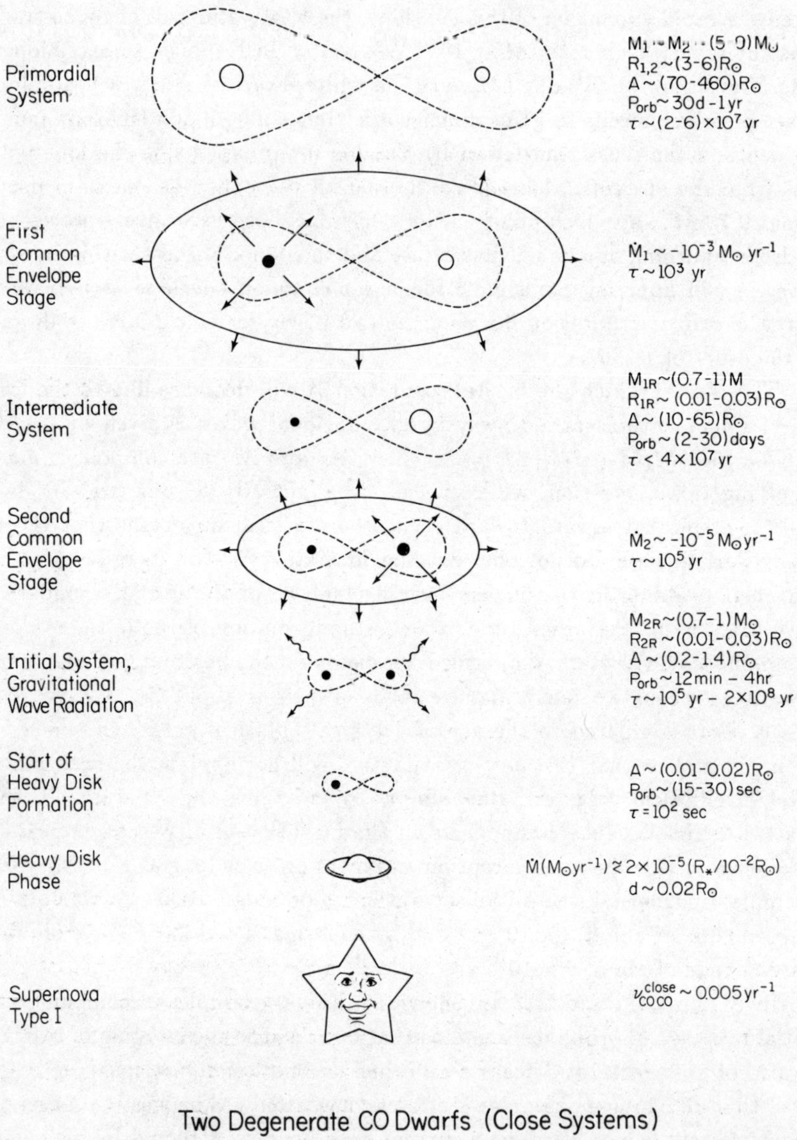

Figure 7.12. Orbital and mass characteristics during several envisioned evolutionary stages of close binaries which may eventually become single carbon-oxygen white dwarfs of mass comparable to or larger than 1.4 $M_\odot$. In most instances both components experience Roche lobe overflow of the case B variety (after exhaustion of central hydrogen but before ignition of central helium), and the total time which elapses between the formation of the initial main-sequence pair and the occurrence of the final merging episode is much less than the age of the universe (Iben and Tutukov 1985).

decrease achieved during the preceding common envelope phase. Furthermore, if the first mass loss episode is of the case B variety, then the second mass loss episode is also likely to be of the case B variety and, since the lifetime of the core helium-burning phase of the first remnant of case B mass loss may be larger than the remaining main-sequence lifetime of the secondary, the secondary may expand to fill its Roche lobe before the primary exhausts its central helium and expands for a second time. Thus intermediate mass components that are initially of about the same mass can alternate as mass donors, and this complicates the situation beyond that employed in constructing the estimates shown in Figures 7.12 and 7.13.

Estimates of the frequency of formation of potentially explosive systems can also be made. From Popova, Tutukov, and Yungelson (1982), the rate at which binary systems of various types are formed in the Galaxy may be approximated by

$$\nu \sim 0.2\ \Delta \log A \int_{M_A}^{M_B} (dM/M^{2.5}) ,$$

where ν is the formation frequency (systems yr^{-1}), A is the orbital separation of the primordial pair ($R_\odot$), and M is the mass of the initial primary ($M_\odot$, $M_A \leq M \leq M_B$). Choosing from Figure 7.12 $A \sim 70 \rightarrow 460$, $M_A = 5$, and $M_B = 9$, we have that potentially explosive systems of the "close" variety are formed at the rate of $\nu_{\rm close} \sim 0.005\ \mathrm{yr}^{-1}$. Choosing from Figure 7.13 $A \sim$ 460–1500, $M_A = 5$, and $M_B = 9$, we have that systems of the "wide" variety are formed at the rate of about $\nu_{\rm wide} \sim 0.003\ \mathrm{yr}^{-1}$. The sum $\nu_{\rm tot} \sim 0.008\ \mathrm{yr}^{-1}$ is suggestively close to the rate $\nu_{\rm SNI} \sim 0.01\ \mathrm{yr}^{-1}$ at which SNeI are thought to occur in our Galaxy.

7.5. Disk Formation, Merging, and Ignition

Just as the details of the transformation from a primordial main-sequence pair into a final close pair of degenerate dwarfs are not yet understood, so the events that take place once the lighter of the dwarfs fills its Roche lobe are far from being understood. We are not yet even at the point that we can say with certainty that a pair of merging white dwarfs whose total mass is larger than the Chandrasekhar mass will disrupt completely, leaving no compact remnant, or collapse totally to become a neutron star, expelling no matter.

One thing is certain. Since the radius of a white dwarf increases as its mass decreases, the lighter of the white dwarfs must continue to lose mass through its Roche lobe at least until the interior of the remnant is no longer supported against gravity by the pressure of degenerate electrons, that is, at least until the mass of the light component drops below that of Jupiter. Another certainty is that angular momentum must be conserved and, unless this conservation is achieved in some unexpected way by significant flow of (angular momentum transporting) matter outward from the system, the implication is that the matter leaving the "dissolving" white dwarf must flow into a "thick disk" (Tutukov and Yungelson 1979) of dimensions comparable to those of the binary orbit when Roche lobe filling first occurs. The matter in the thick disk will at first be considerably less electron degenerate than the matter in the parent dwarf; most of the degeneracy energy in the parent material is used up in expanding

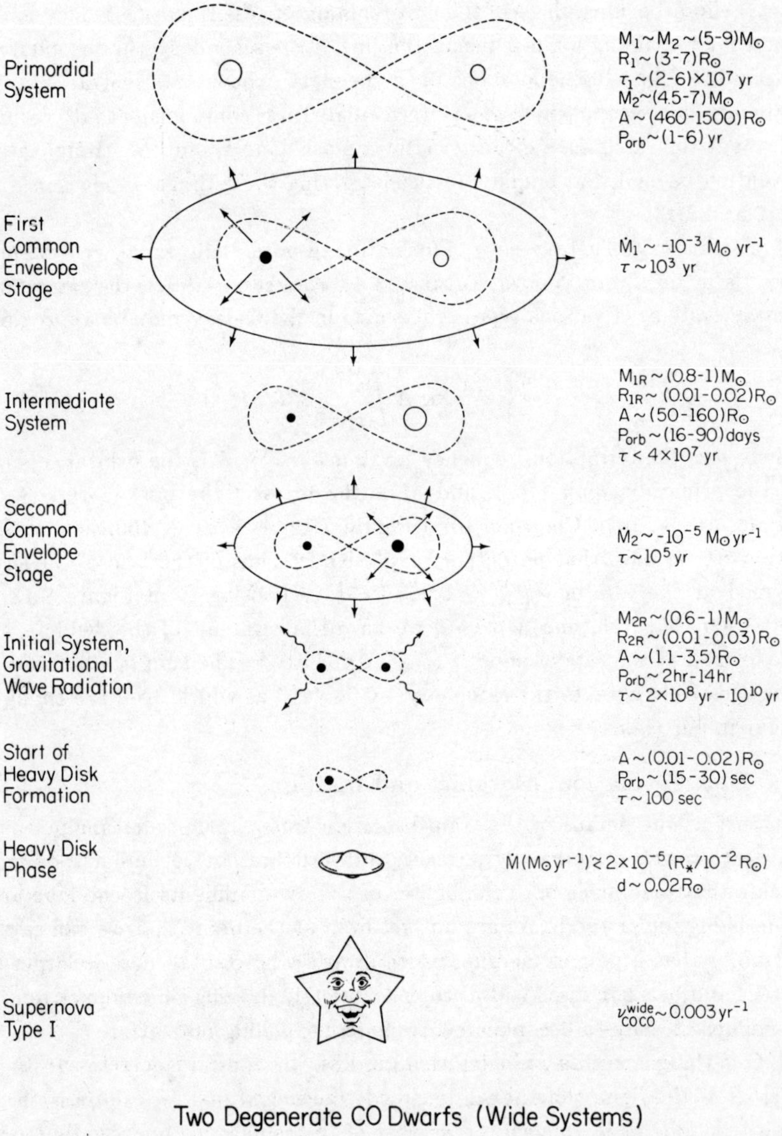

Figure 7.13. Same as Figure 7.12 except that the initial orbital separations are large enough that the more massive of the initial components fills its Roche lobe for the first time only after having exhausted helium at its center. Such systems can require as long or longer than the Hubble time to evolve to explosive conditions (Iben and Tutukov 1985).

against gravity, but additional energy becomes available as the matter flowing through the Roche lobe is rearranged into the developing thick disk and releases gravitational potential energy.

Another effect which must be taken into account is the energy which is radiated back outward by the disk material itself and by the central white dwarf

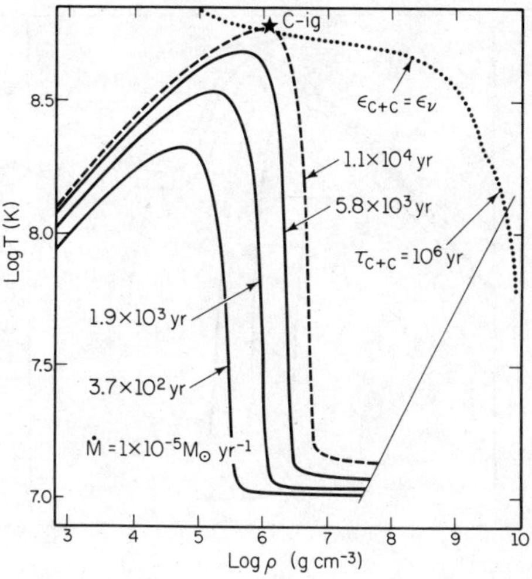

Figure 7.14. Evolution in the density-temperature plane of matter in the interior of a 1 $M_\odot$ white dwarf accreting mass at nearly the Eddington rate for a cold white dwarf ($\dot{M} \sim 10^{-5}\ M_\odot\ \mathrm{yr}^{-1}$) (Nomoto and Iben 1985).

in response to accretion. From experiments assuming spherical symmetry (Iben 1984, unpublished), it is clear that the energy diffusing radiatively outward will tend to buoy up the material falling inward and act in the same sense as angular momentum in slowing down the effective rate of accretion onto the central dwarf.

Unfortunately, a quantitatively reliable estimate of the effective rate of accretion will probably require a full three-dimensional hydrodynamic calculation with radiative flow. In the meantime, however, one can explore possible outcomes of accretion at various rates. The results of two experiments (Nomoto and Iben 1985) are shown in Figures 7.14 and 7.15. In both cases, the central white dwarf is initially of mass 1 $M_\odot$ and has a temperature structure that would be achieved after approximately 5×10^8 yr of cooling. Both the white dwarf and the accreted material are assumed to be half ^{12}C and half ^{16}O.

From these results we infer that, unless the effective accretion rate $\dot{M}_{\mathrm{eff}}$ is over 4 times less than the Eddington rate defined by $GM_{\mathrm{WD}}\dot{M}_{\mathrm{eff}}/R_{\mathrm{WD}} < L_{\mathrm{EDD}} = 4\pi c G M_{\mathrm{WD}}/\kappa_e$, where κ_e is the electron-scattering opacity, then carbon is ignited at a considerable distance from the center. Further study shows that, if carbon is ignited off center, burning proceeds inward and outward by a series of weak shell flashes and all of the carbon in the white dwarf is eventually converted completely into ^{16}O and ^{20}Ne and into ^{24}Mg and other trace elements (Saio and Nomoto 1985). If accretion continues until the dwarf reaches a mass of about 1.37 $M_\odot$, electron captures on ^{23}Na, ^{24}Mg, and ^{20}Ne lead to a core contraction which cannot be halted by oxygen burning, and the central dwarf collapses into a neutron star (Nomoto et al. 1979; Miyaji et al. 1980).

On the other hand, if carbon is ignited at the center, a thermal runaway occurs (Hoyle and Fowler 1960; Arnett 1969; Nomoto, Thielemann, and Yokoi

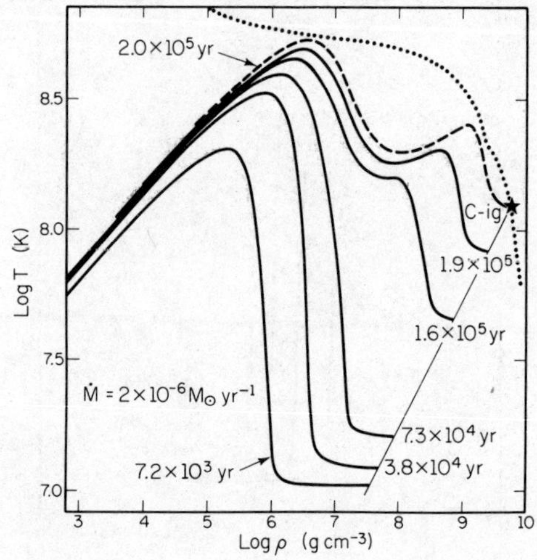

Figure 7.15. Same as Figure 7.14 but for $\dot{M} = 2 \times 10^{-6}\ M_\odot\ \mathrm{yr}^{-1}$ (Nomoto and Iben 1985).

1984), leading to an incineration of perhaps half the original material of the dwarf and to a complete disruption of the star. Detailed models of the nucleosynthesis occurring (Nomoto, Thielemann, and Yokoi 1984) and of the light curve expected (Woosley, Axelrod, and Weaver 1984) give encouraging consistency with the observations (Branch 1984). Since the immediate pre-supernova system contains no hydrogen or helium, and since one expects neither of these elements to be produced out of carbon and oxygen, one does not expect either hydrogen or helium to appear in the ejecta of the explosion. Thus the double degenerate scenario readily explains one of the most striking characteristics of a type I spectrum at maximum light: there is a total absence of any feature produced by hydrogen or helium.

Another aspect of carbon ignition which deserves further study is the precise extent to which Urca processes can delay the thermal runaway. This is important because several crucial isotopic ratios in the final nucleosynthetic product depend critically on the density at which the thermal runaway begins (Woosley, Axelrod, and Weaver 1984). Exploration of the role of Urca processes in white dwarfs began with Tsuruta and Cameron (1970), who pointed out that in a vibrating white dwarf, density variations in matter at any given Lagrangian mass point, especially near a critical density (where the electron Fermi energy is comparable to the threshold for electron capture on an abundant terrestrially stable isotope), would cause the equilibrium electron capture and decay rates to alternate and enhance the net Urca loss rate relative to that occurring in the absence of vibrations. This idea was further elaborated by Bruenn (1973), Paczyński (1972, 1973), and Tsuruta and Cameron (1973), who examined the degree of cooling which might be associated with mass motions at the edge of the convective core which is formed at the beginning of the carbon-burning phase.

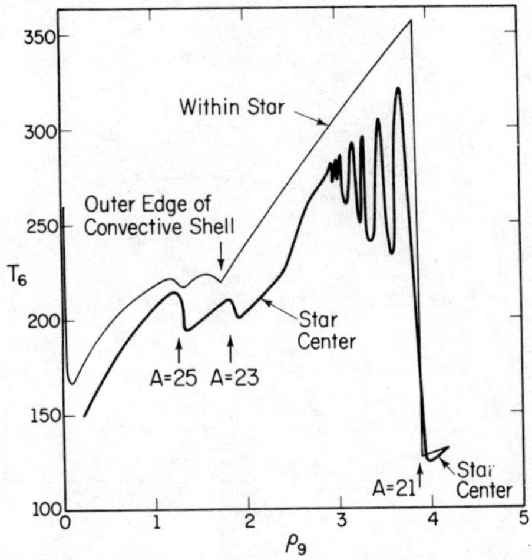

Figure 7.16. The effect of the Urca process on the structure and evolution of a white dwarf merger product if carbon ignites at the center at a density less than about 4×10^9 g cm^{-3}. The heavy solid curve describes the path in the density-temperature plane of matter at the stellar center, and the thin curve gives the temperature-density distribution within the star when the final thermal runaway begins. The heavy arrows mark the density at which the electron Fermi energy equals the threshold for electron capture (at zero temperature) on ^{25}Mg, ^{23}Na, and ^{21}Ne (Iben 1982).

The global Urca process, in which electron decay predominates at the edge of the convective core and electron capture predominates at the base, with overall conservation of charge achieved by an electron current driven inward by convective motions, was explored by Ergma and Paczyński (1974), Couch and Arnett (1974, 1975), and Iben (1978*a*, *b*, 1982).

Concrete studies of the influence of Urca processes in preexplosive white dwarfs have thus far been restricted to (1) the case of a hot white dwarf in the interior of a theoretical AGB star which, when its mass approaches 1.4 $M_\odot$, may be thought of as a central white dwarf accreting matter from a thick disk at a rate approximately equal to 10^{-6} $M_\odot$ yr^{-1}, and (2) the case of a white dwarf accreting at $\sim\dot{M} = 10^{-8}$ $M_\odot$ yr^{-1}.

The path in the density-temperature plane of matter at the center of the model which accretes at the first rate is shown in Figure 7.16. For the choice of screening adopted, carbon "ignites" (begins to produce energy more rapidly than this energy can radiatively diffuse outward) when the central density just exceeds 2.4×10^9 g cm^{-3}, and this is when a convective core is first formed. Prior to carbon ignition, matter in the vicinity of those points where the Fermi energy ε_F equals the threshold ε_{THRES} for electron capture on a particular nuclear species experiences cooling by the local equilibrium Urca process. This is the reason for the decrease in temperature at densities of about 1.2×10^9 g cm^{-3} (^{25}Mg-^{25}Na) and about 1.9×10^9 g cm^{-3} (^{23}Na-^{23}Ne). The largest contributions to the abundances of both the $A = 25$ and the $A = 23$ isotopes

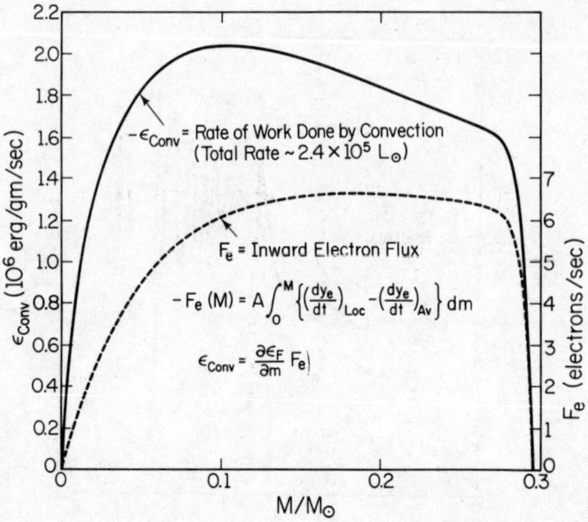

Figure 7.17. The rate at which work is done by convection in driving electrons inward against an adverse gradient in the Fermi energy and the flux of electrons as a function of position in the convective core (Iben 1978*b*).

in the Urca peaks come from carbon-burning reactions within the star prior to full-fledged ignition.

When convection sets in, mass motions act in the sense of homogenizing within the convective region the abundances of those constituents, including electrons, which are created and/or destroyed locally on time scales which are large compared to the "turnover" time for matter in the convective region. In the particular case at hand, this means that, at the base of the convective region where densities are highest, electron captures on ^{23}Na occur more frequently than do electron decays by ^{23}Ne. At the outer edge of the convective region where densities are lowest, the reverse is true. An overall or "global" equilibrium is achieved by means of a convectively driven current of electrons and of ^{23}Na nuclei inward. An example of the magnitude of this current is shown in Figure 7.17. The energy required to raise the average energy of the inward moving electrons to that appropriate to the local Fermi energy is abstracted, presumably, from the local energy of turbulent motions.

Although the global Urca process produces a net cooling effect, thanks to the irretrievable loss of energy by neutrinos, locally there can be heating as well as cooling associated with the process. This is because there is a thermal component to the Fermi distribution so that, for example, some of the energy released in an electron capture can go into thermal motions of ions as well as of electrons. Both the rate of heating, L_E, and the rate of cooling, L_{URCA}, associated with the global Urca process are highly sensitive to the density and temperature distributions within the convective zone where the process is operating. The overall rates are particularly sensitive to the density contrast between the base and the edge of the zone. Of course, the rate of heating by carbon-burning reactions,

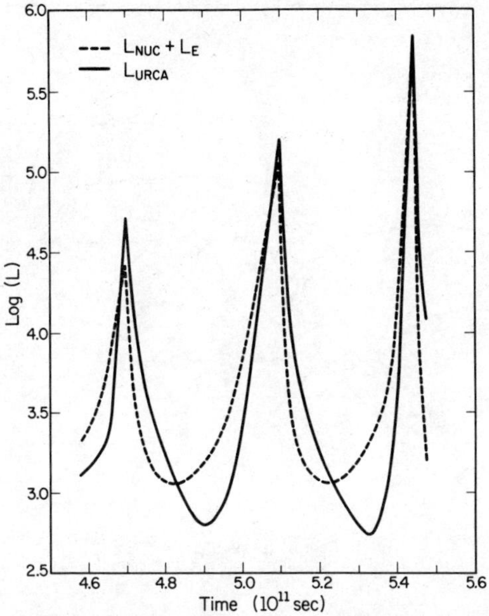

Figure 7.18. The net rate of heating ($L_{\mathrm{NUC}} + L_E$) and of Urca cooling (L_{URCA}) during several thermal oscillations (Iben 1978*b*).

L_{NUC}, is also very temperature-sensitive.

The numerical experiments show that a stable global equilibrium between the total rate of heating, $L_{\mathrm{NUC}} + L_E$, and the total rate of cooling, L_{URCA}, is never achieved. Instead, there is an oscillation back and forth such that heating first wins out, and then cooling. During the heating phase, the outer edge of the convective zone is pushed outward in mass to lower densities and the increasing density contrast between edge and base leads to an increasing cooling rate. Eventually the cooling rate overtakes the heating rate, and temperature in the convective zone begin to decrease, causing a reduction in the rate of nuclear burning. The outer edge of the convective zone moves inward in mass, thanks to a decreasing flux of energy outward, and the cooling rate begins to decrease. Aspects of this oscillatory behavior are described in Figures 7.16 and 7.18. The amplitude of the oscillations increases with time because the central density of the star continues to increase in response to accretion.

As an aside, but also as a salute to the tremendous contributions Al Cameron has made to our understanding of stellar nucleosynthesis, two examples of the flow through reaction networks that have been used in achieving the results just described are shown in Figures 7.19 and 7.20.

Eventually, when the density at the stellar center approaches 3.9×10^9 g cm^{-3}, local Urca cooling by the pair ^{21}Ne-^{1}F causes temperatures near the center to drop precipitously. This dramatically reduces the rate of nuclear burning and turns off convection near the stellar center. The base of the convective zone is thereafter constrained to occur at a point away from the center where

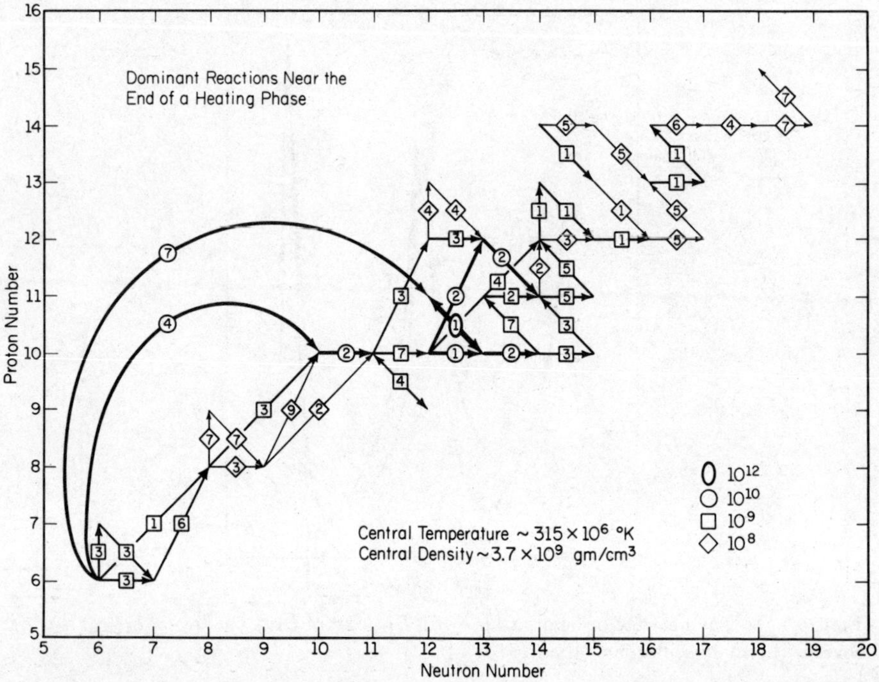

Figure 7.19. Dominant reaction channels in the convective core during the peak of a heating phase. By far the most important processes are electron capture on ^{23}Na and the (p, n) reaction on ^{23}Ne, both of which prevail near the base of the convective core (Iben 1978*b*).

the electron Fermi energy is approximately equal to the threshold energy for electron capture on ^{21}Ne. Thus, as accretion progresses and the central density increases, the maximum density contrast that can be achieved between the base and the edge of the convective zone is fixed. For fixed density, nuclear burning rates are much more temperature-sensitive than are Urca rates, so cooling by the global Urca process involving ^{23}Na and ^{23}Ne can no longer keep up with the heating due to nuclear reactions. A thermal runaway ensues. The thin curve in Figure 7.16 labeled "Within Star" shows the density-temperature distribution within the star when this runaway commences.

There is a "technical detail" that deserves mention, not because it necessarily affects the outcome, but because advance consideration thereof may save the next person who wishes to explore the relevant parameter space considerable grief. In Figure 7.21 is shown the luminosity profile in two adjacent models that differ very slightly with respect to the location in mass of the outer edge, M_{mix}, of the region within which the electrons and the $A = 23$ Urca pair are assumed to be homogenized. Note that, for the choice of smaller M_{mix}, matter is formally unstable against convection out to a point, M_{cc}, considerably larger in mass than is the formally mixed region. The implication is that the mixing zone should have been extended out to the point M_{cc}. However, if the value of

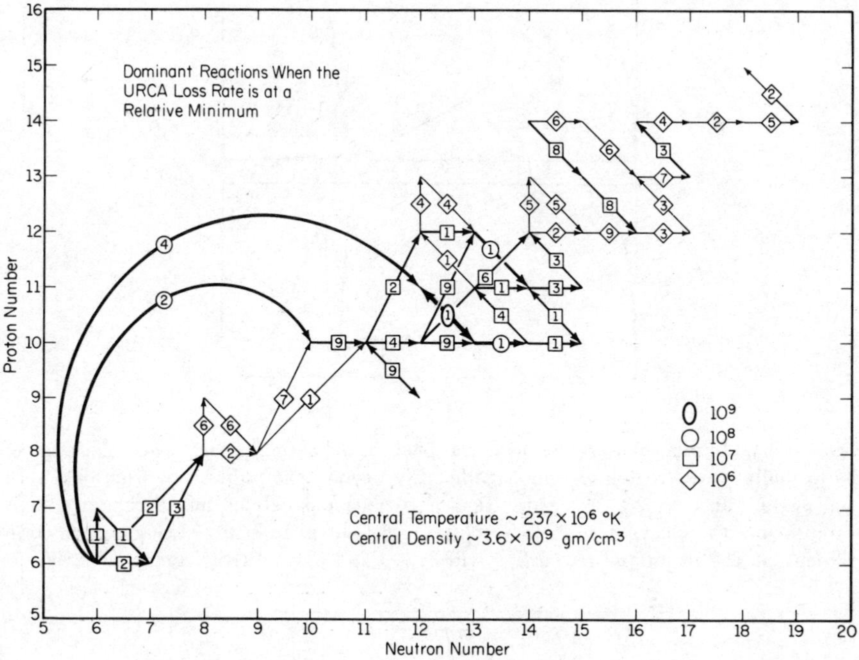

Figure 7.20. Dominant reaction channels in the convective core when the Urca neutrino loss rate is at a relative minimum (Iben 1978*b*).

$M_{\rm mix}$ is increased only very slightly (*lower panel*), the luminosity profile becomes negative within a narrow region interior to $M_{\rm mix}$. This means, of course, that the assumption of homogeneous mixing out to $M_{\rm mix}$ has led to a contradiction. The only way to avoid the contradiction is to give up the assumption of homogeneous mixing of electrons and of the Urca nuclei over the entire region between $M_{\rm mix}$ and M_{cc}. That is, it is necessary to construct a variable distribution of these constituents in such a way as to produce marginal stability against convection between the outer edge $M_{\rm mix}$ of a homogeneously mixed region and a point M_{cc} beyond which matter is in any case stable against convection.

The culprit which requires this complexity is the heating associated with electron decay in a region just outside the point where $\varepsilon_{\rm F} = \varepsilon_{\rm THRES}$. This is illustrated in Figure 7.22.

In summary, if carbon ignition occurs when the central density is less than the threshold density for electron capture on ^{21}Ne, the Urca process involving the $A = 23$ Urca pair will delay a thermal runaway until the central density reaches the threshold for the full operation of the $A = 21$ Urca pair. By the same token, if carbon ignition is delayed until the density is larger than about 4×10^9 g cm^{-3} but less than about 9×10^9 g cm^{-3} (the threshold for the

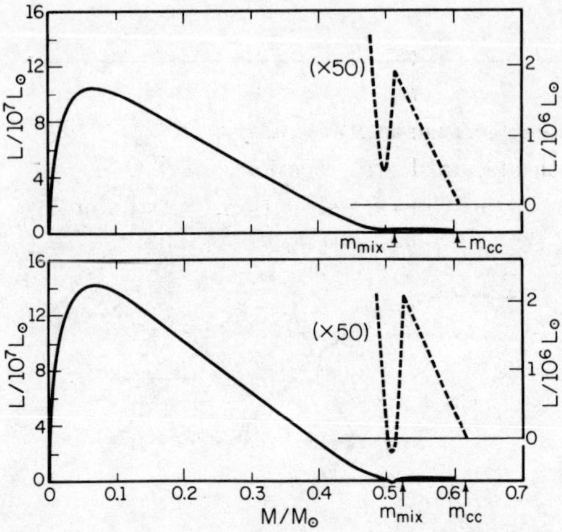

Figure 7.21. Luminosity profiles near the peak of a heating phase, when the edge of the formally mixed region extends significantly beyond the position at which $\varepsilon_F(0) = \varepsilon_{THRES}$ for the $A = 23$ Urca pair. Heating occurs beyond the mixed region, and it is impossible to achieve a consistent solution without introducing strong composition gradients in the region between M_{mix} and M_{conv} and beyond (Iben 1982).

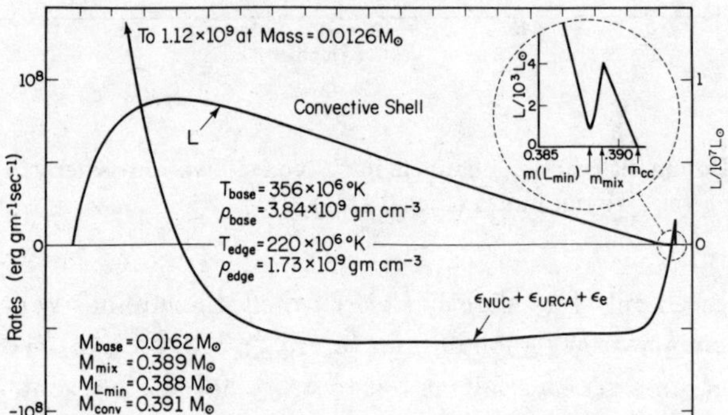

Figure 7.22. Local luminosity and the net local rate of heating and cooling as a function of mass at the beginning of the final thermal runaway (Iben 1982).

operation of the ^{25}Na-^{25}Ne pair), a thermal runaway is delayed until central density reaches 9×10^9 g cm^{-3} (Iben 1982). Such will be the case if the effective accretion from the thick disk onto the central dwarf is less than about 10^{-7} $M_\odot$ yr^{-1}.

There are many additional complexities and uncertainties that enter into a complete understanding of the ignition problem, but they can only be touched upon here. Variations in the treatment of electron screening over the past decade have led to variations of a factor of 2 in the density at which carbon ignition may occur. Improvement in the experimental and theoretical understanding of the ^{12}C$(\alpha, \gamma)^{16}$O rate have in recent years made it quite possible that the final product of complete helium burning is mostly ^{12}C. If this is actually the case, then one must rely on oxygen burning to initiate a star-disrupting explosion. Experiments in the approximation of spherical symmetry show that, with reasonable estimates of electron-capture rates on ^{23}Na, ^{24}Mg, and ^{20}Ne, and other elements and with standard cross sections for $^{16}\mathrm{O} + {}^{16}\mathrm{O}$ initiated reactions, oxygen burning develops into a detonation, but the energy input is not sufficient to halt eventual complete collapse of the inner core (of mass $\sim 1.37\ M_{\odot}$) to neutron star dimensions.

And yet the double degenerate scenario, coupled with the deflagration models which begin with carbon-containing degenerate matter, is so appealing that one hopes this result is not the final word. One possible saving feature is rotation. Considerable shearing at the interface between the inner edge of the thick disk and outer edge of the central accreting white dwarf must occur and this must cause the central dwarf to spin up. In fact, if the central dwarf is not to spin up to breakup velocities, the anticipated shearing action must be rather inefficient. That is, suppose matter were to be accreted onto the dwarf at velocities comparable to the Keplerian velocity at the inner edge of the thick disk; then, after accretion of only about 0.1 $M_{\odot}$, the white dwarf would be severely distorted due to the near balancing of gravity by centripetal forces. The net effect might be that, even if there is not enough ^{12}C in the dwarf to produce a deflagration driven by carbon burning, rotation may permit oxygen to ignite at sufficiently low densities that the tendency toward collapse due to electron captures cannot prevent a star-disrupting deflagration driven by oxygen burning.

Whatever the final result, the problem of how SNeI occur has many facets and it is to be hoped that Al Cameron will continue to contribute new ideas which help in its solution.

Happy 60th birthday, Al.

Acknowledgment

This research has been supported in part by National Science Foundation grant AST 84-13371.

References

Arnett, W. D. 1969, *Ap. Space Sci.,* **5,** 180.

Becker, S., and Iben, I., Jr. 1979, *Ap. J.,* **232,** 831.

Branch, D. 1984, in *Stellar Nucleosynthesis,* ed. C. Chiosi and A. Renzini, p. 19 (Dordrecht: Reidel).

Bruenn, S. W. 1973, *Ap. J. (Letters),* **183,** L125.

Castellani, V.; Chieffi, A.; Pulone, L.; and Tournambe, A. 1985, *Ap. J. (Letters),* **294,** L31.

Couch, R. G., and Arnett, W. D. 1974, *Ap. J.,* **194,** 537.

——. 1975, *Ap. J.,* **196,** 791.

Ergma, E., and Paczyński, B. 1974, *Acta Astr.,* **24,** 1.

Faulkner, J., Flannery, B. P., and Warner, B. 1972, *Ap. J. (Letters)*, **175**, L79.

Greggio, L., and Renzini, A. 1983, *Astr. Ap.*, **118**, 217.

Hoyle, F., and Fowler, W. A. 1960, *Ap. J.*, **132**, 565.

Iben, I., Jr. 1974, *Ann. Rev. Astr. Ap.*, **12**, 215.

——. 1978*a*, *Ap. J.*, **219**, 213.

——. 1978*b*, *Ap. J.*, **226**, 996.

——. 1981, in *Effects of Mass Loss on Stellar Evolution*, ed. C. Chiosi and R. Stalio, p. 373 (Dordrecht: Reidel).

——. 1982, *Ap. J.*, **253**, 248.

——. 1985, *Quart. J.R.A.S.*, **26**, 1.

——. 1986, *Ap. J.*, **304**, in press.

Iben, I., Jr., and Truran, J. W. 1978, *Ap. J.*, **220**, 980.

Iben, I., Jr., and Tutukov, A. V. 1984, *Ap. J. Suppl.*, **54**, 335.

——. 1985, *Ap. J. Suppl.*, **58**, 661.

——. 1986, *Ap. J.*, in press.

Meyer, F., and Meyer-Hofmeister, E. 1979, *Astr. Ap.*, **78**, 167.

Miyaji, S.; Nomoto, K.; Yokoi, K.; and Sugimoto, D. 1980, *Pub. Astr. Soc. Japan*, **32**, 303.

Nather, R. E., Robinson, E. L., and Stover, R. J. 1981, *Ap. J.*, **244**, 269.

Nomoto, K., and Iben, I., Jr. 1985, *Ap. J.*, in press.

Nomoto, K.; Miyaji, S.; Sugimoto, D.; and Yokoi, K. 1979, in *White Dwarfs and Variable Degenerate Stars*, ed. H. M. Van Horn and V. Weidemann, p. 56 (Rochester: University of Rochester).

Nomoto, K., Sugimoto, D., and Neo, S. 1976, *Ap. Space Sci.*, **39**, L37.

Nomoto, K., Thielemann, F.-K., and Yokoi, K. 1984, *Ap. J.*, **286**, 644.

Paczyński, B. 1970, *Acta Astr.*, **20**, 195.

——. 1972, *Ap. Letters*, **11**, 53.

——. 1973, *Ap. Letters*, **15**, 147.

——. 1976, in *The Structure and Evolution of Close Binary Systems*, ed. P. Eggleton, S. Mitton, and J. Whelan, p. 75 (Dordrecht: Reidel).

——. 1985, in *Cataclysmic Variables and Low-Mass X-Ray Binaries*, ed. D. Q. Lamb and J. Patterson, p. xx (Dordrecht: Reidel).

Popova, E. I., Tutukov, A. V., and Yungelson, L. R. 1982, *Ap. Space Sci.*, **88**, 155.

Reimers, D. 1975*a*, *Mém. Soc. R. Sci. Liège*, 6e ser., *8*, 369.

——. 1975*b*, in *Problems in Stellar Atmospheres and Envelopes*, ed. B. Baschek, W. H. Kegel, and G. Traving, p. 229 (Heidelberg: Springer).

Renzini, A., and Voli, M. 1981, *Astr. Ap.*, **94**, 175.

Saio, H., and Nomoto, K. 1985, *Astr. Ap. Letters*, in press.

Sparks, W., and Stecher, T. P. 1974, *Ap. J.*, **188**, 149.

Taam, R. E., Flannery, B. P., and Faulkner, J. 1980, *Ap. J.*, **239**, 1017.

Trimble, V. 1982, *Observatory*, **102**, 133.

Truran, J. W., and Cameron, A. G. W. 1971, *Ap. Space Sci.*, **14**, 179.

Tsuruta, S., and Cameron, A. G. W. 1970, *Ap. Space Sci.*, **7**, 374.

——. 1976, *Ap. Space Sci.*, **39**, 397.

Tutukov, A. V., and Yungelson, L. R. 1979, *Acta Astr.*, **23**, 665.

Webbink, R. F. 1979, in *White Dwarfs and Variable Degenerate Stars*, ed. H. M. Van Horn and V. Weidemann, p. 426 (New York: University of Rochester).

——. 1984, *Ap. J.*, **277**, 355.

Wheeler, J. C. 1982, in *Supernovae: A Survey of Current Research*, ed. M. J. Rees and R. J. Stoneham, p. 167 (Dordrecht: Reidel).

Whelan, J. C., and Iben, I., Jr. 1973, *Ap. J.*, **186**, 1007.

Wood, P. R., and Cahn, J. H. 1977, *Ap. J.*, **211**, 499.

Woosley, S. E., Axelrod, T. S., and Weaver, T. A. 1984, in *Stellar Nucleosynthesis*, ed. C. Chiosi and A. Renzini, p. 263 (Dordrecht: Reidel).

8

Cosmic Rays, Supernovae, and the Origin of Ultrahigh Energy Particles

Stirling A. Colgate

Prologue

At this conference in honor of Al Cameron—organized by his many and able students—it was remarked how all the organizers, Al's students, avoided giving talks and writing papers such as this. However, the perfect topology was broken because, unaware to them, I have considered myself in the same category as one of Al's students. My education in astrophysics started a bit later in age than the organizers', but Al was no less rigorous. I remember him asking me such student-oriented questions as

1. What nuclear synthesis would happen behind such supernova shocks?
2. How would you get enough light out of a supernova to make a quasar?
3. How could you power a quasar with supernovae?
4. What prompt γ-rays and X-rays would you get from a supernova?

Once I had finished a few of these assignments, I was both a student of Al's and an astrophysicist.

Abstract

The acceleration of ultrahigh energy cosmic rays, 10^{15-20} eV, is still an unsolved problem in high energy astrophysics. The now classical mechanism of stochastic acceleration of cosmic rays in a strong shock in the interstellar or intergalactic medium is limited in time and dimension for all likely acceleration sites, particularly for the highest energies. Acceleration of cosmic rays across a plasma shock of velocity β_s ($\beta_s = v_{\rm shock}/c$) requires a $1/\beta_s$ number of crossings and therefore a $(1/\beta_s)^2$ number of scatterings for doubling the energy of a particle. This requires a space of the order of $1/\beta_s \times$ the scattering length, or a multiple of the Larmor radius, and hence the space requirements for cosmic ray acceleration are very many Larmor orbits in dimension and the times are larger by $(1/\beta_s)^2$ than the synchrotron period. The acceleration of cosmic rays by the shock in the envelope of a Type I supernova is reviewed, and the interaction of the accelerated matter with the nearby ISM is considered. The relativistic ejected matter is preferentially trapped in the ISM. Further acceleration of each energy group should take place in both the near relativistic shock wave and during the compression by the following matter. The acceleration of ultrahigh

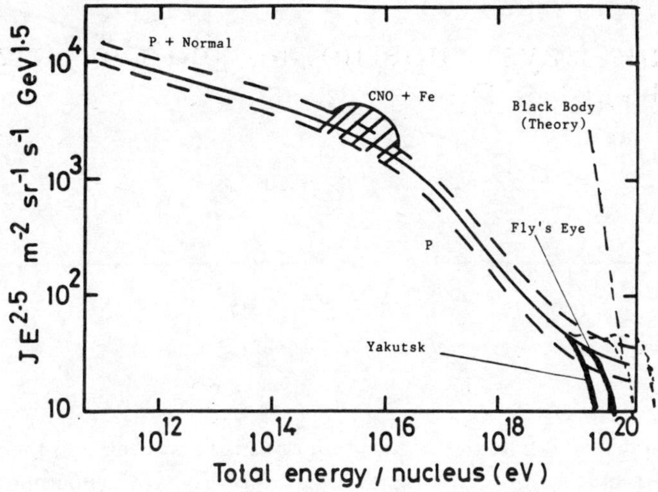

Figure 8.1. The differential energy spectrum of cosmic rays above 10^{11} eV (total energy). The solid curve is an eyeball fit through the data surveyed in /1/. The dashed lines are $\pm 20\%$ from the solid line. It seems certain that the true spectrum lies within these bounds except in the shaded region near 10^{15} eV, where matters may be more complicated, and near 10^{20} eV, where statistical and energy calibration uncertainties may mean that the $\pm 20\%$ bounds are optimistic.

energy cosmic rays in the accretion disk of a nearly stationary magnetic neutron star such as Cyg X-3 is another strong possibility. Here the diffusion of magnetic flux radially outward opposite to the viscous diffusion of mass radially inward is considered as a likely topology for unipolar generator acceleration of ultrahigh energy particles.

8.1. Introduction

The highest thermal temperature observed in astrophysical phenomena is probably that of the March 5 γ burst where $kT \cong 20$ keV. The highest nonthermal phenomenon is certainly the ultrahigh energy cosmic ray spectrum extending up to $\sim 10^{20}$ eV. The highest energies are the most difficult to explain. In Figure 8.1, taken from Hillas (1984), the distribution of cosmic ray energy as a function of flux is shown as a band describing the limits of error from the lowest energy to 10^{20} eV or $\Gamma \equiv 1/\sqrt{1} - \beta^2 = 10^{11}$. The high energy region, above 10^{19} eV, shows several possible modifications from the original curve. In particular, results from the Yakutsk group (Atrashkevich et al. 1983) and Fly's Eye (Baltrusaitis et al. 1985) conflict with the extensive earlier data from Haverah Park. In addition the spectrum expected from progressive blackbody photon attenuation (Stecker 1968; Schramm and Hill 1983) is shown as would be expected if the cosmic rays are stored in the Metagalaxy and attenuated by interaction with

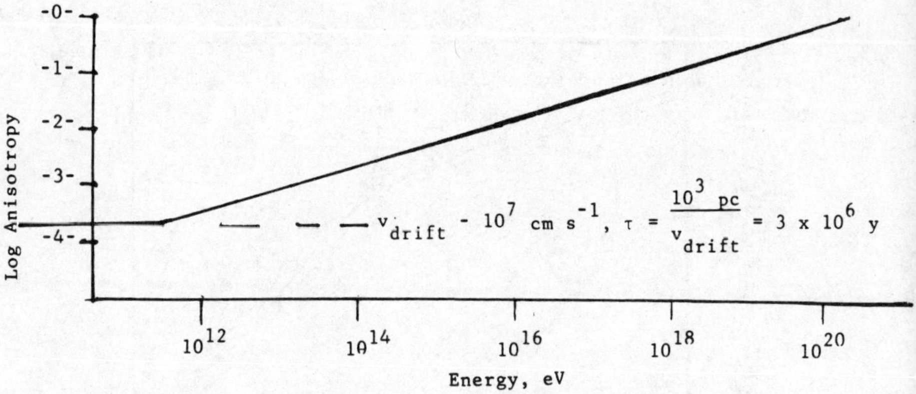

Figure 8.2

the 3 K blackbody photons. This produces a modest peak and then a reduced or steeper spectrum progressing with age. The Fly's Eye experiment seems to confirm the more rapid falloff with energy previously indicated by the Yakutsk data, so that presently there is some uncertainty concerning the ultrahigh energy spectrum, i.e., whether it is flat or falls off as expected due to the blackbody radiation. If indeed the more extensive Fly's Eye data confirm the roll off of the spectrum in the neighborhood of 10^{20} eV, as expected from interaction with the blackbody radiation, then we will begin to have a consistent picture of the highest energy cosmic rays interacting in the Metagalaxy with the primordial photons. This in turn is only partially consistent with the anisotropy shown as a function of energy in Figure 8.2. The relatively large anisotropy at the very highest energies is consistent with galactic as well as extragalactic sources. On the other hand there is another reason to expect an upper limit of 10^{20} eV, namely the Cyg X-3 machine as explained in the last section of this paper.

8.2. Some Mechanisms of Astronomical Acceleration

The one characteristic of astronomical accelerators compared to human-made accelerators is the former have high probability of particle loss during acceleration. This frequently leads to power law spectra.

A few of the mechanisms that have been considered to produce high energy particles are:

1. The Fermi mechanism in the interstellar medium depends upon stochastic particle interaction with turbulent clouds. In a sense this process is a thermalization of the low mass particles, the cosmic rays, with the supermassive particles, the magnetic clouds. A problem is the likelihood of charge-dependent acceleration. The fact that the cosmic ray spectrum is nearly independent of the charge of the nuclei has led to the consideration of other mechanisms.

2. By far the most popular explanation of intermediate energy cosmic rays is acceleration in the plasma shock produced by bulk motion of supernova ejecta in the interstellar medium (Bell 1978; Axford, Leer, and Skadron 1977; Blandford

and Ostriker 1978; Krymsky 1977). A finite compression across the shock is stochastically sampled by cosmic ray particles diffusing back and forth across it. This stochasticity and the attendant distribution of loss lead to a power law spectrum very close to the observed. However, as Legage and Cesarsky (1983) have shown, the maximum energy likely to be achieved in a shock wave in the interstellar medium is only 10^{13-14} eV. This limit is determined by the finite time and dimension (magnetic flux) of the supernova shock wave before it decays.

3. A third mechanism is the ejection of material from the outer envelope of a compact supernova, presumably a Type I supernova, by the hydrodynamic shock (Colgate and Johnson 1960; Johnson and McKee 1971; McKee and Colgate 1973; Colgate and Petschek 1979; Colgate 1984). This mechanism, in contrast to the previous two, is not stochastic in the usual sense, but instead depends upon a purely hydrodynamic property, namely that a shock wave progressing in a density gradient strengthens as a function of decreasing density independent of its origin. A shock wave initially formed from a thermonuclear explosion, or a standing shock on a collapsing neutron star core, experiences some 14 orders of magnitude change in density and a somewhat larger change in mass fraction before reaching the surface layer of the supernova envelope. It is these very large changes of density that allow the hydrodynamic acceleration of cosmic rays. However, current analytical and numerical modeling indicates a power law spectrum steeper than observed by one power of E. In addition, the escape of cosmic rays after interaction with the nearby interstellar medium is problematic. The energy may be either drastically decreased or increased by this interaction.

4. A rotating magnetic neutron star allows acceleration at the singularity at the light cylinder of a corotating magnetic field (Berenzinsky 1983). The stress of a relativistic velocity plasma at the light cylinder modifies the field geometry in a way that is still not agreed upon. The accumulation of mass at the singularity tends to break the topology such that acceleration to extremely high energies appears unlikely without a highly contrived distribution of plasma.

5. A pulsar rotating off axis will excite Alfvén waves (Gunn and Ostriker 1969), which in turn give rise to strong wave-particle in-phase acceleration and very high cosmic ray energies. This phenomenon, unfortunately, is sensitive to a phase destruction due to plasma loading of the waves (Kegel 1971). Strong Alfvén waves from an off-axis pulsar may be seen in the Crab Nebula. The preservation of the phase relationship necessary for ultrahigh energy particle acceleration is still uncertain.

6. Reconnection in helical, force-free magnetic fields of the form $\vec{B}_\theta + \vec{B}_z$ leads to acceleration by the parallel electric field $E_{||}$ due to the interruption of the $J_{||}$ current. An acceleration parallel to B is attractive because at the very highest energies, 10^{20} eV, synchrotron radiation even by protons is immense. For example, a 10^{20} eV proton will radiate its own energy in a pulsar magnetic field of 10^{12} G within a distance of a fraction of a micron. Hence $\vec{E} \cdot \vec{B}$ acceleration

is attractive for 10^{20} eV particles whenever magnetic fields are stronger than a few gauss. The problem with reconnection, or the enhanced resistivity produced in force-free field configurations, is that there is only now emerging a consensus in plasma physics of how it occurs. The creation of the turbulence necessary for reconnection is probably best described as being driven by an anisotropic velocity distribution (Meyerhofer and Perkins 1984). Probably the most complex physical phenomenon that has ever been understood is the plasma loss from a Tokamak fusion toroidal confinement experiment. Perhaps the understanding of this turbulence-driven reconnection phenomenon will be applied throughout astrophysics, but currently it is too complex and too uncertain to form the basis of a theory of the origin of the ultra-energetic particles. On the other hand, there is no confidence in any other mechanism.

7. An emerging possibility for the acceleration of cosmic rays in this Galaxy, as well as all others, is the possibility that the Cyg X-3 machine is universal. The startling γ-rays of greater than 10^{15} eV believed to have been observed from this binary star X-ray source (as well as several other sources) imply an accelerator of (probably) protons of 10^{15} to 10^{17} eV. There is a suggestion by Chanmugan and Brecher (1985) that this accelerator is similar to the accretion disks postulated for black holes and quasars (Lovelace 1976; Blandford and Vnajek 1977). I shall suggest, at the end of this paper, that a more likely explanation resides in the unipolar generator associated with a nearly stationary slowly rotating magnetized neutron star and an accretion disk. The one major problem with such an acceleration mechanism is finding a rational explanation for the effective electrical insulation.

The general property of all these mechanisms with the possible exception of the supernova envelope shock is that a time-dependent loss during a time-dependent acceleration leads to a power law distribution of accelerated particles.

8.3. Stochastic Shock Acceleration

I discuss next the stochastic acceleration in the interstellar medium with the purpose of extending this discussion to the very difficult problem of finding a possible site for acceleration of ultrahigh energy particles in the intergalactic environment. This discussion follows mostly that of Hillas (1984). The fundamental assumption of a cosmic ray acceleration shock is that most of the energy of the shock remains local and does not diffuse to places where the conditions of the medium are significantly different; i.e., the local Hugoniot relations apply. One generally assumes a parallel shock; that is, $\vec{B}$ is roughly parallel to the direction of shock propagation because this represents most of the solid angle. A very small departure from a perpendicular shock allows relativistic particles to escape along lines of force. One further assumes that the shock velocity is small, or $\beta_s \ll 1$. The energy gained by a particle crossing such a strong shock is of the order of $E\beta_s$, and the particle loss by convection downstream is determined by the number of crossings $\cong 1/\beta_s$. Particles are lost when they are convected downstream at the fluid velocity $\cong \frac{1}{4}\beta_s$. Analogous to photon

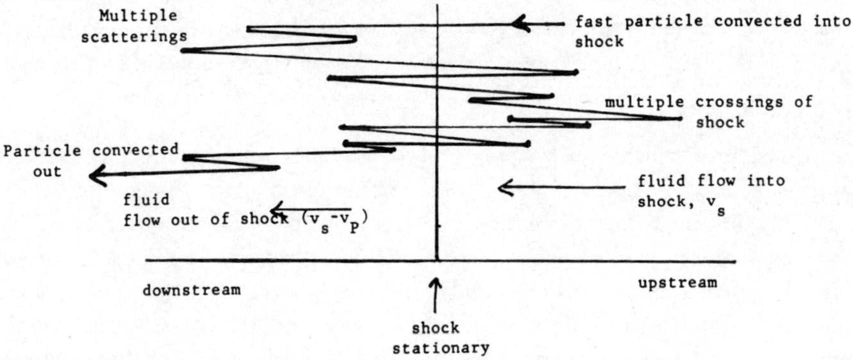

Figure 8.3

diffusion, this occurs at an "optical depth" $\tau = 1/\beta_s$ from the front. On the average $\tau^2 = (1/\beta_s)^2$ scatterings will take place before the particle will get lost downstream, during which time a particle will cross the shock front $\tau = 1/\beta_s$ times, roughly doubling its energy. Therefore there are two requirements: space $\gg (1/\beta_s)$ scattering lengths and time $\gtrsim (1/\beta_s)^2$ scattering times.

These results when combined with statistics result in a spectrum $dN/dE \cong E^{-2}$, and hence the cosmic ray spectrum. This requires that particles be scattered on either side of the shock as shown in Figure 8.3. There will be two different mean free paths, upstream and downstream, λ_1 and λ_2. Typically the shock speed is 10^8 cm s^{-1}, or $\beta_s = 1/300$, and therefore 4×10^5 scatterings are required for a tenfold energy gain. It is this very large number of scatterings that requires that no other phenomena interfere. In particular, energy lost by ultrahigh energy particles is suggested by Eichler (1984). This requires a very delicate balance between gain and loss at high energy so as not to alter the Hugoniot relations based on the local assumption. The rate of energy gain from such a shock is then $\cong c\beta_s^2/(\lambda_1 + 4\lambda_2)$ as given by Legage and Cesarsky (1983).

The scattering on either side of the shock must depend on Alfvén wave turbulence, which in turn must be excited by an anisotropic distribution of the upstream flux of particles. A reasonable estimate of the distance required to scatter a particle in nonsaturated Alfvén wave turbulence is $\lambda_1 \cong 10\ R_{\mathrm{Larmor}}$. This assumes a distribution from saturated turbulence at the shock front to negligible turbulence in the ISM. The growth rate of the relevant velocity space instability is energy density and isotropy dependent. Then the space or distance ahead of the shock required for acceleration is $L = \lambda_1/\beta_s$, or $3 \times 10^3\ R_{\mathrm{Larmor}}$. A supernova collides with roughly a thousand times its mass in 30 to 50 pc radius, and hence this is the radius of what might be called the strong shock in the ISM where $\beta_s = 3 \times 10^{-3}$. For an energy gain of tenfold for each particle, the energy at which 3×10^3 Larmor radii equal the radius is 3×10^{13} eV. This is an upper energy limit for supernova shock acceleration in the ISM. By way of comparison, a 10^{20} eV proton has a Larmor radius of 30 kpc or roughly one Galaxy diameter in a 3 μG field. An extension of these same arguments restricts the energy of cosmic rays in all known extragalactic sites.

8.4. Sites for Possible Extragalactic Ultrahigh Energy Cosmic Ray Acceleration

Consider a typical shock strength of $\beta_s = 1/300$ and a required acceleration dimension of 3000 Larmor radii. For a typical extragalactic magnetic field of 10^{-8} G, this requires a dimension of 3×10^4 Mpc, a dimension greater than the size of the universe. Even a relativistic shock requires a dimension of 100 Mpc, but such a shock would be highly luminous in a region larger than the local supercluster, to say nothing of requiring an energy density at least 10^7 times the typical ambient medium energy density of 100 eV. Radio galactic lobes of typical dimension several times 10 kpc and possibly several microgauss average field are no better than our own Galaxy, which is inadequate by 10 to 10^3, depending upon the shock strength. The hot spots within these lobes, radio galactic hot spots, may be 1/10 the lobe dimension and have 10 times the magnetic field strength. This is still inadequate in size and field unless the shocks are relativistic and the accelerated particles high Z like iron nuclei. The frequent occurrence of both these circumstances seems unlikely as discussed by Hillas (1984). Recently Jokipii and Morfill (1985) have suggested acceleration in the galactic wind shock at $R_{\text{galactic}} = 100$ kpc $\beta_s = 1.7 \times 10^{-3}$ and $B_{\text{wind}} = 1.8 \times 10^{-7}$ G. Even assuming an Archimedes spiral field, Bohm diffusion, i.e., $\lambda_s = R_{\text{Larmor}}$, the space is too small by 3.6×10^3 for a 10^{20} eV proton. The alternative suggested by Jokipii (1985) is to assume a perpendicular shock.

8.5. Time of Acceleration

In addition, the field strength must not be too great; otherwise, synchrotron radiation becomes too large and the particles are decelerated by radiation emission in a time shorter than required for acceleration. The time for emission of the kinetic energy of a 10^{20} eV proton by synchrotron radiation is $t_s = 1.4/(E_{20}B^2)$ yr. Consequently, a typical presumed quasar magnetic field of 100 G is too large, and if we let $R_{\text{Larmor}}/c = t_s$, $B \leq 4/E_{20}$ G. This excludes many other possible sites for acceleration unless the acceleration is parallel to B.

8.6. Active Galactic Nuclei Photon Losses

There has been frequent reference (Brecher and Burbidge 1972) to possible acceleration of ultrahigh energy particles in active galactic nuclei, quasars, etc. There is one major problem with acceleration in active galactic nuclei (AGN): if a particle is accelerated from within such an object, it must encounter a fraction of the photons by which we observe the object. A large energy density of photons is implied by the short fluctuation times and large luminosities observed. An optical or infrared photon interacting with a proton of 10^{20} eV is transformed by its energy factor, $2\Gamma \cong 10^{11}$, to a photon whose energy, $\gtrsim 10^{11}$ eV, is sufficient to produce π's as well as a zoo of other particles. The energy loss rate is immense even for Compton scattering, and meson interactions will increase this by several orders of magnitude. A 10^{20} eV proton traversing the radiation field of the quasar 3C 273 would radiate by Compton scattering alone some thousand times its kinetic energy. Hence the traversal of a small fraction of such an object in any acceleration process would cause a radiation damping that far exceeds any plausible acceleration (Colgate 1984). The infrared emission is much worse than the optical because of the higher photon energy density and the still greater photon number density. X-ray sources that have shorter fluctuation times imply smaller sizes and comparable or even greater photon energy densities, and hence are even more unlikely to produce ultrahigh energy particles without overwhelming radiation loss. As a general rule, if we can observe any object by radiation outside the radio spectrum, its photon density will be so great as to prohibit the acceleration of ultrahigh energy particles.

8.7. Supernova Envelope Shock

If the supernova envelope shock propagates into a tenuous magnetosphere of the pre-supernova star, it could create ultrahigh energy particles and avoid the problem of radiation and finite Larmor orbit. In this case acceleration occurs in the bulk matter by a pressure gradient. A strong relativistic shock divides energy equally between a kinetic and internal energy density as does a nonrelativistic shock. The mass density of this internal energy density means that the fluid velocity, and hence the kinetic energy per original particle, is initially relatively small. The subsequent expansion (and acceleration) results in a final energy of expanded relativistic matter that far exceeds the initial kinetic energy of the same matter immediately behind a relativistic shock. This acceleration is then relatively free of the constraints of Larmor radius and dimension.

A relativistic supernova envelope shock requires a small compact white dwarf as a pre-supernova star where the surface radius $\cong 10^8$ cm. The central density is 10^9 to 10^{10} g cm^{-3}, and the surface density is 10^{-4} g cm^{-3}. The ratio is then $\rho_{\rm cen}/\rho_{\rm surf} \cong 10^{14}$, and the mass fraction of the surface is 10^{-16}. An explosion, either thermonuclear or from neutron star collapse inside this mass distribution, will always lead to a strong shock because of high sound speed inside and a low sound speed outside. A nonrelativistic shock in a density gradient gives rise to $\varepsilon_{\rm shock} \cong \rho v_f^2/2 \propto F^{-0.4}$. The external mass fraction $F \propto N(>E) \propto E^{-2.5}$. This applies to an envelope structure similar to a polytrope of index 3. Unfortunately, this integral spectrum $E^{-2.5}$ is too steep to give a cosmic ray spectrum by roughly one power of E. There is, however, lots of energy from the supernova, so that the total energy with a relativistic factor $\Gamma > 2$, i.e., the kinetic energy equals the rest mass, is $c^2 F_{(\Gamma=2)} M_\odot/2 \cong 5 \times 10^{48}$ ergs for $F_{\Gamma=2} = 10^{-6}$. One needs roughly double this, or 10^{49} ergs, for Type I supernovae to maintain the cosmic rays in the Galaxy.

8.8. Relativistic Shocks

A plane-parallel relativistic shock gives rise to an energy versus mass fraction

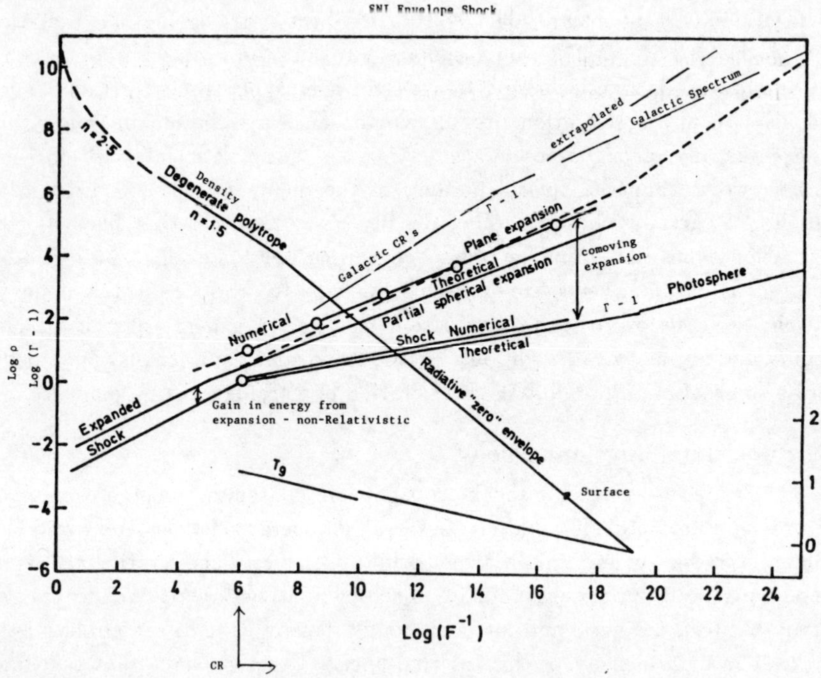

Figure 8.4

$E \propto F^{-0.48}$, or $N(>E) \propto E^{-2}$ and for a spherical shock wave possibly $\propto E^{-2.5}$. The maximum energy when such a shock wave breaks through the stellar surface layer ($\rho r \cong 1$ g cm^{-2}) is of the order $\Gamma \cong 10^5$ to 10^6, $\cong 10^{14-15}$ eV. This limit is similar to the stochastic shock acceleration within our Galaxy, but in this case there are significantly fewer particles than are necessary for cosmic rays. The composition of such matter is determined by the original composition (presumably recently accreted) and the fact that immediately behind the shock the lepton density, i.e., from pairs, is roughly 10^4 per nucleon. These additional leptons supply the cushion for acceleration of the heavy nuclei in the shock transition so that heavy nuclei can be accelerated in the shocked fluid without decomposition due to nuclear spallation. Hence the shocked material should carry the same composition as the unshocked material of the pre-supernova envelope, which in turn should correspond to the composition of the material recently accreted from the companion binary star. This mass fraction is so small ($<10^{-6}$) that it has not been altered by nuclear processes.

Figure 8.4 shows the energy versus mass fraction expected from a shock in the envelope of a compact white dwarf for a SNI. By turning the graph 90° one obtains a cosmic ray spectrum where log F^{-1} becomes equivalent to log $N(>E)$ and log $(\Gamma-1)$ becomes the energy scale. In addition, the density of a degenerate polytrope is superimposed along with the temperature behind the shock. One notes that in the relativistic region the shock temperature is of the order of several 10^9 degrees, enough to cause some nuclear synthesis in the relatively short time before expansion. One also notes the large difference between the energy of the shocked matter immediately postshock and the energy after expansion. This large factor at high energies, for ultrarelativistic shocks, is due to the fact that the energy behind the shock resides in mass energy density. Upon expansion this mass energy density, photons and pairs, recombines and is converted into kinetic energy of nucleons. Hence a modest relativistic shock at the surface of the pre-supernova star with an energy factor $\Gamma_s = 80$ corresponds to a postexpansion energy factor of $\cong 10^6$. A nucleon is accelerated in the comoving relativistic fluid frame by the pressure gradient associated with the original heat deposited by the relativistic shock. One notes in this figure the line drawn to indicate what would be required to produce the cosmic ray spectrum if no subsequent loss or gain mechanisms apply. One observes the large difference between the observations and the expected spectrum. Various effects of interaction within the Galaxy occur that may make this worse or better.

8.9. The High Energy Limit

When a shock breaks through the surface of a supernova star, $\Gamma_s \cong 80$, the initial surface conditions will be scale height $h_{\text{surface}} \cong 100$ cm, $\rho_{\text{surface}} = 10^{-2}$ g cm^{-3}, $h\rho = 1$ g cm^{-2}, $F \cong 10^{-16}$. The relativistic shock, $\Gamma_s = 80$, will compress the fluid in the comoving frame by a factor of $4\Gamma_s$ to a nucleon density of 3 g cm^{-3} and a mass energy density of 240 g cm^{-3}. This relativistic fluid with a few embedded nucleons ultimately expands to a Γ of roughly 10^5 to 10^6 or 10^{14-15} eV per nucleon, depending upon the details of spherical expansion.

One next observes that if the local magnetic field were 10^6 G, a modest value for such a star, the flux within one scale height (hB) would be 10^8 G cm, which is sufficient to contain a proton, local to the scale height, with $\Gamma_{\text{fluid}} = 80$. Hence a relatively modest embedded magnetic field corresponding to the corona of such a star holds the possibility of propagating the relativistic shock to much lower densities and much higher energies. The limiting case corresponds to a final scale height of 5×10^3 cm and density 10^{-14} g cm^{-3} and coronal density n_e of 10^{10} cm^{-3}. This corresponds to a coronal external mass fraction of 5×10^{-20} of the star or Γ_{final} of 10^{11-12} or an energy $E = 10^{20}$ eV. The relativistic shock in the magnetized plasma is dominated by its energy density, namely pairs, photons, and a modest magnetic field. The Larmor radius of a proton is always small compared to the local scale height in the comoving frame. Hence ultrarelativistic acceleration is feasible in the expansion of such a relativistic fluid, but it has other major problems like running into the interstellar medium.

8.10. Expansion in the ISM

The usual picture of the expansion of the mass ejected from the supernova is that it blows a "bubble" in the interstellar medium causing the matter internal to the bubble to undergo adiabatic expansion by PdV work (Kulsrud and Zweibel 1975). Figure 8.5 shows a picture of the expansion of such a bubble with cosmic

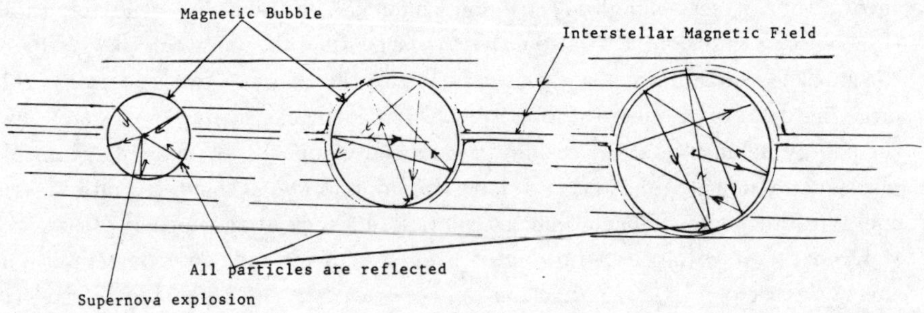

Figure 8.5

rays presumably reflecting from the magnetic boundary back into the expanding debris. The reflection of such particles back into the expanding bubble takes place because of hydromagnetic waves or Alfvén wave scattering. This is the same scattering as is invoked for the return of particles back and forth across the shock in the stochastic shock acceleration mechanism. The excitation of such wave turbulence requires a progression of a relatively large anisotropic flux (energy density $\leq (B^2/8\pi)$) and a distance of $\cong$ 10 Larmor orbits before the necessary turbulence is excited; see Legage and Cesarsky (1983) for an analysis of the gradient of hydromagnetic turbulence intensity ahead of the shock.

8.11. Relativistic Piston and Shock Interaction with the ISM

An alternative view of the supernova expansion into the ISM is that the relativistic ejected matter is trapped in the magnetic boundary and stochastically shock accelerated (Figure 8.6). The cosmic rays ejected from a relativistic supernova shock first expand as a relativistic piston. The leading cosmic rays are of course the most energetic, but immediately following is a larger mass fraction of lower energy matter. The distribution in space of this distributed-in-energy piston is surprisingly compact. For example, the radius at which the relativistic cosmic rays, $\Gamma = 2$, collide with equal mass of the ISM, $n_{i,e} = 0.1\ \mathrm{cm}^{-3}$, is determined by

$$M_{\Gamma=2}A = n_{i,e}R^3 4\pi/3.$$

where A is Avogadro's number and the mass of cosmic rays, $\Gamma > 2$, is $FM_\odot = 10^{27}$ g. Then the radius of equal mass is $R = 2\times10^{17}$ cm. The linear separation between particles of energy Γ and the highest energy at the light front will be $\Delta R = R(1-\beta)$, and so, since $(1-\beta) = \Gamma^{-2}/2$, cosmic rays of $\Gamma \cong 20$ will correspond to ΔR of $10^{-3}R = 2\times10^{14}$ cm. The distance necessary to excite hydromagnetic turbulence for scattering 10 Larmor radii is also 2×10^{14} cm for a proton of $\Gamma = 20$ in an ISM magnetic field of 3×10^{-6} G. Hence the point where the interaction of the supernova ejecta first becomes subrelativistic is close in energy to where the dispersion in ejecta velocities first allows the excitation of the turbulence necessary to scatter the cosmic rays back into the expanding bubble. On the other hand, the cosmic rays are now well embedded

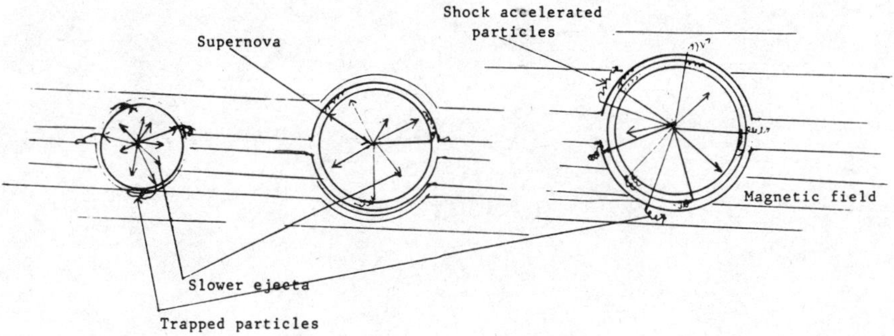

Figure 8.6

in the galactic magnetic field. In order to return to the vacuum bubble, the particles must scatter upstream in a near nonrelativistic medium where the flux of particles is greater by the ratio β^2c^2M/R^2. Since we are now concerned with a relativistic piston driving a relativistic shock in the ISM, $v^2 \propto M^{-0.4}$, the flux is proportional to $M^{0.6}/R^2$. Therefore the hydromagnetic turbulence at any given time will be greater at smaller radii, and the cosmic rays will tend to be trapped ahead of the shock. If cosmic rays are later accelerated from small energy in the ISM by a relatively weak shock of $\beta_s \cong 1/300$, then we would expect an initial relativistic distribution to be far more energized by a relativistic shock. However, the effectiveness of the shock scales as (shock time/time for acceleration) $= (R/\beta_s)/\beta_s^2$ or $\beta_s R$, but the nonrelativistic shock velocity scales as

$$\beta_s = (M^{0.6}/R^3)^{1/2} = M^{0.3}R^{-3/2}.$$

Then the effectiveness scales as $M^{0.3}R^{-1/2}$, and since the radius where equal masses interact gives $M \propto R^3$, the final effectiveness of the shock acceleration scales as $R^{0.4}$, which weakly favors a large radius weak shock. The radius ratio favors large radii by the ratio $(R_{\rm initial}/R_{\rm final})^{0.4}$, or a factor of 10. The relativistic shock spectrum is already seeded with particles up to at least 10^{14-15} eV, but smaller in number by roughly $E^{-0.8}$. The subsequent postshock expansion is comparable for both regions of the supernova ISM shock.

The energy multiplication required for the supernova-injected spectrum is so much smaller than that starting from thermal in the ISM that the injected spectrum and the ISM relativistic shock postacceleration would seem to be more likely to produce the observed cosmic rays. Certainly the previously proposed large adiabatic expansion deceleration is unlikely and a postrelativistic ISM shock acceleration more likely, but the final outcome is not evident.

8.12. Cygnus X-3 Cosmic Ray Acceleration

It has been well recognized that the ultrahigh energy γ-rays greater than 10^{15} eV and up to 10^{16} eV observed from Cyg X-3 are a strong indication of an extraordinary accelerator (Eichler and Vestrand 1984). The cascade models

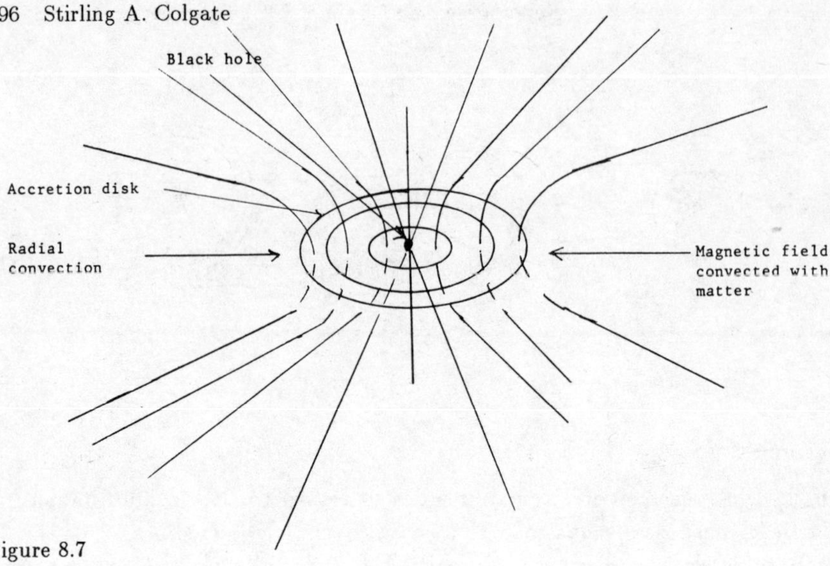

Figure 8.7

of γ-ray production from relativistic ultrahigh energy protons would indicate that the most likely source of the γ-rays is protons of 10 to 100 times higher energy, in this case a minimum of 10^{17} eV. Cyg X-3 is evidently a binary neutron star in a star in a system with an accretion disk, since this model fits the X-ray source. With this as the starting point, Eichler and Vestrand (1984) and recently Chanmugan and Brecher (1985) have proposed models for the acceleration of particles. These models involve either a rotating magnetized neutron star as Eichler (1984) considered or an accretion disk threaded with a magnetic field compressed and convected with the accreting mass flux, as Chanmugan and Brecher (1985) considered (Figure 8.7). The model by Brecher and Chanmugan is essentially that proposed for quasars and active galactic nuclei by Lovelace (1976) and independently by Blandford (1976). I would like to consider a slightly different view of such an accelerator. The magnetic field in the Chanmugan and Brecher (1985) model is presumed convected with the accretion disk, as indeed it must be in the case of black hole accretion in an active galactic nucleus. The reason is that there is no central object in the case of a black hole to pin and support the magnetic flux needed to produce the high voltages required for acceleration. However, we consider a magnetized neutron star, essentially stationary, with a surrounding accretion disk, as shown in Figure 8.8. The magnetic flux of the neutron star should diffuse radially outward into the convectively evolving disk despite the opposite and implied radially inward mass flow. The magnetic flux in the neutron star should diffuse radially outward in a steady-state vacuum situation simply because the angular momentum of the disk must be flowing radially outward by whatever diffusion mechanism is leading to the viscosity necessary for the radial accretion of the matter in the first place, namely the α-viscosity of the disk. Presumably the diffusion that leads to the α-viscosity must allow diffusion of the vector field in both directions so that a gradient of magnetic flux, steeper than the vacuum field configuration,

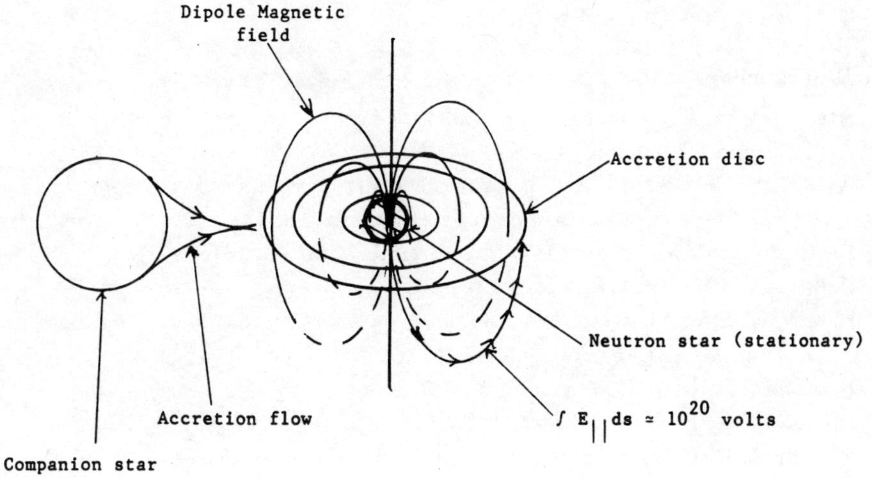

Figure 8.8

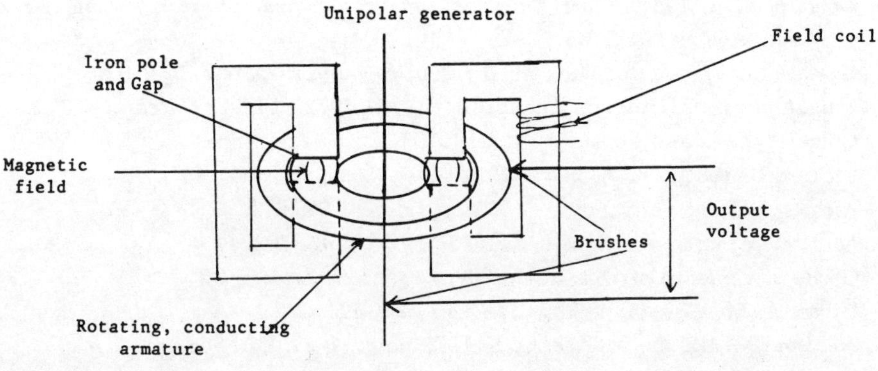

Figure 8.9

should lead to a relaxed state of magnetic flux threading the disk. Under these circumstances we have a field threading the disk much like a unipolar generator (Figure 8.9). It is a simple matter to integrate the equations to give the electric field to obtain a total potential of 10^{20} eV. The catch in this picture is that the electric field generated by the unipolar generator must exist along the lines of force between the disk and the neutron star. Just why such a large electric field should not lead to breakdown by pairs and gammas and then the current necessary to maintain the magnetic hydrodynamic conducting limit is a puzzle. However, the circumstance of a binary accretion disk and neutron star X-ray source, and a necessary α-viscosity, leads one to suspect that there is a pony somewhere. A conclusion is that there is as yet no conclusion to the problem of how to accelerate the ultrahigh energy cosmic rays.

Acknowledgments

I am indebted to Albert Petschek for discussions, understanding, and doubt.

This work was supported by DOE.

References

Atrashkevich, E., et al. 1983, *8th European Symposium,* ed. N. Iucci et al., p. 67 (Bologna: Tecnoprint).

Axford, W. I., Leer, E., and Skadron, G. 1977, *Proceedings of the 15th International Conference on Cosmic Rays* (Plovdiv), **11,** 132.

Baltrusaitis, R. M., et al. 1985, *Phys. Rev. Letters,* **16,** 1875.

Bell, R. A. 1978, *M.N.R.A.S.,* **182,** 147.

Berenzinsky, V. S. 1983, *Proceedings of the 18th International Conference on Cosmic Rays* (Bangalore, India), **2,** 275.

Blandford, R. D. 1976, *M.N.R.A.S.,* **176,** 465.

Blandford, R. D., and Ostriker, J. P. 1978, *Ap. J. (Letters),* **221,** L29.

Blandford, R. D., and Vnajek, R. L. 1977, *M.N.R.A.S.,* **179,** 433.

Brecher, K., and Burbidge, G. R. 1972, *Ap. J.,* **174,** 253.

Chanmugan, G., and Brecher, K. 1985, *Nature,* **313,** 767.

Colgate, S. A. 1984, *Super Envelope Advances in Space Research* (Conference Graz, Austria), **4,** 367.

Colgate, S. A., and Johnson, M. H. 1960, *Phys. Rev. Letters,* **5,** 235.

Colgate, S. A., McKee, C. R., and Blevins, B. 1972, *Ap. J. (Letters),* **173,** L87.

Colgate, S. A., and Petschek, A. G. 1979, *Ap. J.,* **229,** 682.

Eichler, D. 1984, *Ap. J.,* **277,** 429.

Eichler, D., and Vestrand, E. T. 1984, *Nature,* **307,** 613.

Foot, E. A., and Kulsrud, R. M. 1979, *Ap. J.,* **233,** 302.

Gunn, J. E., and Ostriker, J. P. 1969, *Phys. Rev. Letters,* **22,** 728.

Hillas, A. M. 1984, *Ann. Rev. Astr. Ap.,* **22,** 425.

Johnson, M. H., and McKee, C. F. 1971, *Phys. Rev.,* **D3,** 858.

Jokipii, J. R. 1985, private communication.

Jokipii, J. R., and Morfill, G. E. 1985, *Ap. J. (Letters),* **290,** L1.

Kegel, W. H. 1971, *Astr. Ap.,* **12,** 452.

Krymsky, G. F. 1977, *Dokl. Akad. Nauk. SSSR,* **234,** 1306.

Kulsrud, R. M., and Zweibel, E. G. 1975, *Proceedings of the 14th International Conference on Cosmic Rays* (Munich), **2,** 465.

Legage, P. O., and Cesarsky, C. J. 1983, *Astr. Ap.,* **125,** 249.

Lovelace, R. V. E. 1976, *Nature,* **262,** 649.

McKee, (Chester) R., and Colgate, S. A. 1973, *Ap. J.,* **181,** 903.

Meyerhofer, D. D., and Perkins, F. W. 1984, *Phys. Fluids,* **27,** 2483.

Schramm, D. N., and Hill, C. T. 1983, *Proceedings of the 18th International Conference on Cosmic Rays* (Bangalore, India), **2,** 393.

Stecker, F. 1968, *Phys. Rev.,* **21,** 1016.

9

High Energy Radiation from Neutron Stars

M. Ruderman

9.1. Introduction

Neutron stars are observed in two families:

i) Isolated, mainly from their nonthermal emission as radiopulsars; these are generally observed as relatively young objects ($\lesssim 10^7$ yr).

ii) In binaries, mainly as X-ray sources; these are usually, but not always, old ($\gtrsim 10^7$ yr). They are usually observed from the thermal X-ray emission associated with accretion from their companions or from near surface nuclear explosions of the accreted matter.

Isolated radiopulsars from Crab-like supernova explosions have a birthrate of one every few centuries (approximately one for each 5 or so explosions). Their population in our Galaxy $\sim 10^8$. They turn off as radiopulsars after 10^7 yr so that only about 10^5 are currently observable. Of these almost 400 have now been observed. Neutron stars in binaries appear to have a birthrate of order 10^{-5} yr^{-1}. These can be formed when a supernova explosion does not disrupt the binary. (Some may be formed in relatively mild events when certain white dwarfs grow beyond their Chandrasekhar limits because of accretion from their companions.) The present galactic population of neutron stars in binaries $\sim 10^5$. Of order 30–300 of these have been observed as accreting X-ray binaries. If, as in many currently popular models, γ-ray burst sources are also a local population of neutron stars in binaries, at least another 10^2 such neutron star binaries have been detected. The numbers of observed (and possibly of observable) isolated and binary neutron stars are comparable.

Neutron star surface dipole magnetic fields (B_0) are inferred in a somewhat model-dependent way ($B_0^2 \propto P\dot{P}$) from measured radiopulsar stellar spin-down rates ($\dot{P}$) and periods (P). The evolutionary track ($\dot{P}$ versus P) has been measured for only two pulsars, the very young Crab and PSR 1509–58. They give $P\dot{P} \propto P^{\alpha}$ with $\alpha = 0.5$ and 0.2, respectively (Groth 1975; Manchester, Durdin, and Newton 1985), and are consistent with $B_0 \sim 3 \cdot 10^{12}$ and 10^{13} G which grow slightly with time. Most radiopulsars are much older and have smaller $P\dot{P}$, interpreted as implying $B_0 \sim 10^{12}$ G; many of these have very much smaller B_0. For the above two very young pulsars $P\dot{P}$ and (perhaps) by implication B_0 will ultimately drop if they are to evolve into typical older pulsars. This is a popular interpretation because there are a number of ways for B_0 to be diminished strongly after several million years (an age which is probably comparable to the decay time for crustal currents). We shall consider separately below high energy radiation from young isolated radiopulsars with large B_0 and old ones spun up in accreting binaries which usually have very small B_0.

9.2. Young Rapidly Spinning Pulsars

Young pulsars with spin-down energy loss rates $\gtrsim 10^{34}$ ergs s^{-1} are embedded in X-ray synchrotron nebulae (Helfand 1982). The nebula radiation is presumably sustained by the injection of energetic e^-/e^+ into the circumstellar magnetic field. For the Vela pulsar this injection rate is 10^{36} s^{-1}; for the Crab pulsar it is 10^{38} s^{-1}. These are both several orders of magnitude greater than the maximum radio of charge separated particle emission from the pulsar itself, limited by the magnetic field of the electric current out of and into the star. This suggests, or is at least consistent with, a currentless flux of $e^{\pm}$ produced near the star as the injected particle source. The assumed production of such an $e^{\pm}$ wind from the near vicinity of the pulsar leads to a successful quantitative description of the optical and X-ray emission of the Crab Nebula (Kennel and Coroniti 1984).

Production of $e^{\pm}$ pairs in the magnetosphere of a young pulsar can be sustained by high energy γ-rays in several ways:

1) conversion on strong magnetic fields (Sturrock 1971), important in the very strong magnetic field ($\sim 10^{12}$ G) very near the stellar surface;
2) conversion to $e^{\pm}$ pairs by collisions with X-rays or other γ-rays;
3) conversion by collisions with soft photons (optical or IR), important for 10^{12} eV γ-rays.

The energetic γ-rays can be radiated, often by the very e^+/e^- particles they create, by three mechanisms:

a) curvature radiation from accelerated e^-/e^+ moving along curved magnetic field lines;
b) inverse Compton scattering of extreme relativistic e^-/e^+ on dense soft photon fluxes;
c) synchrotron radiation by extreme relativistic electrons moving in tight spirals along $\vec{B}$.

A test of a young pulsar model is its ability to show how one of the above 3×3 combinations supplies the required $e^{\pm}$ wind from the neutron star.

A second feature of observations of the youngest pulsars is the double pulse structure of the energetic radiation emitted by them. The Crab pulsar gives two cusp-shaped subpulses ($\sim 140°$ separation) at energies from optical through GeV (Figure 9.1) as well as at radio frequencies. The spectrum extends to 10^{12} eV in at least the "main" pulse (Dowthwaite et al. 1984) (Figure 9.2). The radiated power is $\sim 10^{-3}$ the total spin-down energy loss rate of the neutron star. A double pulse structure ($\sim 70°$ separation) is also present in the observed optical and X-ray pulse structure from LMC 0540−693, a young pulsar in the Large

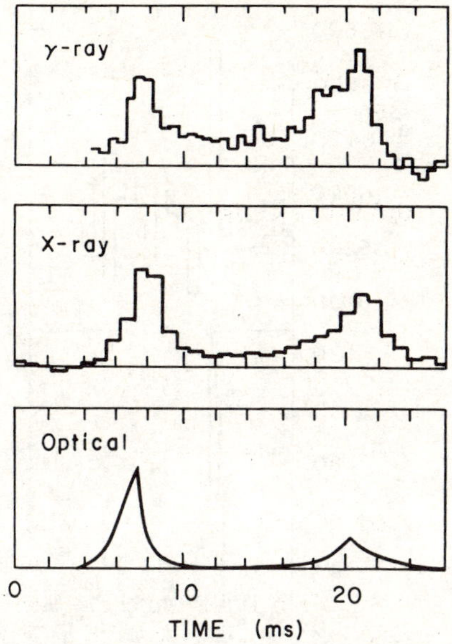

Figure 9.1. Average optical, X-ray, and γ-ray pulse profiles of the Crab pulsar (after Manchester and Taylor 1977).

Magellanic Cloud similar in age to the Crab. The Vela pulsar has a double-pulsed γ-ray spectrum similar in structure to that from the Crab pulsar ($\sim 140°$ subpulse separation), no, so far observed, X-ray intensity, and only a very weak double pulse optical emission, whose separation ($\sim 90°$) differs from that of its γ-rays (Figure 9.3).

The γ-rays needed to create an $e^{\pm}$ wind and those that are observed directly in double pulses indicate that associated with young high magnetic field rapidly spinning neutron stars are strong and efficient charged particle accelerators. There is no consensus on their structure (or even their location), or on whether they are "static" accelerators ($\vec{E}$ parallel to $\vec{B}$ and static in corotating frame) or dynamic (e.g., acceleration by electromagnetic waves).

From observations and models we take the following constraints:

i) The $e^{\pm}$ production mechanisms and, presumably, the accelerator of the particles which produce the γ-rays needed for them are effective only at distances from the star $r \lesssim c\Omega^{-1}$ ($\Omega \equiv 2\pi/P$). This follows because all of the above $e^{\pm}$ production mechanisms (a, b, or c) need either large $\vec{B}$ or large fluxes of IR, optical, or X-rays which exist only within such distances of the neutron star.

ii) The 10^{12} eV Crab (and Vela) γ-rays imply an emission distance $r \gtrsim c/3\Omega$; otherwise pair conversion by mechanism (1) in the local $\vec{B}$ will absorb the emitted γ-rays before they can propagate away from the star.

Then *if* $e^{\pm}$ production and γ-ray emission are from the same, or closely contiguous, regions the needed particle accelerator is in the outer magnetosphere ($r \lesssim c\Omega^{-1} \equiv$ light cylinder radius) or the trans-light-cylinder region. (For other

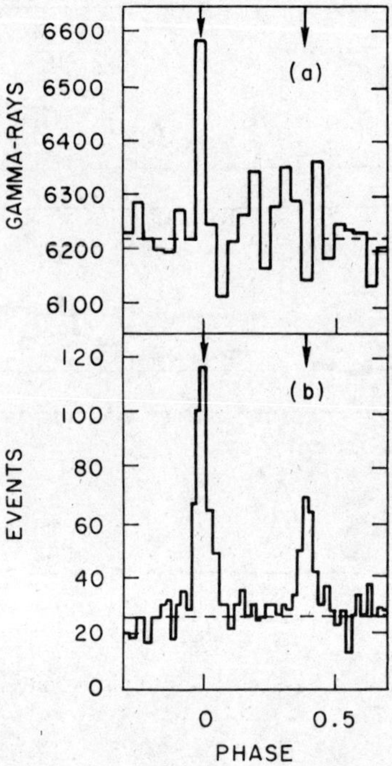

Figure 9.2. γ-ray pulse profile of the Crab pulsar at 10^{12} eV (Dowthwaite et al. 1984) and 10^8 eV (Wills et al. 1982).

proposed sites for the origin of γ-rays see Tademaru 1973; Hardee 1977; Harding, Tademaru, and Esposito 1978; Daugherty and Harding 1982).

iii) The double beams of radiation are rotating fan beams (latitude coverage $\gg$ observed width) rather than cones. If they were cones, the probability for having detected both the Crab and Vela pulsars in searches for pulsars in young supernova remnants would be only about $(1/5) \times (1/5) = 1/25$.

We turn now to models which appear to offer an explanation of these properties and also a quantitative description of the emitted spectra which agrees rather well with the observed ones.

9.3. "Static Gaps" in Outer Magnetospheres

In the laboratory a solid conductor rotating with angular speed $\vec{\Omega}$ has an interior velocity

$$\vec{v} = \vec{\Omega} \times \vec{r}, \tag{1}$$

an electric field

$$\vec{E} = \frac{\vec{v}}{c} \times \vec{B}, \tag{2}$$

and a charge density

$$\rho = \rho_0 \equiv \frac{\vec{\nabla} \cdot \vec{E}}{4\pi} = \frac{\vec{\Omega} \cdot \vec{B}}{2\pi c} + \text{terms of order } \vec{\nabla} \times \vec{B}. \tag{3}$$

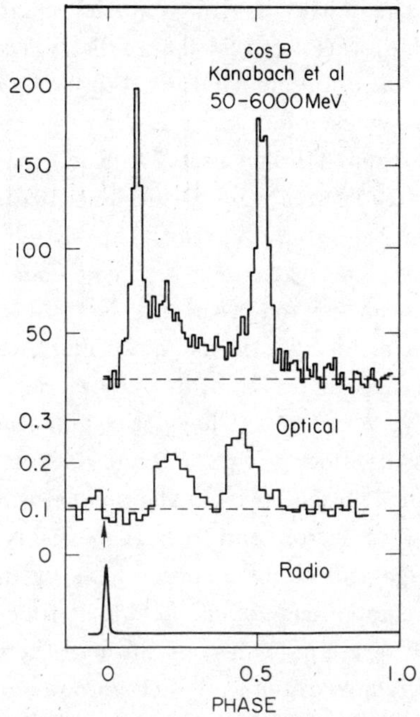

Figure 9.3. Average γ-ray, optical, and radio profiles of the Vela pulsar (after Manchester and Taylor 1977 and Kanbach et al. 1980).

From equation (2) it follows that, where there is no current,

$$\vec{E} \cdot \vec{B} = 0 \tag{4}$$

in both the laboratory and corotating frames. In the corotating frame the interior potential (V) satisfies

$$\nabla^2 V = 4\pi(\rho - \rho_0). \tag{5}$$

Outside of the spinning solid, $\rho = 0$, $\vec{E} \cdot \vec{B} \neq 0$, and, between distant points near the star, the potential difference $\Delta V \sim \Omega R^2 B c^{-1}$. In the laboratory, maximum rotation speeds for a 10^2 cm 10^4 G rotator give $\vec{E} \cdot \hat{B}$ outside the rotator $\sim$10 volts cm^{-1}, and $\Delta V \sim 10^3$ volts, insufficient to pull electrons or ions from the surface. For the Crab pulsar, with $B_0 \sim 3 \cdot 10^{12}$ G, the exterior $\vec{E} \cdot \hat{B}$ would be $3 \cdot 10^{12}$ volts cm^{-1} if the exterior region were to remain empty. This is quite impossible in the presence of such strong $\vec{E}$ along $\vec{B}$. Charge will be pulled from the star until equation (4) is achieved. This in turn would, if achieved, imply equations (1), (2), and (3) for the corotating plasma outside of the star whenever $\rho \neq 0$. Thus the charges in the magnetosphere beyond the star tend to flow until they achieve a rigidly rotating extension of the star (see also Krausse-Polstoroff and Michel 1985). If the magnetosphere were bounded by a thin insulating cylinder which prevented current flow through it (and whose radius were enough less than $c\Omega^{-1}$ that relativistic velocities would not be approached), the resulting "static" magnetosphere would satisfy equations (1)–(5). The "static" model charge distribution is indicated in Figure 9.4. However, this is not a unique solution. Empty ($\rho = 0$) regions can exist in the magnetosphere, on whose boundaries $\vec{E} \cdot \vec{B} = 0$ so that none of the boundary charge is pulled into them (Holloway 1973; Ruderman and Sutherland 1975; Michel 1981; Cheng, Ruderman, and Sutherland 1976; Arons 1981). Some of these are indicated on Figure 9.5. There the charge-depleted slots are bounded by magnetic field lines along their length. Their ends, however, cross through field lines, but only near the magnetosphere's "null surface" where $\rho_0 = \vec{\Omega} \cdot \vec{B}/2\pi c = 0$ and ρ, in a filled static magnetosphere, changes sign. The exact position and shape of these ends cannot be fixed a priori but must be determined self-consistently to give $\vec{E} \cdot \vec{B} = 0$ (Cheng, Ho, and Ruderman 1985*a*). Within the gap, away from its boundary, $\vec{E} \cdot \vec{B} \neq 0$ and particle acceleration is possible. For a gap which begins and ends within the magnetosphere, the total potential drop (in a corotating frame) from end to end along $\vec{B}$ is zero. Such a gap can be easily filled with charge and is not a source for sustained particle acceleration through it. However, a gap along "open" field lines which do not close within the light cylinder may accelerate particles out through the light cylinder and out of the magnetosphere. We now argue that such an open gap might be expected to form in a static magnetosphere if the insulating cylinder around it were removed so that magnetospheric currents could flow through the light cylinder.

9.4. "Dynamic Gaps" in Pulsar Outer Magnetospheres

If the insulating sheet which, in the static magnetosphere approximation, separates the magnetosphere from the rest of the circumstellar region were removed, current would flow along the open magnetic field lines through it. If the initial particle flow were only outward through the light cylinder in the general pattern of Figure 9.6, this flow would cause a charge depletion region to grow around the null surface. Positive (negative) charge moving out through the light cylinder leaves a void ($\rho = 0$) which, according to equation (5) has the electrostatic properties of negative (positive) charge in the depleted region. This pushes negative (positive) charge back toward the star from the starward side of the null surface. As the charge-depleted regions ("voids") grow, so does $\vec{E} \cdot \vec{B}$ within them until pair production processes within a void, of the sort discussed in § 9.2, limit further growth by supplying the charge carried away when current flows through the null surface.

Of course when γ-rays from one void propagate through another one, the latter void may be quenched by $e^{\pm}$ materialization there. However, all of the processes for making high energy γ-rays give rise only to γ-rays initially directed essentially tangent to the local $\vec{B}$. As a result, a copious γ-ray production from voids in certain regions can quench voids in other regions but not vice versa. The only voids which cannot be quenched because, for geometrical reasons, points within them cannot be reached by high energy γ-rays created near their own (upper) open field line boundaries or in other regions are shown in Figure

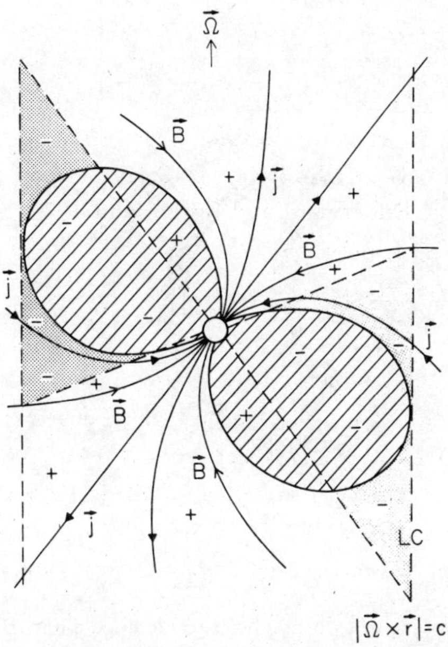

Figure 9.4. Charge distribution in a "static" magnetosphere of a spinning magnetized neutron star with $\vec{\Omega} \cdot \vec{B} < 0$ above the polar cap. An insulating cylindrical sheet surrounds the magnetosphere and prevents the flow of current through it. (Its surface charge is not shown.) The charge distribution is $\rho \sim \rho_0$ with the negatively charged open field line region shown dotted; $\rho_0 = 0$ on the "null surface" $\vec{\Omega} \cdot \vec{B} = 0$. The closed field lines which do not penetrate the light cylinder ($|\vec{r} \times \vec{\Omega}| = c$) are in the shaded area. $\vec{B}$ field lines are represented by those of a static dipole, a bad approximation near the light cylinder and possibly near the star.

9.6. It is the crosshatched region bounded below by the last closed (= first open) magnetic field line, bounded above by an open field line boundary, and ending within the magnetosphere slightly inward of the null surface where the gap $\vec{B} \cdot \vec{B} = 0$. This is an example of the "open" static gap of § 9.3 and Figure 9.5. There are two such gaps, one on each side of the star.

If, now, the possibility of inward flow through the light cylinder is included, such a gap can remain unquenched. It has a roughly constant electric field along $\vec{B}$ through its central region, which falls to zero only at its inner end. Negative (positive) charge sucked into this "outergap" through the light cylinder will be accelerated through the gap and flow down onto the star below. To quench an empty outergap such charges, and oppositely charged ones pulled into the gap from its inner end, would have to stop (or be strongly slowed) within the gap, a behavior quite opposite to their accelerated passage through it. Thus an outergap, as sketched in Figure 9.6, once formed, may be stable and continue to grow in thickness until the $\vec{E} \cdot \vec{B}$ through its interior (which, for thin gaps, is proportional to the square of the gap thickness) sustains enough $e^{\pm}$ production near the gap's (upper) open field line boundary to limit further growth. Details of the structure of such dynamically limited gaps and the energetic radiation

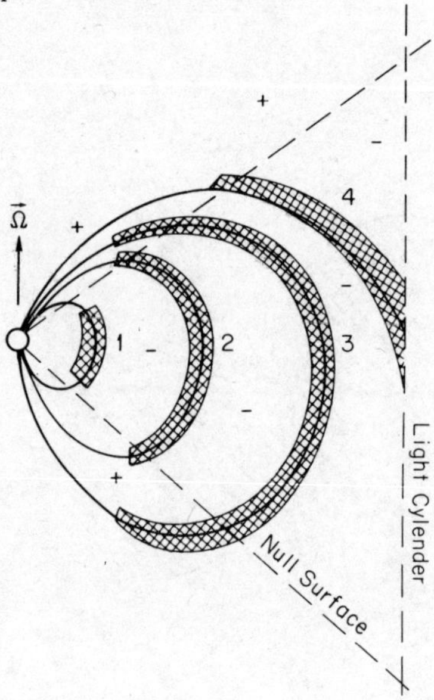

Figure 9.5. Four examples of thin magnetosphere gaps along $\vec{B}$ field lines in a rotating magnetosphere with axial symmetry. All charge has initially been removed from the interior of the gaps (shown crosshatched). In the corotating reference frame these gaps are embedded in an equipotential conducting plasma. In gap 1, $\vec{E} \cdot \hat{B} < 0$ at the gap ends; negative charge is pulled into the gap through its ends, quickly replenishing the negative charge initially removed and thus destroying the gap. (No charge flows across $\vec{B}$.) In gap 3, $\vec{E} \cdot \hat{B} > 0$ at the gap ends (positive charge was initially removed there to form that part of the gap) and positive charge flows into the gap to replenish what was initially removed from that gap region. Neither of these gaps is static. Gap 2 has its end near where $\vec{E} \cdot \hat{B} = 0$ and is static. If $e^{\pm}$ production existed in it, the $\vec{E} \cdot \hat{B}$ in the gap would accelerate e^{+} toward the gap ends but keep e^{-} bound, thus replenishing the gap's charge deficit and annihilating the gap. Gaps of type 4 on open $\vec{B}$ field lines, but whose magnetospheric end is similar to that of gap 2, are the outergaps considered in this paper.

from them have been considered by Cheng, Ho, and Ruderman (1985*a*, *b*).

9.5. Pulse Structure of Energetic Radiation Sustained by Outergap Pair Production

Although details of outergap $e^{\pm}$ production can be complicated and very sensitive to pulsar period, certain features of the gap-related emitted radiation are not. Symmetry of the magnetosphere structure always results in a pair of gaps on opposite sides of the magnetosphere as shown in Figure 9.6. The strongest part of that radiation comes from synchrotron and inverse Compton emission by pairs created just above a gap's open field line boundary. A quite general feature of such radiation is that it is mainly emitted in directions almost tangent to the local $\vec{B}$ and that it is emitted in both tangential directions, reflecting the approximate symmetry between the flow of $e^{-}(e^{+})$ inward and $e^{+}(e^{-})$ outward for all $e^{\pm}$ created within the gap. (This fails, of course, near the gap ends since the total $e^{-}(e^{+})$ current flow there is the sum of all $e^{-}(e^{+})$

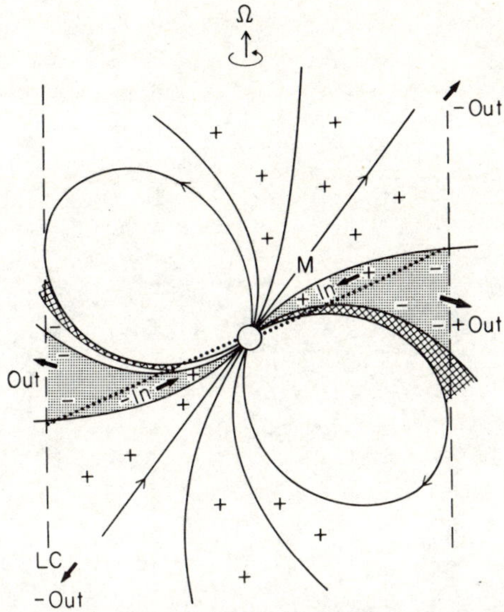

Figure 9.6. A schematic indication of assumed current flow patterns on open field lines. Current flow of positive charge away from the star (mainly via non-charge-separated plasma) is assumed balanced by a return current which is mainly through the null surface. The inward-directed arrow indicates current flow direction only: negative electrons will be *out*flowing in the dotted region. Negative charge is assumed to flow out leaving behind a partial void ($\rho \sim 0$) in the outer magnetosphere near the null surface. The Coulomb field of $\rho - \rho_0 \sim -\rho_0$ pushes positive charge on the other side of the null surface toward the star. Pair production processes can fill all voids except the two outergaps shown crosshatched.

creation within the gap and there has not yet been space enough for a large amount of $e^+(e^-)$ production.) The directions of emitted beams are shown in Figure 9.7. These are fan beams whose longitudinal extent is, for thin gaps, almost 90°. An observer would see two of them whose arrival time separation comes from differences in time of flight of the radiation across the magnetosphere and from aberration effects since the azimuthal velocity of the local radiating region (Ωr) is typically 0.6 c. The exact pulse time separation depends upon the inclination angle between the stellar spin and magnetic moment and upon the angle between the spin and the observer. Aberration also changes the shape of the emitted beam so that an observer should see a different (cusped) beam shape from that emitted in the corotating frame (Figure 9.8) at varying r through the beam. The resulting double beam tilted cusp resembles that of the well-measured Crab optical profile of Figure 9.1. For Vela (but not the Crab) the weak optical radiation originated from closer to the star, where aberration is less, than do the γ-rays. This results in closer optical pulse spacing, but the midpoint between pulses (corresponding physically to radiation from $r = 0$) should be the same for all radiation in Vela, as observed.

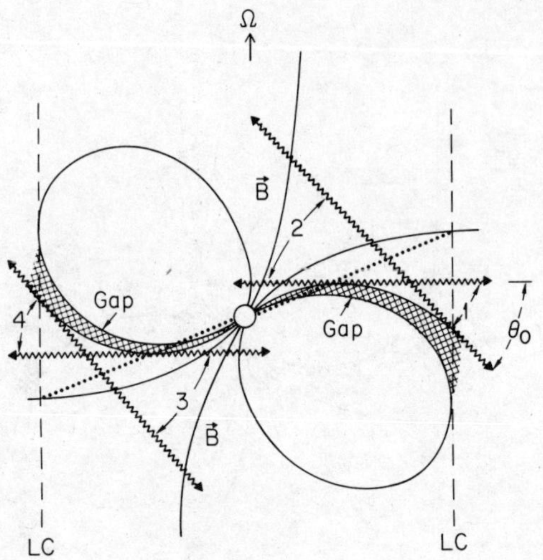

Figure 9.7. Model outer magnetosphere with indicated radiation beaming of secondary radiation from four emission regions near the outergap boundary.

9.6. Outergap Radiation: Crab Pulsar

A mechanism which is found to limit the growth of the Crab pulsar outergap and to supply enough $e^{\pm}$ pairs to prevent charge depletion elsewhere in the trans-null-surface magnetosphere is the following (Cheng, Ho, and Ruderman 1985*b*). An e^{+}/e^{-} accelerated through the outergap reaches an energy (limited by radiation reaction) at which it radiates $\sim 10^5$ multi-GeV curvature γ-rays per gap passage. These convert to $e^{\pm}$ pairs from collisions with soft X-rays (§ 9.2, mechanism 2*a*). When such pairs are created within the gap, the e^{+}/e^{-} are separated, are accelerated, and repeat the process. Beyond the gap they are unaccelerated and lose their energy by synchrotron radiation (MeV to optical), which supplies the needed X-ray flux through the gap. Inverse Compton scattering of the outside gap $e^{\pm}$ on these same X-ray photons boosts some of them into the MeV–GeV regime. The Crab outergap is limited by this combination of processes when the particle flow through it is $\sim 10^{33}$ s^{-1} and the related total potential drop in it along $\vec{B}$ is $\sim 10^{15}$ volts. The total double fan beam power radiated from the gap transboundary region, which is the source of most of the gap sustained observed radiation, is $\sim 10^{36}$ ergs s^{-1}. The total number of created $e^{\pm}$ pairs $\sim$ the total number of multi-GeV curvature γ-rays coming out of the gap (almost all completely converted to $e^{\pm}$ pairs) $\sim 10^5 \times 10^{33}$ s$^{-1} = 10^{38}$ s^{-1}. This radiated power and $e^{\pm}$ flux are consistent with those which, apparently, come directly from the Crab pulsar. A computed emission spectrum (Cheng, Ho, and Ruderman 1985*b*), arbitrarily adjusted to fit the integrated intensity, is given in Figure 9.9. (The above mechanisms will not give 10^{12} eV γ-rays. These will be generated in this model by a process which begins with inverse Compton scattering of gap e^{-}/e^{+}. The resulting 10^{13} eV γ-rays will convert in

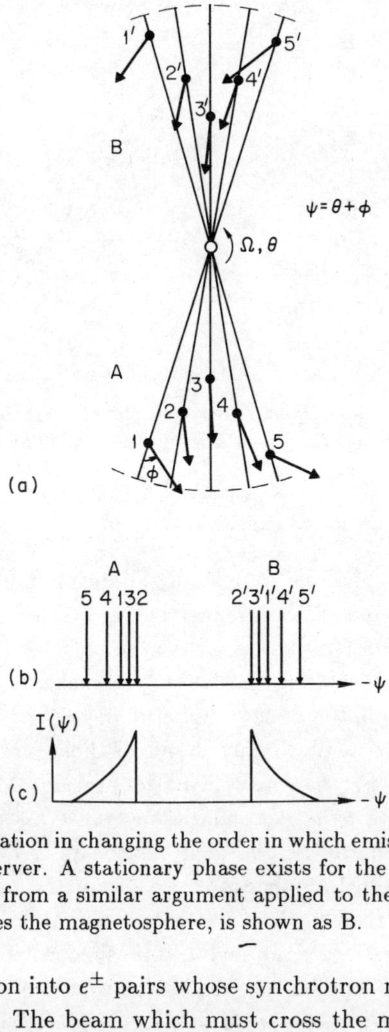

Figure 9.8. Effect of aberration in changing the order in which emission from neighboring field lines reaches an observer. A stationary phase exists for the beam from point 2 of beam A. The cusp shape from a similar argument applied to the second pulse (Figure 9.7), whose beam traverses the magnetosphere, is shown as B.

the transboundary region into $e^{\pm}$ pairs whose synchrotron radiation is typically in the 10^{12} eV regime. The beam which must cross the magnetosphere to be observed [beam 3 of Figure 9.7] may be absorbed by pair production on the stronger $\vec{B}$ through which it must pass.) In this model the extent of inverse Compton scattering on optical photons deep within the Crab outergap depends upon unknown details of the shape of the heavily plasma loaded magnetic field gap boundary lines near the light cylinder. If such inverse Compton scattering were to be sufficiently large, the Crab pulsar outergap might be controlled in part or even mainly by the Vela mechanism of § 9.7.

9.7. Outergap Radiation: The Vela Pulsar

The mechanism ($2a$) which limits the model outergap for Crab parameters is not effective in limiting a similar model outergap for Vela. This is mainly because Vela's outergap synchrotron radiation time scale ($\propto B^{-2} \propto P^6$) is of order 10^3 that of the Crab. There is then insufficient time for Vela $e^{\pm}$ to radiate

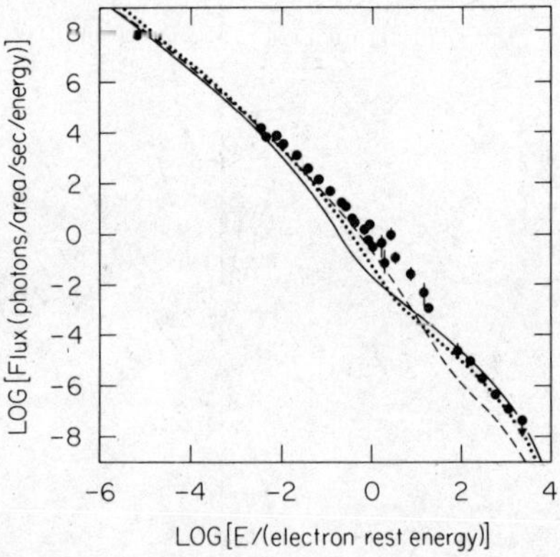

Figure 9.9. Model pulse spectra and observed spectrum for the Crab pulsar. (For details and references, see Cheng, Ho, and Ruderman 1985*b*.) The model integrated intensity is adjusted to fit that observed. The three theoretical curves differ in assumed gap magnetic field and primary e^-/e^+ energy. The solid curve is for outergap $B = 2 \cdot 10^5$ G and $E = 10^{13}$ eV; the dotted curve has $B = 10^6$ G.

enough synchrotron photons to sustain the $2a$ mechanism of § 9.6 before the radiating pairs flow out of the magnetosphere. Vela, instead, limits its gap in the following, different, way.

In Vela's outergap, each accelerated e^-/e^+ makes of order a hundred 10^{13} eV γ-rays by inverse Compton scattering on IR ($\sim 10^{-1}$ eV) photons. These γ-rays then make $e^{\pm}$ on this same IR flux. Some of these pairs are formed in the gap and thus supply e^-/e^+ there to repeat the process. Most convert beyond the gap boundary to give pairs (secondaries) whose crossed beams of synchrotron (secondary) radiation are the observed Vela beams. Additional (tertiary) pairs ($\sim 10^{36}$ s^{-1}) are made by pair production from the passage of low energy γ-rays of one secondary beam through the highest energy X-rays of the other and vice versa. These tertiary pairs give the IR synchrotron radiation which flows through the gap and supplies the IR photons whose inverse Compton scattering by gap e^-/e^+ initiates the coupled sequence of outergap controlling processes. Calculations (Cheng, Ho, and Ruderman 1985*b*) give a self-limited outergap for Vela parameters when the secondaries' radiated crossed fan beam power is of order 10^{35}–10^{36} ergs s^{-1} and the total (tertiary) pair production rate is $\sim 10^{36}$ s^{-1}. The computed emission spectrum, adjusted to fit the total integrated intensity is given in Figure 9.10 and compared there to observations. The spectral break at around 10^2 keV occurs because $e^{\pm}$ pairs leave the magnetosphere before their characteristic synchrotron radiation energy falls much below this value. This model spectrum falls below the very weak observed optical intensity if the low energy (optical and IR) radiation from the tertiaries is not included. The latter gives double fan beams (total power $\sim 10^{30}$ ergs s^{-1}) emitted from regions enough closer to the neutron star that the pulse interpulse separation is only about 90° (see § 9.5). In the Vela outergap model the secondaries lose almost all their energy to synchrotron radiation, but inverse Compton scattering on the

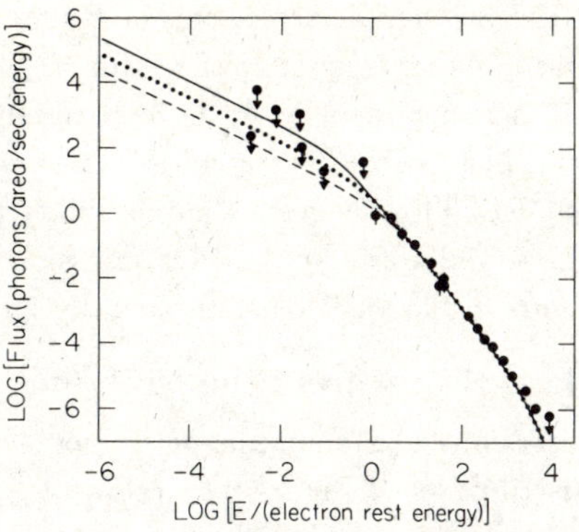

Figure 9.10. Model pulsed spectrum and observed spectrum for the Vela pulsar. (For details and references, see Cheng, Ho, and Ruderman 1985*b*.) The model integrated intensity is adjusted to fit that observed. Tertiary optical emission (which agrees with that observed) is not included in the theoretical curve. The theoretical curve for modulated X-rays around 1 keV does not include an expected contribution between pulses from thermal polar cap emission. Its existence would raise the observationally derived upper limit by diminishing the amount of observed modulation through the pulse. The three curves correspond to different residence times for secondary pairs in the outer magnetosphere and/or different synchrotron radiation decay times for these pairs. These determine a minimum for the characteristic energy of synchrotron radiation where there is a break in the spectrum at 80 keV (*solid curve*), 225 keV (*dotted curve*), and 575 keV (*dashed curve*). The uncertain structure of the outer magnetosphere does not support a definitive prediction for the break energy.

tertiary IR flux does take about 10^{-2} of it. This appears as 10^{12}–10^{13} eV γ-rays with total emitted power about 10^{-2} that in the 10^2–10^3 MeV band.

This Vela outergap model calculation also suggests that an outergap ultimately fills the entire outer magnetosphere region whose $\vec{B}$ field lines are threaded by the null surface when the spin-down energy loss rate of a young pulsar drops to about 10^{34} ergs s^{-1}. At longer pulsar periods the outergap would no longer be effective in controlling gap size by copious outer magnetosphere $e^{\pm}$ production (see § 9.1).

9.8. Radioemission

For the Vela and Crab pulsars, radioemission (RF) is presumed to take place much nearer the star, possibly from polar caps at the stellar surface, through mechanisms entirely unrelated to those which support outergap emission. If the surface field is irregular, so are the surface distribution, shape, and magnetic field directions of polar caps. Then the observed radiopulse arrival time is unrelated to those of the higher energy outergap emitted pulse and interpulse. But if a somewhat irregular outergap secondary $e^{\pm}$ flow is illuminated by an RF beam which originated elsewhere, the beam can be coherently inverse Compton boosted by the motion along $\vec{B}$ of the moving "mirror" formed by the dense plasma of secondaries, even though the relatively strong local magnetic field strongly suppresses the incoherent scattering of RF by single e^-/e^+. The resulting RF would have a higher frequency than that of the incident beam and be beamed in two directions coincident with those of the higher energy incoherent secondary synchrotron beams from the same $e^\pm$ as those which coherently upscattered the incident RF. This suggests the possibility that the Crab outergap does not create the observed double-pulsed RF emission of the Crab pulsar, but merely diverts a precursor like an RF beam partially incident upon it.

9.9. 10^{15} eV Nucleon Beams from the Crab and Vela Pulsars?

If the relative directions of stellar magnetic moment and spin are such as to give the charge distribution of Figure 9.4, a flow of gap-accelerated e^- down onto the neutron star surface would pull protons (or positive ions) out of the stellar surface to balance the negative charge of the e^- column. When these protons reach the outergap, they will be accelerated through it with negligible energy loss from inverse Compton scattering or curvature radiation. They will therefore achieve the full 10^{15} eV gap potential drop. If emitted from the Vela pulsar, such a beam would give no easily observable effect. If emitted from the Crab pulsar, such a beam might be a strong source of 10^{15} eV γ-rays since 300 eV soft X-rays in the X-ray flux through which the proton passes are at the 300 MeV (center of mass) peak for the $\gamma + p \to \pi^0 + p$ neutral photomeson production cross section. The calculated superhigh energy γ-ray emission rate from the $\pi^0 \to 2\gamma$ decay ($\sim 10^{33}$ s^{-1}) would give a flux at the Earth of 3×10^{-12} cm^{-2} s^{-1}, near that which has been reported from the direction of the Crab (Boone et al. 1983; Wdowczyk 1984). However, 10^{15} eV γ-rays would not escape from within the Crab light cylinder, so the π^0 production cannot be that close to the star.

9.10. High Energy Radiation during the Accretion Spin-up of Older Neutron Stars

Older isolated neutron stars with smaller $\vec{B}$ and $\vec{\Omega}$ would no longer be expected to have Vela- (or Crab-) type outergaps. However, older accreting neutron stars in binaries may, especially during their spin-up phase when the inner part of the accretion disk which feeds them is spinning much more rapidly than the neutron star magnetosphere. If the Crab/Vela "open" outergap exists, it is because corotation breaks down: matter outside the light cylinder cannot corotate with the magnetosphere within it and the potential drop between open field lines causes current to flow. If, instead, the outside environment were to rotate faster than the stellar magnetosphere, the potential differences and current flow directions would reverse. This is, of course, what happens between the inner accretion disk and the star when there is a large mismatch in angular speed between them. The maximum potential drop which can be developed across the stellar magnetosphere of a slowly rotating neutron star whose magnetic field is penetrated by a surrounding accretion disk is largest when the stellar field is weak enough to permit the inner edge of the accretion disk to just reach the stellar surface, about 10^8 gauss for an Eddington limit accretion rate ($\dot{m}$) of 10^{18}

gs^{-1} (Brecher and Chanmugam 1985). In general the maximum magnetosphere potential drop achievable in this way is $\sim B_0(BMR)^{1/2}c^{-1} \sim 10^{16}$ volts. For smaller $\dot{m}$ and/or larger B_0

$$\Delta V \sim (10^{-8}B_0)^{-10/7}(\dot{m}_{18})^{5/7}10^{16} \text{ volts.} \tag{6}$$

This ΔV is comparable to that achieved across the Crab of Vela radiopulsar outergaps only if $B_0 < 10^9$ G (assuming a sub-Eddington accretion $\dot{m}_{18} < 1$). Thus potential drops similar to those discussed above for the Crab or Vela may be possible in very low magnetic field accreting neutron stars (presumably of age $\gtrsim 10^7$ yr) long before their spins have been spun up to a steady state with their accretion disks.

Because B_0 is small and the accreting stellar surface is hot ($kT \sim$ keV), the electrical conductivity across $\vec{B}$ in the star is high enough that stellar ohmic resistivity cannot account for a significant part of ΔV. Despite the large plasma accretion rate, the ΔV of equation (6) may then be found in the magnetosphere. How and where it will give particle acceleration is not clear. When $B_0 \sim 10^9$ G, the interface between a slowly rotating neutron star (Ω^*) and the rapidly rotating inner edge of its accretion disk (Ω_a) will probably be the seat of rapid magnetic recombination $[\dot{\vec{B}} \sim (\Omega - \Omega^*)\vec{B}]$ and thus of $\Delta V \sim R^2(\Omega_a - \Omega^*)B_0c^{-1} \sim 10^{16}$ volts.

Several galactic objects (in addition to the Crab) have been reported as strong sources of 10^{15} eV γ-rays. If confirmed, this would imply accelerators with at least this potential drop. At least two, Vela X-1 (Protheroe, Clay, and Gerhardy 1984) and Her X-1 (Baltrusaitis et al. 1985), are accreting neutron stars. A third, Cyg X-3, probably is, but there are also other theoretical suggestions. Since the nature and period of the condensed object in Cyg X-3 are not yet determined, various combinations of stellar P and B_0 can be proposed to give such a ΔV. Her X-1 is a $P = 1.24$ s X-ray pulsar which is being spun up by an accretion disk fed by Roche lobe overflow by its companion. Vela X-1 has a stellar rotation period $P = 282$ s. Its accretion appears to be fed by the stellar wind of its companion (Henricks 1983), and the large but alternating stellar $\dot{P}$ suggests that the accretion disk formed from the wind close to the neutron star fluctuates in rotation direction. It is difficult to see how Her X-1 with its 1 s period and especially how the very slowly rotating Vela X-1 could generate $\Delta V \gtrsim 10^{15}$ volts unless the accreting disk motion cutting through the stellar B is the potential drop source and B_0 is $\lesssim 10^{10}$ G, characteristic of many (and perhaps all) old neutron stars (see Brecher and Chanmugam 1985). In such cases "outergap accelerators" may be formed by magnetospheric current flow in old, low B_0, long period neutron stars.

9.11. Postscript

Seventeen years ago when the initial observations of radiopulsars promoted the theory of neutron stars to a respectable role in astrophysics, Al Cameron was among the first to study their internal structure. He introduced several of our colleagues at this conference into a lasting affair with this subject. He has long since moved on to other problems, but uncharacteristically, he forgot to tell those who remain involved with neutron stars what the answers are. Clearly we miss him and wish he were with us.

It is a pleasure to thank Professor S. Drell and the Theoretical Group at SLAC for their very kind hospitality.

This research was supported in part by the Department of Energy under contract DE-AC03-76SF00515 and a grant from the National Science Foundation.

References

Arons, J. 1981, in *IAU Symposium 74, Origin of Cosmic Rays,* p. 175.

Baltrusaitis, R.; Cassiday, G.; Cooper, R.; Elbert, J.; Gerhardy, P.; Loh, E.; Mizumoto, Y.; Sokosky, P.; and Steck, D. 1985, University of Utah preprint.

Boone, J.; Cady, R.; Cassiday, G.; Elbert, J.; Loh, E.; Sokosky, P.; Steck, D.; and Waserbaech, S. 1983, in *Proceedings of the Cosmic Ray Workshop,* p. 268 (University of Utah).

Brecher, K., and Chanmugam, G. 1985, *Nature,* **313,** 767.

Cheng, A., Ruderman, M., and Sutherland, P. 1976, *Ap. J.,* **212,** 808.

Cheng, K.-S., Ho, C., and Ruderman, M. 1985*a,* preprint.

——. 1985*b,* preprint.

Daugherty, J., and Harding, A. 1982, *Ap. J.,* **252,** 337.

Dowthwaite, J.; Harrison, A.; Kirkman, I.; Macrae, H.; McComb, T.; Oxford, K.; Turner, K.; and Walmsley, M. 1984, *Ap. J. (Letters),* **286,** L35.

Groth, E. 1975, *Ap. J. Suppl.* **29,** 431.

Hardee, P. 1977, *Ap. J.,* **216,** 873.

Harding, A., Tademaru, E., and Esposito, L. 1978, *Ap. J.,* **225,** 226.

Helfand, D. 1982, in *IAU Symposium 101, Supernova Remnants and Their X-Ray Emission.*

Henricks, H. 1983, in *Accretion-driven Stellar X-Ray Sources,* ed. W. Lewin and E. van den Heuvel (Cambridge University Press).

Ho, C. 1984, Ph.D. thesis, Columbia University Department of Physics.

Holloway, N. 1973, *Nature,* **246,** 6.

Kanbach, G.; Bennet, K.; Bignami, G.; Buccheri, R.; Caraveo, P.; D'Amico, N.; Hermson, W.; Lichti, G.; Masnou, J.; Mayer-Hasselwander, H.; Paul, J.; Sacco, B.; Swanenberg, B.; and Wills, R. 1980, *Astr. Ap.,* **90,** 163.

Kennel, C., and Coroniti, F. 1984, *Ap. J.,* **283,** 710.

Krausse-Polstoroff, J., and Michel, R. C. 1985, *Astr. Ap.,* **144,** 72.

Manchester, R., Durdin, J., and Newton, L. 1985, *Nature,* **313,** 374.

Manchester, R., and Taylor, J. 1977, *Pulsars* (San Francisco: Freeman).

Michel, R. C. 1981, in *IAU Symposium 95, Pulsars,* ed. W. Sieber and R. Wielebinski.

Protheroe, R., Clay, R., and Gerhardy, P. 1984, *Ap. J. (Letters),* **280,** L47.

Ruderman, M., and Sutherland, P. 1975, *Ap. J.,* **196,** 51.

Sturrock, P. 1971, *Ap. J.*, **164**, 529.

Tademaru, E. 1973, *Ap. J.*, **183**, 625.

Wdowczyk, J. 1984, in *Rencontre de Moriond: High Energy Astrophysics*, ed. J. Tranthanh Van.

Wills, R., et al. 1982, *Nature*, **296**, 723.

Part IV

Star Formation

10

Molecular Cloud Cores and Star Formation

Frank H. Shu

Abstract

We examine the problem of how stars form from the great reservoirs of cold matter in giant molecular clouds. A fundamental issue any rational discussion must address is why the classical concept of a protostar as an accreting body of gas and dust has to date eluded definitive discovery by radio and infrared astronomers. A viable picture, at least for the incipient sites of the birth of low mass stars like the Sun, is the quasi-static condensation, by the process of ambipolar diffusion, of a dense molecular cloud core from its magnetically (and turbulently) supported envelope. The gravitational collapse of such a core is likely to take place from "inside-out," with the region of appreciable velocities being too small to be detectable by single-dish radio instruments. Moreover, the optical accretion luminosity which is generated by the collapse process is systematically degraded to successively longer wavelengths by the infalling envelope of dust, with the extended dusty "atmosphere" making the color temperature much lower than the effective temperature so that the emergent spectral energy distribution peaks at far-infrared wavelengths ($\sim$100 μm) instead of the mid-infrared wavelengths ($\sim$10 μm) where most searches for protostars have previously been conducted. From the viewpoint that stars form in an inefficient manner from massive molecular clouds, the masses of the resulting stars must be set by processes which terminate the accretion. Observations suggest that one such process is the onset of a powerful stellar wind. The energy source for driving these winds may be the release of the energy of differential rotation through the mediation of magnetic activity. In a low mass star, such magnetic activity would presumably commence once an outer convection zone (probably driven in an accreting protostar by the onset of deuterium burning) breaks through to the surface. Breakout of the incipient corona, smothered initially by the rotating infalling material, is likely to occur in a bipolar manner. When the accretion flow has been completely reversed and the newly born star becomes optically visible over 4π steradians, its location in the Hertzsprung-Russell diagram will correspond to those observed for young T Tauri stars (the classical pre-main-sequence phase of stellar evolution), providing the solar mass or so of material was accumulated in a total accretion time of 10^5–10^6 yr.

10.1. Introduction

It is a privilege and an honor to join in the celebration of Al Cameron's 60th birthday. Judging from the participant list at this symposium, I draw the obvious conclusion that Al is a pioneering scientist whom we all greatly admire and respect. What has always impressed me most about his exploration is its global perspective, the unifying view with which he scans any landscape fortunate enough to attract his attention and energy. Because the territory he travels is so large, with huge areas left unmapped by the investigations of others, the end point of his journey is sometimes not the destination he originally sought. But a deep philosophic drive underlies all his efforts, and he has erected enduring landmarks inside and outside the boundaries of our science. Let me cite just one example, the bridge he has built to span the divide between the studies of the formation of stars and planets. I am paraphrasing his sentiments; if he were to speak for himself, he would undoubtedly express it much more expansively and comprehensively.

a) When contemplating theories of the origin of the solar system, one should make sure one can account for its most important member, which is not Jupiter, but the Sun (Cameron 1962).

b) When constructing models for the birth of stars, one had better include realistic levels of angular momentum, not only in the surrounding nebular disk (Cameron 1973, 1978), but also in the central object (Mercer-Smith, Cameron, and Epstein 1984).

By the end of this review, I hope it will transpire how truly prescient was his insistence on these points.

10.2. Molecular Clouds

Star formation at the present epoch of the Galaxy occurs almost exclusively in giant molecular cloud complexes (Zuckerman and Palmer 1974; Burton 1976). The immense amount of raw material, 10^5–10^6 $M_\odot$ (Solomon, Sanders, and Scoville 1979; Blitz and Thaddeus 1980) in each such aggregate of gas and dust, combined with its relatively dense, cold, and clumpy state (Sargent 1977; Evans 1978; Blitz 1978; Rowan-Robinson 1979; Ho and Townes 1983), make giant molecular clouds ideal factories for the production of new stars. Indeed, it is surprising that they do not make stars even more efficiently than they do. Ever since this realization gained general acceptance, radio astronomers have diligently tried to find evidence for the collapse motions which should be attendant on star formation, and infrared astronomers have searched widely for the "holy grail" of the true protostar (Wynn-Williams 1982). Alas, so far the quest has been in vain: in every young stellar object examined in sufficient detail, outflows have been found to predominate over inflows (Welch et al. 1985), and while point sources abound in the infrared sky, all of the ones identified from ground-based surveys correspond to a later stage of stellar evolution than that of a star in the process of being made (Cohen 1985).

Does this imply that the classical picture (Hayashi 1966, 1970; Larson 1969,

1973; Appenzeller and Tscharnuter 1975; Winkler and Newman 1980*a, b*) of star formation as the gravitational collapse of an extended gas cloud, followed by a phase of prolonged accretion by a central object, has to be abandoned? No, but clearly the premise underlying the earlier calculations needs to be reexamined. The most important modification, I believe, lies in the outmoded concept that each incipient star forms from the infall of a separate parcel of interstellar gas which itself fragments dynamically from a larger collapsing cloud (see also Tohline 1982). I would prefer to interpret the message of the last decade of observation of giant molecular clouds to imply that if they are fragmenting, the process is quasi-static and not dynamic, at least to the stage of molecular cloud core formation. From this point of view, discussions of *triggers for star formation within molecular clouds* represent overkill. Once giant complexes have been assembled (Mouschovias, Shu, and Woodward 1974; Elmegreen 1979; Blitz and Shu 1980; Cowie 1981; Jog and Solomon 1984; Shu 1985), the central problem of their mechanics is how to support the individual clumps of mass $\sim 10^3\ M_\odot$ (envelope plus cores), not how to trigger their free-fall collapse. As we shall argue, because giant molecular clouds contain many Jeans masses of material, the collapse of cores is inevitable; it will come naturally and inexorably without any external inducement.

Since the thermal pressure in molecular clouds is generally quite small, some other agent must be responsible for the overall support. Three viable candidates have been proposed at one time or another: rotation (Field 1978), turbulence (Norman and Silk 1980; Larson 1981), and magnetic fields (Chandrasekhar and Fermi 1953; Mestel 1965; Spitzer 1968; Mouschovias 1976). Observationally, the rotation rates of molecular clouds may be sufficient, in some instances, to provide centrifugal equilibrium for the envelopes, but the rates are too low for the cores (Goldsmith and Arquilla 1985). Similarly, CO line widths suggest that turbulent velocities approach virial values for the complex as a whole (Sanders, Scoville, and Solomon 1985), so they probably also play an important role in the mechanical structure of the individual clumps. On the other hand, the observations of Vrba, Strom, and Strom (1976) indicate that turbulence cannot be totally dominant over magnetic fields because maps of the interstellar polarization vectors in dark clouds, which presumably mark the local (projected) direction of the magnetic field, show this direction to be ordered over the dimension of the cloud. If the energy of a turbulent velocity field, in the form of a cascade of eddies (see, e.g., Landau and Lifshitz 1959), were much greater than the energy of the magnetic field, then we should expect to see a much more tangled configuration for the latter. The fact that we don't suggests that the observed "turbulent" motions are probably more wavelike than eddylike (cf. Arons and Max 1975) and that the fluid velocities associated with the waves, while often supersonic, are usually sub-Alfvénic. The ultimate source for the waves may be outflows driven by young stellar objects, but the finite ability of such sources to maintain a truly chaotic velocity field (Bally and Lada 1983) may imply that the velocities will automatically decay to sub-Alfvénic (or sub-magnetosonic) levels, where they are more easily sustained for the life of the cloud.

10.3. Cloud Cores and Star Formation

A molecular cloud which is composed of a lightly ionized gas and supported in large part against its self-gravity by magnetic fields will suffer ambipolar diffusion (Mestel and Spitzer 1956; Nakano 1981; Mouschovias 1981; Black and Scott 1982; Shu 1983; Mouschovias, Paleologou, and Fiedler 1985). The gradual leakage of magnetic and "turbulent" support from every initial density enhancement of sufficient scale can be expected to lead to the quasi-static fragmentation of the envelope of a typical cloud clump into many dense cores (Lizano and Shu 1986). Moreover, the long contraction and diffusion time scale for this process, 10^6–10^7 yr, implies that the forming cores will tend to brake magnetically (Gillis, Mestel, and Paris 1974, 1979) to the (slow) rotation rates characteristic of their envelopes (Mouschovias and Paleologou 1979, 1980). The quiet (i.e., subsonic), slowly rotating, molecular cloud cores observed by Myers and Benson (1983) apparently represent an advanced stage of such evolution, and they seem to be the sites of the birth of low mass stars (Benson, Myers, and Wright 1984). The spread in ages of the newly born stars in T associations (Strom, Strom, and Grasdalen 1975; Cohen and Kuhi 1979) of $\sim 10^7$ yr is consistent with the slow formation of many independent cores in a given cloud clump and their subsequent rapid collapse when they become sufficiently centrally condensed to pass the brink of stability.

If the unstable cores so produced have $1/r^2$ density profiles, which is the natural state for subsonic gravitational evolution under isothermal conditions (Bodenheimer and Sweigart 1968), and which seems indicated by certain observational studies (Bally and Lada 1983; Loren, Sandqvist, and Wootten 1983; but see Arquilla and Goldsmith 1985), then the collapse phase proceeds in a self-similar manner (Shu 1977). The infall starts from "inside-out," with the total amount of matter accessible for accretion in typical molecular cloud cores (plus their envelopes) potentially much larger than the star which will eventually emerge. However, substantial velocities are generated only close to the central accreting protostar. Since the regions of molecular inflow are too small to have been resolved by single-dish radio observations, this probably explains why such studies have only found the bipolar outflows of larger scale (cf. Snell, Loren, and Plambeck 1980; Canto et al. 1981) that come at a later stage of the star formation process. Moreover, until such outflows occur and sweep up the surrounding shell of material, the continued infall enshrouds the central object in a thick veil of gas and dust which obscures the protostar from optical view (e.g., Yorke and Krugel 1977). In fact, detailed calculations show that the "extended atmosphere," combined with the fairly steep wavelength dependence of the dust opacity, makes the "color temperature" of the emergent radiation much lower than the "effective temperature," so that practically all of the photons generated by the star or the accretion shock are degraded into the far-infrared (Adams and Shu 1985). Even mid-infrared radiation has difficulty escaping from the protostellar configuration, which probably explains why ground-based surveys at 10 μm have failed to find genuine protostars. Presumably, this situation will improve now that the *IRAS* satellite has turned up numerous far-infrared sources associated with dense cloud cores (C. Beichman and P. C. Myers, private communication).

10.4. Termination of the Accretion Process

T Tauri stars are pre-main-sequence objects with strong mass loss (Herbig 1962; Kuhi 1964). It is tempting to suppose that a T Tauri star represents the post-clearing stage which follows the buildup of a protostar. The locations of T tauri stars in the H-R diagram (Cohen and Kuhi 1979) are consistent with spherically symmetric calculations of protostellar buildup at mass accretion rates of 10^{-6}–10^{-5} $M_{\odot}$ yr^{-1} (Stahler, Shu, and Taam 1980; Stahler 1983; Adams and Shu 1985). These rates also follow if gravitational collapse occurs from initial conditions characteristic of quiet molecular cloud cores. However, the spherically symmetric models give little indication why the accretion phase should terminate when it does, i.e., they offer no explanation why a star ends up with the mass it does (see the discussion of Shu and Terebey 1984). A more dramatic way to phrase the same issue is to ask, How does a giant molecular cloud of 10^5–10^6 $M_{\odot}$ know how to form stars of 1 $M_{\odot}$? Or even more fundamentally, How does the interstellar medium know that roughly $(\hbar c/Gm_{\mathrm{H}}^2)^{3/2}$ hydrogen atoms of mass m_{H} are needed to produce a self-gravitating object capable of thermonuclear fusion?

I have proposed (Shu 1985) that, in fact, the interstellar medium has no inkling of this combination of fundamental constants (basically, Chandrasekhar's limit); it is stars which know this number. The idea is that winds from young stellar objects derive their power from the release of energy stored in differential rotation. In protostars of low mass (comparable to the Sun), this release is triggered when the onset of deuterium burning drives the outer envelope of the star completely convective. The combination of convection and differential rotation is imagined to lead to dynamo action and magnetic activity, which will produce the observed strong chromospheres of T Tauri stars if the magnetic fields can buoy out of the interior (as they can when the convection zone extends to the surface). This proposal predicts that the "birthline" where T Tauri stars first appear as visible objects in the H-R diagram, along a classical pre-main-sequence track (Hayashi, Hoshi, and Sugimoto 1962; Ezer and Cameron 1962, 1967; Iben 1965), should approximately correspond to a mass to radius ratio (i.e., a central temperature) that gives the ignition of deuterium. This condition will probably not be very sensitive to the exact way the star is built up, so it may help to reconcile the controversy between Stahler (1983) and Mercer-Smith, Cameron, and Epstein (1984) concerning the relative importance of direct infall and disk accretion. More work is needed to sort out the precise timing of events.

The geometry by which a stellar wind will break out into an overlying accretion flow is also interesting. On the basis of some analytical calculations of rotating gravitational collapses, we have speculated that breakout takes a bipolar form, along the rotation axis of the system (Cassen, Shu, and Terebey 1985). A useful analogy is a pressure cooker, where the internal steam accumulates until it can break through at the weakest point of the lid (the safety valve). In the present circumstance, the steam is the hot magnetized gas of the smothered corona, and the lid is the ram pressure of the infalling matter, with the weakest points in the lid lying at the rotational poles of the swirling inflow (see Figure 6*e* of Terebey, Shu, and Cassen 1984). It is important to note that both hot gas and magnetic fields are needed: to sustain winds having the heavy mass loss rates typical of young stars by thermal effects alone would entail enormous radiative losses (Bisnovatyi-Kogan and Lamzin 1977; DeCampli 1981; S. Terebey and K. MacGregor, private communication); to achieve breakout through magnetospheric effects alone is probably impossible (for a similar problem in a different context, see Arons and Lea 1976). On the other hand, with a combination of coronal gas and magnetic fields (plus, perhaps, Alfvén waves), breakout can occur because the amount of material which infalls directly on the star is a decreasing function of time when the collapsing matter contains angular momentum.

Our proposal that the outflow will initially be channeled along the rotation axis is hardly controversial given the many observations relating to this point (e.g., Cohen 1985; Welch et al. 1985); its novel aspect is the prediction that well-collimated bipolar-flow sources should correspond to systems where inflow and outflow occur simultaneously. Because of the dust embedded in the inflow, unless one is looking right down the length of the outflow jet, the associated star should be fairly obscured, a condition which can be checked observationally for sources with the most highly directed outflows. In any case, as time proceeds, the opening angle of the cone of the outflow should widen, until eventually the infall has been reversed over 4π steradians of solid angle. The protostellar phase of stellar evolution will then have been irrevocably terminated by a complex interaction between deuterium burning and rotation, two processes whose importance for the evolution of young stars was noted very early by Al Cameron. The star will begin its contraction down one of the pre-main-sequence tracks computed by Ezer and Cameron, surrounded by planets made from a nebular disk like one of Cameron's description. This dissipation of the cocoon of surrounding dust will allow astronomers to peer in or, in the case of the mature solar system, allow Al Cameron to peer out and help deduce the marvelous intricacy of how stars and planets are made.

References

Adams, F., and Shu, F. H. 1985, *Ap. J.*, in press.

Appenzeller, I., and Tscharnuter, W. 1975, *Astr. Ap.*, **40**, 397.

Arons, J., and Lea, S. M. 1976, *Ap. J.*, **207**, 914.

Arons, J., and Max, C. E. 1975, *Ap. J. (Letters),* **196,** L77.

Arquilla, R., and Goldsmith, P. F. 1985, *Ap. J.,* **279,** 664.

Bally, J., and Lada, C. J. 1983, *Ap. J.,* **265,** 824.

Benson, P. J., Myers, P. C., and Wright, E. L. 1984, *Ap. J. (Letters),* **279,** L27.

Bisnovatyi-Kogan, G. S., and Lamzin, S. A. 1977, *Astr. Zh.,* **54,** 1268.

Black, D. C., and Scott, E. H. 1982, *Ap. J.,* **263,** 696.

Blitz, L. 1978, Ph.D. thesis, published as NASA Tech. Memo. 79708.

Blitz, L., and Shu, F. H. 1980, *Ap. J.,* **238,** 148.

Blitz, L., and Thaddeus, P. 1980, *Ap. J.,* **241,** 676.

Bodenheimer, P., and Sweigart, A. 1968, *Ap. J.,* **152,** 515.

Burton, W. B. 1976, *Ann. Rev. Astr. Ap.,* **14,** 306.

Cameron, A. G. W. 1962, *Icarus,* **1,** 13.

——. 1973, *Icarus,* **18,** 407.

——. 1978, *Moon and Planets,* **18,** 5.

Canto, J.; Rodriguez, L. F.; Barral, J. F.; and Carral, P. 1981, *Ap. J.,* **244,** 102.

Cassen, P., Shu, F. H., and Terebey, S. 1985, in *Protostars and Planets II,* ed. D. C. Black and M. S. Matthews (University of Arizona Press), in press.

Chandrasekhar, S., and Fermi, E. 1953, *Ap. J.,* **118,** 116.

Cohen, M. 1985, *Phys. Reports,* in press.

Cohen, M., and Kuhi, L. V. 1979, *Ap. J. Suppl.,* **41,** 743.

Cowie, L. L. 1981, *Ap. J.,* **245,** 66.

DeCampli, W. M. 1981, *Ap. J.,* **244,** 124.

Elmegreen, B. G. 1979, *Ap. J.,* **231,** 371.

Evans, N. J. 1978, in *Protostars and Planets,* ed. T. Gehrels, p. 158 (University of Arizona Press).

Ezer, D., and Cameron, A. G. W. 1962, *Icarus,* **1,** 422.

——. 1967, *Canadian J. Phys.,* **45,** 2315.

Field, G. B. 1978, in *Protostars and Planets,* ed. T. Gehrels, p. 243 (University of Arizona Press).

Gillis, J., Mestel, L., and Paris, R. B. 1974, *Ap. Space Sci.,* **27,** 167.

——. 1979, *M.N.R.A.S.,* **187,** 311.

Goldsmith, P. F., and Arquilla, R. 1985, in *Protostars and Planets II,* ed. D. C. Black and M. S. Matthews (University of Arizona Press), in press.

Hayashi, C. 1966, *Ann. Rev. Astr. Ap.,* **4,** 171.

——. 1970, in *Evolution stellaire avant la séquence principale, 16me Liège Symposium,* p. 127 (Université de Liège).

Hayashi, C., Hoshi, R., and Sugimoto, D. 1962, *Progr. Theoret. Phys. Suppl.,* no. 22.

Herbig, G. 1962, *Adv. Astr. Ap.,* **1,** 47.

Ho, P. T. P., and Townes, C. H. 1983, *Ann. Rev. Astr. Ap.,* **21,** 239.

Iben, I. 1965, *Ap. J.,* **141,** 993.

Jog, C. J., and Solomon, P. M. 1984, *Ap. J.,* **276,** 127.

Kuhi, L. V. 1964, *Ap. J.,* **140,** 409.

Landau, L. D., and Lifshitz, E. M. 1959, *Fluid Mechanics* (London: Pergamon Press).

Larson, R. B. 1969, *M.N.R.A.S.,* **145,** 271.

——. 1973, *Ann. Rev. Astr. Ap.,* **11,** 219.

——. 1981, *M.N.R.A.S.,* **194,** 809.

Lizano, S., and Shu, F. H. 1986, in preparation.

Loren, R. B., Sandqvist, Aa., and Wootten, A. 1983, *Ap. J.,* **270,** 620.

Mercer-Smith, J. A., Cameron, A. G. W., and Epstein, R. I. 1984, *Ap. J.,* **287,** 445.

Mestel, L. 1965, *Quart. J.R.A.S.,* **6,** 161.

Mestel, L., and Spitzer, L. 1956, *M.N.R.A.S.,* **116,** 503.

Mouschovias, T. Ch. 1976, *Ap. J.,* **207,** 141.

——. 1981, in *IAU Symposium 91, Fundamental Problems in the Theory of Stellar Evolution,* ed. D. Sugimoto, D. Q. Lamb, and D. N. Schramm, p. 27 (Dordrecht: Reidel).

Mouschovias, T. Ch., and Paleologou, E. V. 1979, *Ap. J.,* **230,** 204.

——. 1980, *Ap. J.,* **237,** 877.

Mouschovias, T. Ch., Paleologou, E. V., and Fiedler, R. A. 1985, *Ap. J.,* **291,** 772.

Mouschovias, T. Ch., Shu, F. H., and Woodward. 1974, *Astr. Ap.,* **33,** 73.

Myers, P. C., and Benson, P. J. 1983, *Ap. J.,* **266,** 309.

Nakano, T. 1981, *Progr. Theoret. Phys. Suppl.,* no. 70, 54.

Norman, C., and Silk, J. 1980, *Ap. J.,* **238,** 158.

Rowan-Robinson, M. 1979, *Ap. J.,* **234,** 111.

Sanders, D., Scoville, N. Z., and Solomon, P. M. 1985, *Ap. J.,* **289,** 373.

Sargent, A. I. 1977, *Ap. J.,* **218,** 736.

Shu, F. H. 1977, *Ap. J.,* **214,** 488.

——. 1983, *Ap. J.,* **273,** 202.

——. 1985, in *IAU Symposium 106, The Milky Way,* ed. H. van Woerden, R. J. Allen, and W. B. Burton, p. 561 (Dordrecht: Reidel).

Shu, F. H., and Terebey, S. 1984, in *Cool Stars, Stellar Systems, and the Sun,* ed. S. Baliunas and L. Hartmann, p. 78 (Berlin: Springer-Verlag).

Snell, R. L., Loren, R. B., and Plambeck, R. L. 1980, *Ap. J. (Letters),* **239,** L17.

Solomon, P. M., Sanders, D. B., and Scoville, N. Z. 1979, in *IAU Symposium 84, The Large Scale Characteristics of the Galaxy,* ed. W. B. Burton, p. 35 (Dordrecht: Reidel).

Spitzer, L. 1968, in *Nebulae and Interstellar Matter,* ed. B. M. Middlehurst and L. H. Aller, p. 1 (Chicago: University of Chicago Press).

Stahler, S. W. 1983, *Ap. J.,* **274,** 822.

Stahler, S. W., Shu, F. H., and Taam, R. E. 1980, *Ap. J.,* **241,** 637.

Strom, S. E., Strom, K. M., and Grasdalen, G. L. 1975, *Ann. Rev. Astr. Ap.,* **13,** 187.

Terebey, S., Shu, F. H., and Cassen, P. 1984, *Ap. J.,* **286,** 529.

Tohline, J. E. 1982, *Fund. Cosm. Phys.,* **8,** 1.

Vrba, F. J., Strom, K. M., and Strom, S. E. 1976, *Astr. J.,* **81,** 958.

Welch, W. J.; Vogel, S. N.; Plambeck, R. L.; Wright, M. C. H.; and Bieging, J. H. 1985, *Science,* in press.

Winkler, K. H., and Newman, M. J. 1981*a, Ap. J.,* **236,** 201.

——. 1981*b, Ap. J.,* **238,** 311.

Wynn-Williams, C. G. 1982, *Ann. Rev. Astr. Ap.,* **20,** 587.

Yorke, H. W., and Krugel, E. 1977, *Astr. Ap.,* **54,** 183.

Zuckerman, B., and Palmer, P. 1974, *Ann. Rev. Astr. Ap.,* **12,** 279.

11

Dynamical Properties of the Primordial Solar Nebula

D. N. C. Lin and J. Papaloizou

The Sun and its planetary system formed about 4.6 billion years ago. Revolving around the proto-Sun was a primordial solar nebula out of which the planetary system was formed. The basic physical properties of the primordial solar nebula are very similar to those of an accretion disk. The structure, stability, and evolution of the nebula are primarily determined by the efficiency of the angular momentum and heat transport processes. The dominant transport mechanisms may be turbulence and convection. Owing to constraints on (1) grain condensation temperatures, (2) the minimum nebular mass required to allow accretion of the protoplanets, and (3) the magnitude of nebular viscosity in the vicinity of the protogiant planets' orbit around the proto-Sun, the final epoch, when the protoplanets acquired most of their masses, must have lasted less than $\sim 10^6$ yr.

11.1. Introduction

Studies of the origin of the primordial solar nebula date back to the work of Descartes (1644). Kant (1755) and Laplace (1796) proposed that the Sun and its planets were formed from a rotationally flattened gaseous nebula with the Sun at the center and the planets at various positions in the nebula. According to Laplace's original model, the nebula may have broken up into several discrete rings, due to rotational instabilities, out of which the planets formed. This aspect of the Laplacian model was later challenged by Jeffreys (1924), who showed that in a differentially rotating gaseous disk, such as the primordial solar nebula, viscous stress can lead to angular momentum transfer and matter diffusion. Nevertheless, if molecular viscosity provides the only momentum transport mechanism, the nebula's mass distribution would not be significantly changed, during the course of its evolution, from its initial conditions. But if the nebula is turbulent, the effective Reynolds stress may induce efficient momentum transport and cause the nebula to evolve on a relatively short time scale (von Weizsäcker 1943, 1948).

If the nebula is turbulent, the energy associated with chaotic motion is continually transferred to small eddies and dissipated on some "viscous" length scale (Tennekes and Lumley 1972). In order for the nebula to sustain a turbulent flow pattern, it must be driven to do so by some intrinsic instabilities. There are many potential instabilities which may cause the gas flow to become turbulent in the nebula. For example, for some time after the formation of the nebula, infall of material may continue. Since the infalling material may have velocity and angular momentum substantially different from those in the nebula at the arrival locations, considerable shear in the interface regions can be induced. Such an intense shear may cause the flow near the interface to become Kelvin-Helmholtz unstable. The mixing of angular momentum between the infalling and nebular material can provide additional stirring (Cameron 1976). Alternatively, if a sufficient fraction of the nebula gas is ionized, a seed magnetic field may be dragged by the ionized particles. Collisions between the ionized and the neutral particles induce them to follow the similar Keplerian differential rotation around the proto-Sun. If the field is coupled to the flow over an extended radial range, it may be stretched and its strength may grow until field lines break and reconnect. If the amplified random field can be effectively coupled to the nebular material, gas from different regions of the nebula would be connected and angular momentum would be transferred between these regions (Hoyle 1960; Hayashi 1981). The nebula may also be convectively unstable (Lin and Papaloizou 1980). We shall discuss the convective stability in §11.2.

Turbulent motion in the solar nebula induces an effective internal friction, which is normally referred to as viscous stress. In a differentially rotating disk such as the solar nebula, the viscous stress induces a continuous outward transfer of angular momentum and a net inward flux of matter. In this respect the solar nebula is an accretion disk. The mathematical foundation of viscous accretion-disk flow was basically established by Lüst (1952). Important theoretical understanding has been gained (Lynden-Bell 1969; Prendergast and Burbidge 1968; Shakura and Sunyaev 1973; Lynden-Bell and Pringle 1974) in an attempt to understand the underlying physical processes that govern mass transfer flows in binary systems and around massive black holes. This new information and knowledge has been applied to models of solar system formation (Cameron 1976, 1978). Today, accretion-disk theory has become an integral part of quantitative models for the origin and evolution of the solar nebula (Lin and Papaloizou 1980; Lin 1981; Lin and Bodenheimer 1982). The physical nature and the mathematical foundation of accretion-disk theory is extensively reviewed by Pringle (1981) for a wide range of astrophysical contexts. In §11.3 we discuss the evolution of the primordial solar nebula according to accretion disk theories.

Despite the application of the accretion-disk theories, the detailed dynamics of the solar nebula are still subject to considerable uncertainty. Due to the lack of an accurate, non–ad hoc prescription for turbulent viscosity, we cannot assess with confidence the effectiveness of angular momentum transport in the nebula. There is an indirect argument with which we can set limits on the magnitude of viscosity in the nebula. This argument is based on the protoplanet-nebula tidal interaction. When a protoplanet has gained sufficient mass, it induces tidal transfer of angular momentum with the nebula in a manner similar to tidal interaction between Saturn's satellites and rings. Unless the magnitude of kinematic viscosity is sufficiently high, the protoplanet can tidally truncate the nebula by opening up gaps in the vicinity of its orbit. Thereafter, the nebular material can no longer be accreted by the protoplanet and the protoplanet's growth is terminated. Thus, from the present mass of the giant planets, we may estimate the magnitude of viscosity at the time of protoplanetary formation. In §11.4 we discuss protoplanet-nebula tidal interaction in detail. Finally, we summarize our results in §11.5.

11.2. Local Analysis and the Vertical Structure of the Nebula

In principle, the evolution of an axisymmetric nebula should be analyzed in a two-dimensional time-dependent framework. However, if the pressure scale in the vertical direction H is compared with the radius r, the radial and vertical dependences in the momentum and energy transport equations can be essentially decoupled. For example, the evolution of the mass distribution in a thin axisymmetric nebula can be derived from the continuity equation, which can be reduced to

$$\frac{\partial \Sigma}{\partial t} + \frac{1}{r}\frac{\partial}{\partial r}\Sigma U_r r - S_\Sigma(r,t) = 0, \tag{1}$$

where $S_\Sigma(r,t) = -\rho(r,z,t)U_z(r,z,t)|_{z=-\infty}^{z=\infty}$ is the infall flux onto the nebula per unit area, U_r and U_z are the velocities in the radial and vertical directions, and $\Sigma = \int_{-\infty}^{\infty} \rho dz$ is the surface density. Provided $|\partial \ln \Sigma / \partial \ln r| \sim 1$, $|\partial \ln T/\partial \ln r| \sim 1$, and $|(1/\rho)(\partial p/\partial r)| \approx c_s^2/r \ll GM/r^2$, the r component of the equation of motion is reduced to

$$\Omega^2 r = \frac{U_\phi^2}{r} = \frac{GM}{r^2}, \tag{2}$$

which is the condition for Keplerian flow. After integrating over z, the ϕ component of equation of motion becomes

$$\Sigma U_r r \frac{d\Omega r^2}{dr} = \frac{\partial}{\partial r}\Sigma \nu r^3 \frac{d\Omega}{dr} + S_\Sigma(r,t)J(r,t)r, \tag{3}$$

where $J(r,t)$ is the excess specific angular momentum carried by the infalling material. Combining equations (1) and (3), we deduce the standard mass diffusion equation

$$\frac{\partial \Sigma}{\partial t} - \frac{1}{r}\frac{\partial}{\partial r}\left[3r^{1/2}\frac{\partial}{\partial r}\left(\Sigma \nu r^{1/2}\right) - \frac{2S_\Sigma(r,t)J(r,t)}{\Omega}\right] - S_\Sigma(r,t) = 0. \tag{4}$$

The evolution of any initial Σ distribution can be calculated provided a prescription for the coefficient of kinematic viscosity ν is specified. The most uncertain and important aspect of any accretion-disk theory is the determination of the magnitude and functional dependence of ν. If $\nu = \nu(\Sigma,T,r)$, we need to consider the thermal properties of the nebula. In the absence of infall, the equation of motion in the vertical direction of a geometrically thin, non-self-gravitating accretion disk is reduced to

$$\frac{1}{\rho}\frac{\partial p}{\partial z} = -\Omega^2 z. \tag{5}$$

Similarly, the energy equation is reduced to

$$C_v\rho\left(\frac{DT}{Dt} - (\Gamma_3 - 1)\frac{T}{\rho}\frac{D\rho}{Dt}\right) = D - \underline{\nabla} \cdot \underline{\mathcal{F}}, \tag{6}$$

where C_v and Γ_3 are the specific heat and adiabatic index and $\underline{\mathcal{F}}$ is the total heat flux. If the nebula is geometrically thin and is in local thermal equilibrium, equation (6) reduces to

$$\frac{\partial \mathcal{F}_z}{\partial z} = \frac{9}{4}\rho\nu\Omega^2. \tag{7}$$

The vertical structure of the nebula is determined from the solutions of equations (5) and (7) provided the equations of state and radiative transfer, the opacity law, and the viscosity prescription are specified.

11.2.1 Convective Stability

If the nebula is not turbulent initially so that ν and D are essentially zero, it will undergo quasi-stationary subsonic contraction toward the mid-plane as heat was lost from the nebula's surface. If the opacity is a power law function of T only, say $\kappa = \kappa_0 T^\beta$, then it is possible to construct a set of homologously contracting solutions (Lin and Papaloizou 1980) and show that a sufficient criterion for the onset of convective instability is

$$1/(4 - \beta) \geq (\gamma - 1)/\gamma. \tag{8}$$

In most regions of the nebula, the temperature at the mid-plane is $T_c < 2000$ K so that the hydrogen gas is primarily in the form of molecules and the metals condense into sub-micron-size dust grains. In this case $\gamma \approx 1.4$ so that the critical β for convective instability is ~0.6. At these temperatures, if the dust grains are well mixed with the gas, they provide the dominant sources of opacity. The magnitude of the grain opacity roughly varies as the square of the temperature (Kellman and Gaustad 1969; DeCampli and Cameron 1979) so that *the nebula is convectively unstable.*

The above results may be interpreted in the following physical manner. Since κ increases with T, a cooler surface region of the nebula has a lower opacity and therefore can cool off more quickly than the hotter regions near the mid-plane. Eventually the thermal structure of the nebula adjusts to a new equilibrium in which the top surface layers of the nebula are considerably cooler than the mid-plane. Owing to the large temperature and pressure gradients in the vertical direction, the nebula may become superadiabatic and convectively unstable in that direction. In some parts of the inner region, the grain population changes rapidly with temperature due to evaporation and condensation processes. In this temperature range, the opacity also undergoes rapid variations. The stability analysis is more complicated, but similar arguments can still be applied to show that the nebula is convectively unstable.

The above arguments are only relevant for an optically thick nebula. When the optical depth becomes of order unity, we can no longer assume $p_0 \gg p_s$

and $T_0 \gg T_s$. The temperature gradient may become so small that the vertical structure is essentially isothermal. An isothermal nebula is convectively stable since ∇ $(= \partial \ln T/\partial \ln p) = 0$. When the optical depth becomes less than unity, the thermal structure of the nebula may be analyzed by detailed radiative transfer calculations in a manner analogous to stellar atmosphere calculations. Our numerical solutions indicate that the nebula is convectively unstable even in the limit that τ is of order unity. For $\tau \lesssim 1$, the nebula is likely to be convectively stable.

The stability of the nebula may also be modified by a substantial rate of infall in the vertical direction. In this case, as external material arrives onto the nebula, the infall velocity is reduced from supersonic to subsonic flow through a shock wave. In the shock front, intense energy dissipation acts to reduce the temperature gradient in the vertical direction. If the infall flux is sufficiently large, ∇ in most regions of the nebula may be reduced below $(\gamma - 1)/\gamma$ so that the nebula is stabilized against convection. Consequently, if convection were the *only* mechanism for generating non-negligible effective viscous stress in the solar nebula, the infalling material would accumulate in the vicinity of the infall radius and Σ there would increase. Below the shock surface near the mid-plane, the nebula material would adjust into a quasi-hydrostatic equilibrium. Due to the absence of viscous dissipation of energy, any heat loss would lead to a temperature decrease and a contraction in the vertical direction. If there were sufficiently large Σ, the adiabatic work resulting from such contraction would become comparable to or larger than the shock dissipation rate and the nebula might once more become convectively unstable. Thus, for any given mass infall rate S_Σ, there is a critical surface Σ_c above which the nebula is convectively unstable.

11.2.2 The Vertical Structure of a Convectively Unstable Nebula

If the nebula is convectively unstable, turbulent motion will be induced. Convectively driven turbulence has three major consequences for the evolution of the nebula. (1) Convection increases the efficiency of thermal energy transfer from the mid-plane to the nebula surface layers. (2) Turbulent eddies induce mechanical mixing of material over a radial extent comparable to the size of the largest convective eddies and thereby increase the efficiency of angular momentum transport. (3) Turbulent motion induces a continual transfer of kinetic energy from the differential rotation in the main flow to the largest eddies. The energy associated with the chaotic motion is then dissipated into heat. These effects can be quantitatively analyzed with a prescription for the effective turbulent eddy viscosity. For example, a widely adopted, ad hoc prescription for $\nu_{\rm eff}$ in accretion-disk theories is the α model (Shakura and Sunyaev 1973). In standard α models,

$$\nu_{\rm eff} = \frac{2}{3}\alpha_s c_s H = \frac{2}{3}\alpha_s c_s^2/\Omega, \tag{9}$$

where α_s is a dimensionless parameter which is comparable to or less than unity for transonic or subsonic turbulence.

With this prescription, the vertical structure can be determined from a set of ordinary equations, (5), (7), and (9). If we ignore the energy flux carried by convection, the radiative transfer equation may be approximated by a one-zone model (Shakura and Sunyaev 1973) such that the surface flux

$$F_s = 2\sigma T_e^4 = \frac{8\sigma T_c^4}{3\tau_c} = \frac{16\sigma T_c^4}{3\kappa_c \Sigma}, \tag{10}$$

where κ_c and τ_c are the opacity and optical depth at the mid-plane of the nebula. From equations (7), (9), and (10) we find

$$\frac{3}{2}\alpha\Sigma\frac{\mathcal{R}T_c}{\mu}\Omega = \frac{4acT_c^4}{3\kappa_c\Sigma}, \tag{11}$$

which can be solved once the opacity law is specified. The analytic approximate solutions are in general agreement with the numerically integrated solution when the convective transport of heat is included (Lin and Papaloizou 1985).

In the above analysis it is evident that the α prescription is very useful for constructing thermal equilibrium models for the nebula. Perhaps the weakest assumption in the α prescription is that the origin of turbulence is not specified. In the solar nebula we have already demonstrated that there is *at least one* source of turbulence, i.e., convection. If we assume that convection is the *only* source of turbulence in the solar nebula, we can construct a self-consistent model in which all three major effects of convection are taken into account. In this case we set $\nu_{\text{eff}} = D = 0$ in the radiative region and

$$\nu_{\text{eff}} = \alpha_c v_c l_c = \alpha_c v_c^2/\Omega \tag{12}$$

in the convective region, where v_c is the convective velocity and α_c is a dimensionless parameter comparable to or less than unity. Since typical convective eddies break up on a time scale comparable to $\tau_c \approx \Lambda/c_s \approx \Omega^{-1}$, where Λ is the scale height or the mixing length, we estimate the "mixing length" in the radial direction

$$l_c \approx v_c \tau_{\text{eddy}} \approx v_c \Omega^{-1}. \tag{13}$$

The results of these computations are in good agreement with the solutions deduced from the conventional, radiative, α-prescription, thin disk approximation (Lin and Papaloizou 1980). For example, both treatments led to similar Σ and Ω dependences on various physical parameters such as T_c, ρ_c, and $\Sigma\nu$. The only difference is that α_c is approximately equivalent to 0.005–0.02 α_s. This difference results from the fact that the convection zone only occupies a fraction of the disk, i.e., $l < H$, and that convective motion is subsonic, i.e., $v_c \lesssim 0.2\ c_s$. Nonetheless the general functional form of the solutions is essentially reproduced from the simple standard thin disk approximations for most values of Σ.

11.2.3 A Brief Summary

The result presented in this section indicate that: (1) During its formation stage, the solar nebula adjusts to a "quasi-stationary," local hydrostatic equilibrium on dynamical time scales. (2) From the conditions required for such a local hydrostatic equilibrium, we can deduce the vertical structure of the nebula in the absence of turbulence with or without infall. (3) In the absence of infall, the nebula undergoes homologous contraction. Due to the temperature dependence of the grain opacity, the central portion of the nebula is superadiabatic so that the gas flow in the nebula is convectively unstable. (4) Infall can provide an additional heat source near the surface of the nebula, thereby reducing the temperature gradient in the vertical direction and temporarily stabilizing against convection. (5) Preliminary solutions for the infall case indicate that there is always a critical Σ_c for which the nebula is convectively unstable. If convection is the only source of non-negligible effective viscous stress, infall would induce Σ to increase to Σ_c before the nebula became unstable. (6) The solution for local thermodynamic equilibrium can be rigorously determined from the vertical structure equations with any prescription for viscosity with or without infall. (7) With the standard α prescription, a set of analytic solutions can be obtained. These analytic solutions adequately approximate the numerical solutions and describe the overall properties of the nebula. However, we cannot use such an ad hoc approach to determine the value of α. (8) In detailed numerical solutions for the α model, we find that although the mid-plane region of the nebula is convective, convective flux does not carry much heat. (9) Convection qualitatively changes the rate of viscous dissipation and reduces the magnitude of $\mathcal{R}_{\rm eff}$ to values typically on the order of a few thousand. (10) If convection is the only source of turbulence, the vertical structure equations can also be solved with a "self-consistent" approach. The solutions for various physical parameters obtained therefrom have similar functional dependence on Σ and Ω as those found with the standard α prescription for viscosity. Nevertheless the magnitude of the effective viscosity may be smaller than that deduced from the α prescription by about two orders of magnitude, that $\alpha_c = 1$ corresponds to $\alpha_s \approx 10^{-2}$.

11.3. Evolution of the Nebula

For the convenience of general discussion we divide the evolution of the nebula into three major stages: (1) the infall stage, when the evolution of the nebula is strongly influenced by the influx of external material; (2) the viscous stage, when the external supply of material is exhausted and the evolution of the nebula is primarily determined by its internal viscous frictional force; (3) the clearing stage, when most of the nebular material is accreted by either the proto-Sun or the protoplanets, and through various physical processes the remaining nebular material is driven away.

11.3.1 The Infall Stage

In the early stages of nebular formation, matter from the collapsing cloud continuously arrives at the nebula at a rate faster than that of the redistribution of matter induced by the viscous frictional force, so that the mass of the nebula gradually builds up. The reservoir of cloud material is depleted on a characteristic collapse time scale $\tau_{\rm eff}$, which for a 10 K, 1 $M_\odot$ molecular cloud may be on the order of 10^5 yr (Cameron 1978). The infall stage may be further divided into two substages: (1) the initial buildup stage and (2) the quasi-stationary stage. In the initial buildup stage, shock dissipation reduces the temperature gradient in the vertical direction and may stabilize convection when Σ is relatively small. If convection is the only source of turbulence, viscous diffusion is essentially absent until Σ becomes larger than Σ_c. For $\Sigma > \Sigma_c$, shock dissipation can no longer stabilize the convective instability. Thereafter viscous stress increases with Σ until the viscous diffusion rate in the nebula becomes comparable to the infall rate so that Σ changes slowly with time. Generally, the outer edge of the nebula can expand freely and the entire nebula cannot establish a steady state. If, however, the boundary conditions J, S_Σ, and S_T change slowly with time, at least the inner regions of the nebula may evolve toward a quasi-steady state.

The results of the detailed analysis of the infall stage (Cameron 1978; Cassen and Moosman 1981; Lin and Bodenheimer 1982; Lin and Papaloizou 1985) have several implications worth noting. (1) During the initial buildup phase the typical time scale for the nebula to increase its surface density to a particular value Σ_1 is $\tau_1 \approx \Sigma_1/S_\Sigma$. In order for infall to dominate over the viscous diffusion process, $\tau_1 \lesssim \tau_\nu(\Sigma_1) \approx r^2/\nu(\Sigma_1) = \mathcal{R}_{\rm eff}/\Omega$, where $\mathcal{R}_{\rm eff}$ is the effective Reynolds number. This initial buildup stage is short-lived for relatively small $\mathcal{R}_{\rm eff}$. (2) During the quasi-equilibrium stage, $\Sigma\nu \approx S_\Sigma r^2$. If during this stage the mass of the nebula is comparable to the minimum amount needed to make up for the "missing" gaseous constituents, the infall rate needed to sustain such a quasi-steady flow is $\dot{M} \approx 10^{-2}\mathcal{R}_{\rm eff}^{-1} M_\odot$ yr^{-1}. If $\mathcal{R}_{\rm eff} \approx 10^3$, $\dot{M} \approx 10^{-5} M_\odot$ yr^{-1}, which can hardly last for more than 10^5 yr. (3) In both the initial buildup stage and the quasi-steady stage, the radial distribution of T_c is determined by the radial dependence of S_Σ. Rather restricted functional forms are required for $S_\Sigma(r,t)$ to reproduce the temperature distribution ($T_c \propto r^{-1}$) traced out from the condensation temperature of the dominant rocky material in terrestrial planets and satellites. (4) Since both Σ and T_c increase with the infall rate, the self-gravitating effect of the nebula is a weak function of S_Σ; i.e., a large infall flux does not necessarily imply that the nebula is more prone to gravitational instability.

11.3.2 The Viscous Stage

Although infall may continue throughout the evolution of the nebula, infall flux probably decreases on a time scale comparable to the free fall time for the original cloud. When the infall flux is reduced to a relatively low level, the solar nebula evolves under the action of its own internal friction and enters a viscous stage. During this stage, angular momentum is transported outward to allow most of the matter to fall inward, so that nebular mass decreases and nebular size increases on a characteristic diffusion time scale $\tau_\nu \approx r/U_r \approx r^2/\nu$. One important issue to be addressed is the evolutionary time scale for a minimum mass nebula during the viscous stage. Quantitatively, the evolutionary time scale is model-dependent. In the $\nu = \nu(r)$ case, the characteristic diffusion time scale in the nebula is $\sim\mathcal{R}_{\rm eff}/\Omega$ (Lynden-Bell and Pringle 1974; Cameron 1978). If, however, ν is a function of local thermal variables, the magnitude of ν can change during the evolution of the nebula (Lin and Papaloizou 1980, 1985; Lin and Bodenheimer 1982). In this case, as the mass of the nebula is gradually redistributed, its vertical structure adjusts to new thermal equilibria on a characteristic thermal time scale $\tau_{\rm th} \approx \Sigma c_s^2/F_z$. For a thin disk, $\tau_\nu \gg \tau_{\rm th}$ so that local quasi-stationary thermal equilibrium is always maintained. The results in §11.2 indicate that in these thermal equilibria, T_c, H, and ν decrease as Σ decreases. Thus the evolutionary time scale increases as the nebular material is depleted. The thermal equilibrium solutions can be applied to the diffusion equation to numerically determine the quantitative evolutionary trend of the nebula. The main results of the numerical computations are: (1) The evolution of a minimum mass nebula with a physical dimension comparable to that of the present solar system takes place on a time scale of $\sim 10^6$ yr for $\alpha_c \approx 1$. In principle, the evolutionary time scale can be indefinitely long if $\alpha_c \ll 1$. (2) After an initial transition phase, the evolution of the nebula follows a unique pattern, regardless of how the nebular mass was distributed at the termination of the infall stage. Thus there is a self-adjusting process in the overall nebular structure. (3) The nebular mass, surface density, density, temperature, and thickness at all but the outermost regions of the nebula decrease with time. Since the scale height also decreases with time, angular momentum transport gradually becomes less efficient and the evolutionary time scale gradually lengthens. (4) A minimum mass nebula contains sufficient surface density such that it is optically thick throughout the nebula between the orbital radius of Mercury and that of Neptune, even in the case that α_c is of order unity. For smaller α_c, the surface density may be higher and the nebula has a larger optical depth unless the primary sources of opacity, namely grains, are significantly depleted. (5) In these optically thick regions, T_c may be larger than T_e by a factor of several. It is of interest to note that convective eddies may exert a viscous drag of small-size grains, inducing them to circulate between the relatively cool nebular surface region and the relatively hot nebular interior. For certain types of grains such a circulation provides a mechanism for repeated condensation and evaporation (Morfill 1983; Cameron and Fegley 1982). (6) The inner and intermediate regions of the nebula always evolve into a quasi-steady state in which the mass transfer rate is independent of the distance from the proto-Sun. (7) At any given time, $T_c \propto r^{-3/2}$ and $T_e \propto r^{-3/4}$ near the inner and intermediate regions where the flow is quasi-steady and both decrease slower with r at the outer regions of the nebula. Provided $0.1 \lesssim \alpha_c \lesssim 1$, the temperature distribution in a minimum mass nebula model approximately resembles that deduced from cosmochemical data. For smaller values of α_c, larger nebular surface density, or equivalently nebular mass, is needed to provide sufficient viscous dissipation. These temperature distributions can be maintained over several viscous diffusion time scales at that epoch. (8) In the intermediate region of the nebula, there is a first-order phase-transition zone for water molecules. Inside this zone the temperature of the nebula is sufficiently hot that most of the water molecules are in a gaseous state. At larger radii, most of the water molecules can undergo a first-order phase transition and condense into ice grains. Similar transition zones exist for other molecules. Because iron and silicon grains have much higher phase-transition (condensation) temperature than the ice grains, their transition zones are at much smaller radii. Consequently, metal grains are probably found throughout the nebula whereas ice grains are only found at large radii. (9) Inside the ice grain transition zone, the opacity is somewhat lower, which causes the convective zone to be smaller. If we adopt a self-consistent prescription for viscosity, the nebular viscosity would also be smaller. When a steady-state flow is established in this region, the nebular surface density is higher inside the transition zone by an order of magnitude than that outside the zone. (10) The ice grain transition zone evolves inward in time as the nebula gradually cools off and exhausts most of its material. Melting and recondensation of ice grains are possible as a consequence of convective circulation in the transition zone. (11) At the outer regions of the nebula there is a critical radius R_W at which the viscous stress W reaches a maximum. Inside this radius the inward diffusion velocity decreases with radius whereas the nebular material diffuses outward outside R_W. As the nebular material is depleted, R_W evolves outward. (12) In the outer region of the nebula near R_W, typical fluid elements undergo orbital migration. However, since they cannot diffuse rapidly toward small radii, these fluid elements continually evolve toward a cooler environment. As a result, the gas elements may become supersaturated, creating new nucleation sites which enhance grain condensation. In these regions grain condensation is a continuous process throughout the viscous stage. These continuous condensation regions generally spread inward in time.

11.3.3 The Clearing Stage

Today there is very little gas left between the planets. At some stage of the solar nebula evolution, the gas must have been cleared away. There are four basic models to account for the clearing of the nebula. First, sometime after the formation of the proto-Sun there may have been a very strong stellar wind which swept out the bulk of the nebular material (Horedt 1978; Elmegreen 1978). Although there is no satisfactory model to account for the origin of such a wind, mass outflow has been observed in a class of newly forming stars, T Tauri stars (Herbig 1962; Cohen and Kuhi 1979). With the typical mass outflux and wind velocity in the T Tauri stars, a minimum mass nebula may be cleared in a few thousand to a few million years. A second model is based on the general expectation that there may be a hot coronal layer above the surface layer of the nebula which is established by the dissipation of mechanical waves generated by turbulence (Cameron 1978). Such a model is analogous to one of the two models that account for the solar corona. Just as the presence of a hot corona in the

Sun causes a hydrodynamic flow of mass away from the Sun, coronal layers high in the atmosphere of the nebula may induce mass flow away from the nebula (DeCampli 1981). The quantitative verification of these arguments is still to be carried out.

A third possible mechanism for clearing is associated with the photoionization effect due to the solar UV radiation. Although the Sun's present UV flux is so low that it hardly has any significant effect on the upper atmosphere of the Earth, its UV flux may have been considerably higher during its T Tauri phase. If a small portion of the nebular gas is ionized by the UV photons of the early Sun, it may have sufficient kinetic energy to leave the solar system. Consequently, a radiatively driven ionization "wind" may be induced. Such a process could be particularly efficient for the outer regions of the nebula, say beyond the orbit of Jupiter, where the escape velocity is less than about 10 km s^{-1}. In fact, we would expect a considerable amount of material to accumulate in the outer regions of the nebula for two reasons. First, a fraction of the nebular material must move outward to carry the angular momentum viscously "removed" from the infalling material. If most of the mass in the nebula is initially located near the present orbit of Jupiter, about 30% of the infalling material must expand to the present orbit of Neptune to "absorb" the redistributed angular momentum. Second, near the outer regions of the nebula, the temperature is sufficiently cool that the nebula is likely to become optically thin. In this case, if convection is switched off and there is no other source of turbulence, nebular material would accumulate at these large radii. Note that this process is only effective for removing the gaseous component of the nebula. Small grains may undergo orbital decay under the action of the Poynting-Robertson effect, whereas larger "grains" or planetesimals may be left behind. The disk of large-size grains may have observational properties similar to those which are being attributed to a planetesimal disk around Vega by the *IRAS* team.

A fourth model is based on protoplanet-gas interaction. Protoplanets with very small mass have little effect on the structure and evolution of the nebula. When they have grown to sufficiently large masses, the protoplanets can induce a tidal torque on the nebula and eventually open up gaps in the vicinity of the protoplanets' orbits (Lin and Papaloizou 1980; Cameron 1979). Subsequently, further mass growth by accretion from the nebula is terminated and the nebular flow between the protoplanets becomes unstable so that the remaining nebular gas trapped between the protoplanets may be ejected by a wind or, perhaps more likely, be driven toward the proto-Sun. The advantage of this tidal model is that it allows most of the protoplanets to acquire their mass before the remaining nebular gas is cleared away. It also provides a strong dynamical constraint on the magnitude of ν_{eff} in the nebula.

11.4. Protoplanetary-Nebular Tidal Interaction

Giant planets contain mainly hydrogen and must have been formed in a gas-rich environment. If the gas they contain was accreted from the nebula, their present properties, such as mass, chemical composition, and orbital position, may have been determined by the details of the dynamical interaction with the nebula. For example, a strong tidal interaction may have occurred between a protoplanet and the nebula. Since a protoplanet revolves around the Sun faster than does the nebula outside its orbit, angular momentum is transferred from the protoplanet's orbit to the nebula in order to achieve a lower energy state, namely uniform rotation. Similarly, angular momentum is transferred from the nebula inside a protoplanet's orbits to the protoplanet.

The simplest estimate of the efficiency of tidal transfer of angular momentum is based on an impulse approximation (Lin and Papaloizou 1979). In this treatment we assume that when nebula material orbits at a radius close to that of a protoplanet, the tidal force of the protoplanet can only be felt during the orbital phase when particles have a close approach with the protoplanet since the mass ratio q between typical protoplanets and the Sun is much less than unity. Using this approximation, we can deduce that the rate of total angular momentum transfer is given by

$$\dot{H}_T = \frac{8}{27}\left(\frac{M_p}{M}\right)^2 \Sigma\omega^2 r_p^2 \left(\frac{r_p}{\Delta_0}\right)^3, \tag{14}$$

where M_p, r_p, and ω are the mass, orbital radius, and orbital angular frequency of the protoplanet and Δ_0 is the distance between the protoplanet and the nebula edge. In the limit that the protoplanet influences the nebula strongly, Δ_0 can be taken to be the Roche radius of the protoplanet, i.e., $\Delta_0 = R_L = r_p(q/3)^{1/3}$, so that

$$\dot{H}_T = 8q\Sigma r_p^4 \omega^2/9. \tag{15}$$

The above analysis can be modified to include the influence of the Sun on the close encounter (see appendix A of Lin and Papaloizou 1979).

In the impulse approaches above, we take the equation of motion for a fluid element to be that of a noninteracting particle. The fluid nature of the nebular flow is described by continuity, momentum, and energy equations. In the limit $M_p/M \ll 1$, the protoplanet's tide has little effect on the nebular flow except near orbital resonances where Ω is commensurable with ω. The rate of resonant tidal transfer of angular momentum can be calculated in many different ways (for reviews see Goldreich and Tremaine 1982). For example, the rate of tidal transfer of angular momentum can also be calculated by summing up a series of pointlike contributions from low order Lindblad resonances ranging from the inner 2 : 1 resonance to those near the edge of the disk (Goldreich and Tremaine 1980). The basic functional dependence of dH_T/dt derived from the two different approaches is identical. The numerical coefficient differs from that in equation (14) by a factor of 2.83.

The discussions in §§ 11.2 and 11.3 indicate that the nebular temperature is sufficiently large that $H/r \sim 0.05$–0.1. The effects of pressure gradient can be incorporated into linearized analytic approximations (Papaloizou and Lin 1984).

These analyses show that if the pressure varies on a scale comparable to H, the epicyclic frequency κ and the location of the resonances can be significantly changed from those in the cold disk case where they are determined by Kepler's laws. Moreover, a disk edge at which Σ decreases rapidly to zero on a scale length shorter than H in general implies $\kappa^2 < 0$, so that specific angular momentum decreases with radius. A differentially rotating flow with a negative angular momentum gradient is unstable against axisymmetric perturbations according to the Rayleigh criterion (Landau and Lifshitz 1959). When the Rayleigh criterion is violated, turbulence induced by Taylor vortices would act to smooth out the density gradient (Rayleigh 1917; Greenspan 1969). When a gap is first formed, $\Delta_0 \sim R_L$ so that the necessary condition for gap formation is $R_L \gtrsim H_D$. In the solar nebula where $H_D/r \sim 0.05$–0.1, a protoplanet must have a mass larger than $\sim 10^{-5}$–$10^{-4}\ M_\odot$ before it can truncate the nebula.

What is the sufficient condition for gap formation? The central idea of gap formation is that there is a balance between the rate of angular momentum transport due to the action of viscosity on Keplerian shear and that due to tidal effects, i.e.,

$$\frac{dH_T}{dt} = [3\pi\nu\Sigma r^2\Omega]_{r=r_1} \tag{16}$$

at the edge of the gap, r_1. If the nebula is tidally truncated near r_0 and Σ vanishes over some characteristic length scale comparable to H, the value of dH_T/dt may be modified from equation (14) (Papaloizou and Lin 1984) such that

$$\frac{dH_T}{dt} = 0.23\ q^2\Sigma_\infty r_0^2\omega^2\left(\frac{r_0}{H}\right)^3, \tag{17}$$

where Σ_∞ is the value of Σ at large distance away from the protoplanet. Taking $\Sigma(r_1) = \Sigma_\infty$ and assuming r_1 to be the same order as r_0, we find that, in order to produce a gap, we require

$$\frac{\nu}{r_0^2\omega} \lesssim 0.025\ q^2\left(\frac{r_0}{H_\infty}\right)^3 \tag{18}$$

or equivalently

$$q \gtrsim \left(\frac{40\nu}{r_0^2\omega}\right)^{1/2}\left(\frac{H_\infty}{r_0}\right)^{3/2}. \tag{19}$$

This condition does not depend very much on the detailed dissipation pattern which determines the scale of the gap. However, the time scale to form the gap is determined by the rate at which tides can significantly redistribute the angular momentum of the material originally in the gap region. This clearly lengthens if the gap region is enlarged.

Setting $H_\infty = R_L$ in equation (19), the condition to form a gap becomes

$$q \gtrsim \frac{40\nu}{r_0^2\omega} = \frac{40}{\mathcal{R}} = 40\alpha_s\left(\frac{H_\infty}{r_0}\right)^2. \tag{20}$$

This is to within a numerical factor the same condition as given by Lin and Papaloizou (1979), that the mass ratio should be greater than the inverse Reynolds number.

The above linear treatment provides some useful results. However, in the context of protoplanetary-nebular tidal interaction, what is needed is a full nonlinear solution of the hydrodynamic equations governing the tidal interaction, i.e., the solution of the two-dimensional (r and ϕ) momentum equation, including a non-axisymmetric perturbing potential and the continuity equation. The solutions of these equations are needed whether or not the perturbations of the nebula by the protoplanet can be treated using linear theory because the surface density profile has to be determined self-consistently as a result of the interplay between tidal and viscous processes. When the protoplanet is sufficiently massive, it can induce spiral shock waves in the nebula. Recently, Lin and Papaloizou have initiated a numerical analysis in which we adopt a viscosity law in which $\nu = \nu_0(\Sigma/\Sigma_0)^2$ where ν_0 and Σ_0 are scaling constant. For the equation of state, we use a polytropic gas model. The shock dissipation, when it occurs, is computed with von Neumann's method. The first results of the case $q = 10^{-3}$, $\nu_0 = 10^{-5}$, and $c_s = 10^{-1}$ and $c_s = 10^{-3}$ are presented in Figures 11.1*a* and 11.1*b*. These results are in general agreement with that deduced from the linear approximation despite the fact that we relax the assumptions on the surface density profile, the role of nonlinear shock dissipation, and the variation of Ω. In Figures 11.1*a* and 11.1*b* we plot the surface density profile for the $M_p/M = 10^{-3}$ case. The effect of nonlinear shock dissipation at several important resonances is clearly illustrated. Note that despite the transport of angular momentum wave by the larger sound speed in Figure 11.1*b*, the truncation condition is not significantly modified.

The above results have important implications for the protoplanetary formation process. As the nebula evolves, the mass of the protoplanet grows, so that the rate of tidal transfer of angular momentum is increased. At the same time the nebular material is exhausted, and therefore the rate of the viscous transfer of angular momentum in the nebula is decreased. Eventually, the tidal effect can redistribute angular momentum faster than can the viscous effect and thereby open up gaps in the nebula in the vicinity of the protoplanets' orbits and further mass growth by accretion will be inhibited (Cameron 1979; Lin and Papaloizou 1980). Furthermore, depending on the total mass and structure of the nebula, the net angular momentum exchange rate may cause secular orbital evolution of the protoplanet (Goldreich and Tremaine 1980).

The tidal effect of protoplanet-nebula interaction can be incorporated into a self-consistent calculation of the nebular evolution and protoplanetary orbital migration. We have performed a series of numerical computations for this purpose (Lin and Papaloizou, in preparation). Preliminary results show that a protoplanet's orbital migration does not induce additional constraints on the truncation condition even in the limit that the nebula's mass is considerably larger than that of the protoplanet. These calculations also indicate that the orbital evolution of a protoplanet always occurs on a time scale comparable to or longer than the viscous time scale of the nebula gas after some initial transients.

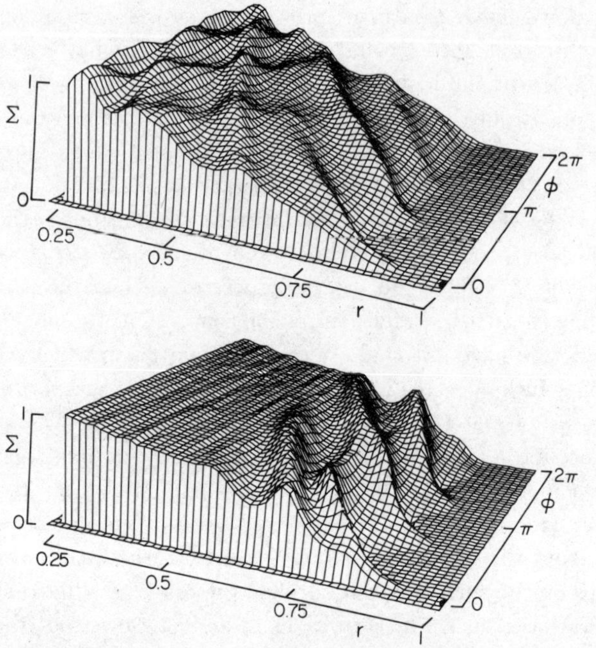

Figure 11.1. Tidal interaction between a protoplanet and the primordial solar nebula. The coordinate r denotes the radius, normalized with respect to the position of the protoplanet. The location of the Sun is at $r = 0$ and that of the protoplanet is marked with a dot. In Figures 11.1*a (top)* and 11.1*b (bottom)* the sound speeds are 10^{-1} and 10^{-3} times the Keplerian velocity of the protoplanet. We adopt $q = 10^{-3}$ and $\nu = 10^{-5}$ in both cases.

If the protoplanet starts at some intermediate position in the disk, a feedback mechanism operates to balance inward and outward angular momentum transfer so that little orbital evolution occurs. Even if the protoplanet is placed on the outside of the disk, after rapidly reaching an equilibrium from the disk, further evolution occurs on the slow viscous time scale.

11.5. Implication of Nebular Models on Protoplanet Formation

In the past few years, theoretical investigations, led by Cameron, on the origin and evolution of the solar nebula have progressed rapidly. Through these investigations we find that the nebular phase of the solar system may be turbulent, of relatively low mass, and short-lived. Convectively driven turbulence not only provides an energy transport process in the vertical direction but also induces viscous energy dissipation and mass transfer in the radial direction. From the nebula models outlined in §§ 11.2 and 11.3, the temperature, mass, density, and pressure distribution can be deduced at any given epoch of solar nebula evolution. From these results we are now in a position to examine the physical conditions in the solar nebula at the time the protoplanets acquired most of their mass and find constraints on and evidence for or against protoplanetary formation scenarios. For example, the tidal constraint in §11.4 implies that the viscous evolution time scale in the nebula must be less than $\sim 10^6$ yr in order to avoid tidal truncation and gap formation. Therefore Jupiter and Saturn must have been formed within a million years after the initial formation of the Sun. Although this argument provides a strong limit on the duration of the protoplanetary formation epoch, there is still considerable uncertainty on just when and how the protoplanets did form (for further details see Bodenheimer, this volume). The diversity of opinion may be divided into three major scenarios: (1) All protoplanets formed with 10^5 to 10^6 yr of the infall stage by gravitational instability which occurred in a very massive and extended nebula (Cameron 1978). (2) The protogiant and terrestrial planets formed at different epochs through different processes. While protogiant planets may have formed through massive gravitational instability, as in the first scenario, the terrestrial planets formed through a series of gradual accretion processes in an essentially gas-free environment (Safronov 1969, 1972; Cameron 1962, 1973; Kusaka, Nakano, and Hayashi 1970; Wetherill 1980). (3) The protogiant as well as prototerrestrial planets formed through a process of gradual accretion, first of dust, then of gases in a gas-rich environment (Hayashi, Nagazawa, and Adachi 1977; Hayashi 1981; Lin 1982; Lin and Papaloizou 1985). It is still somewhat premature to suppose one particular scenario may be more favorable than the others. However, if the protogiant planetary formation proceeds through a series of gradual accumulation processes, most of the protoplanets probably acquired the bulk of their mass within about a million years after the formation of the proto-Sun.

Acknowledgments

We thank P. Bodenheimer, F. Shu, and S. Ruden for useful conversations and helpful comments. This work was supported in part by a grant from the National Science Foundation, AST 83-01223. Parts of the numerical results presented here were generated on an IBM PC-XT 370 on loan from the IBM Palo Alto Scientific Center.

References

Cameron, A. G. W. 1962, *Icarus,* **1,** 13.

——. 1973, *Icarus,* **18,** 407.

——. 1976, in *The Origin of the Solar System,* ed. S. F. Dermott, p. 49 (Chichester, England: Wiley).

——. 1978, in *Protostars and Protoplanets,* ed. T. Gehrels, p. 453 (Tucson: University of Arizona Press).

——. 1979, *Moon and Planets,* **21,** 173.

Cameron, A. G. W., and Fegley, M. B. 1982, *Icarus,* **53,** 1.

Cassen, P., and Moosman, A. 1981, *Icarus,* **48,** 353.

Cohen, M., and Kuhi, L. V. 1979, *Ap. J. Suppl.,* **41,** 743.

DeCampli, W. M. 1981, *Ap. J.,* **244,** 124.

DeCampli, W. M., and Cameron, A. G. W. 1979, *Icarus,* **38,** 367.

Descartes, R. 1644, *Principia Philosophia* (Amsterdam).

Elmegreen, B. G. 1978, *Moon and Planets,* **19**, 262.

Goldreich, P., and Tremaine, S. 1980, *Ap. J.,* **241**, 425.

——. 1982, *Ann. Rev. Astr. Ap.,* **20**, 249.

Greenspan, H. P. 1969, *The Theory of Rotating Fluids* (Cambridge: Cambridge University Press).

Hayashi, C. 1981, *Progr. Theoret. Phys. Suppl. Japan,* **70,** 35.

Hayashi, C., Nakazawa, K., and Adachi, J. 1977, *Pub. Astr. Soc. Japan,* **19**, 163.

Herbig, G. H. 1962, in *Advances in Astronomy and Astrophysics,* ed. Z. Kopal, vol. 1, p. 47 (New York: Academic Press).

Horedt, G. P. 1978, *Astr. Ap.,* **64,** 173.

Hoyle, F. 1960, *Quart. J.R.A.S.,* **1,** 28.

Jeffreys, H. 1924, *The Earth* (Cambridge: Cambridge University Press).

Kant, I. 1755, *Allemeine Naturgeschichte und Theories des Himmels.*

Kellman, S. A., and Gaustad, J. E. 1969, *Ap. J.,* **157,** 1465.

Kusaka, T., Nakano, T., and Hayashi, C. 1970, *Progr. Theoret. Phys. Japan,* **44,** 1580.

Landau, L. D., and Lifshitz, E. M. 1959, *Fluid Mechanics* (New York: Pergamon Press).

Laplace, P. S. de. 1796, *Exposition du système du monde* (Paris).

Lin, D. N. C. 1981, *Ap. J.,* **246,** 972.

——. 1982, in *Rubey Colloquium,* vol. 3, *The Solar System,* ed. M. Kivelson (Englewood, NJ: Prentice-Hall), in press.

Lin, D. N. C., and Bodenheimer, P. H. 1982, *Ap. J.,* **262,** 768.

Lin, D. N. C., and Papaloizou, J. 1979, *M.N.R.A.S.,* **186,** 789.

——. 1980, *M.N.R.A.S.,* **191,** 37.

——. 1985, in *Protostars and Planets,* vol. 2, ed. D. Black and M. Mathews (Tucson: University of Arizona Press).

Lüst, R. 1952, *Zs. Naturforschung,* **7a**, 87.

Lynden-Bell, D. 1969, *Nature,* **223,** 690.

Lynden-Bell, D., and Pringle, J. E. 1974, *M.N.R.A.S.,* **168,** 603.

Morfill, G. E. 1983, *Icarus,* **53,** 41.

Papaloizou, J., and Lin, D. N. C. 1984, *Ap. J.,* **285,** 818.

Prendergast, K. H., and Burbidge, G. R. 1968, *Ap. J. (Letters),* **151,** L83.

Pringle, J. E. P. 1981, *Ann. Rev. Astr. Ap.,* **19,** 137.

Rayleigh, Lord. 1917, *Proc. Roy. Soc. London,* **A93,** 148.

Safronov, V. S. 1969, *Evolution of the Protoplanetary Cloud and Formation of the Earth and Planets* (Moscow: Nauka; translation 1972, Israel Program for Scientific Translation, NASA TTF-677).

——. 1972, in *On the Origin of the Solar System,* ed. H. Reeves, p. 89 (Paris: CNRS).

Shakura, N. I., and Sunyaev, R. A. 1973, *Astr. Ap.,* **24,** 337.

Tennekes, H., and Lumley, J. L. 1972, *A First Course in Turbulence* (Cambridge, MA: MIT Press).

von Weizsäcker, C. F. 1943, *Zs. Ap.*, **22**, 319.

——. 1948, *Zs. Naturforschung,* **3a**, 524.

Wetherill, G. W. 1980, *Ann. Rev. Astr. Ap.*, **18**, 77.

Part V

Solar System Evolution

CP-Q

12

Properties of Giant Gaseous Protoplanets

P. Bodenheimer

12.1. Introduction

Al Cameron approached the problem of the formation of the Sun and planets from a very broad point of view, which is well summarized by the following quotation from his well-known paper in *Icarus,* volume 1 (1962):

> One feature of this approach is that it is necessary to consider a very large number of physical and chemical processes in many fields of astrophysics. We must rely on the theories of nucleosynthesis and of stellar evolution to give us the relative production rates of the extinct radioactivities. We must investigate the mixing mechanisms by means of which the fresh radioactivities, presumably present in the expanding shells of supernovae, can be transferred into the general gas of the interstellar medium and thus become available for incorporation in newly formed stars. We must investigate how this star formation takes place, and ask how a solar nebula can arise that is suitable for the formation of the planets. We must consider how planets can accumulate in such a solar nebula. In particular, we must be concerned with the time required for this entire chain of physical and chemical events to occur, and we are interested to see whether this total time is consistent with the evidence provided by the amounts of the extinct radioactivities present in the meteorite parent bodies.

Al Cameron's research encompassed many of the above-mentioned topics. Before turning to the question of the origin of the giant planets, I should mention briefly a few of his early and pioneering contributions on two very closely related problems—the early evolution of the Sun and the structure of the solar nebula. In 1963, Ezer and Cameron published one of the first evolutionary tracks for the Sun during its pre-main-sequence contraction (Figure 12.1). The models were assumed to be homologously contracting, and they were constructed by the fitting method (Schwarzschild 1958) on the supercomputer of the time, an IBM 7090. The question of the depletion of lithium by convective mixing to the warm interior was also investigated, a problem that was to receive much attention during subsequent years. This calculation was one of the first confirmations of Hayashi's (1961) theory of pre-main-sequence evolution. The sharp bend in the track at 10^6 yr is associated with the transition from a largely convective to a largely radiative structure and thus cannot be modeled accurately by homologous contraction. However, Ezer and Cameron improved their calculation and

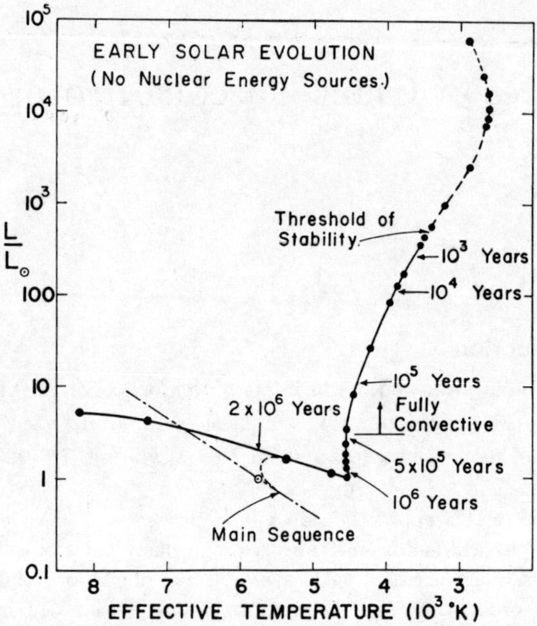

Figure 12.1. An early calculation of the pre-main-sequence evolution of the Sun (Ezer and Cameron 1963). Reprinted by permission. ©Academic Press.

in 1966 published a revised track (Figure 12.2) based on the Henyey method of solving the full time-dependent equations of stellar structure (Henyey, Forbes, and Gould 1964). The parameters in the evolutionary calculation were adjusted so as to provide a fit to the radius and luminosity of the present Sun at an age of 4.5×10^9 yr.

A very early model of the solar nebula was calculated by Cameron in his *Icarus* paper of 1962 and plotted in an article published by *Sky and Telescope* in 1962 (Figure 12.3). The calculation of the nebular surface density is based on the assumption that an initial uniformly rotating polytrope collapses, conserving the angular momentum of each mass element, to form an equilibrium disk. The collapse of the initial configuration is induced by the dissociation of molecular hydrogen. The assumption regarding the initial configuration was later modified (Cameron 1963) to correspond to a uniformly rotating, uniform density sphere. A more elaborate model, in equilibrium in both the radial and vertical directions, was calculated by Cameron and Pine (1973). The distribution of angular momentum in the nebula was derived under the assumption of collapse with conservation of angular momentum from an initially uniformly rotating cloud with a density distribution that decreased linearly with radius. The self-gravity of the nebula was included, as were detailed thermodynamics and opacities from dust grains. The two-dimensional structure of one of these models is illustrated in Figure 12.4. Of particular interest are the two convection zones, the inner one corresponding to the region where molecular hydrogen dissociates and the outer

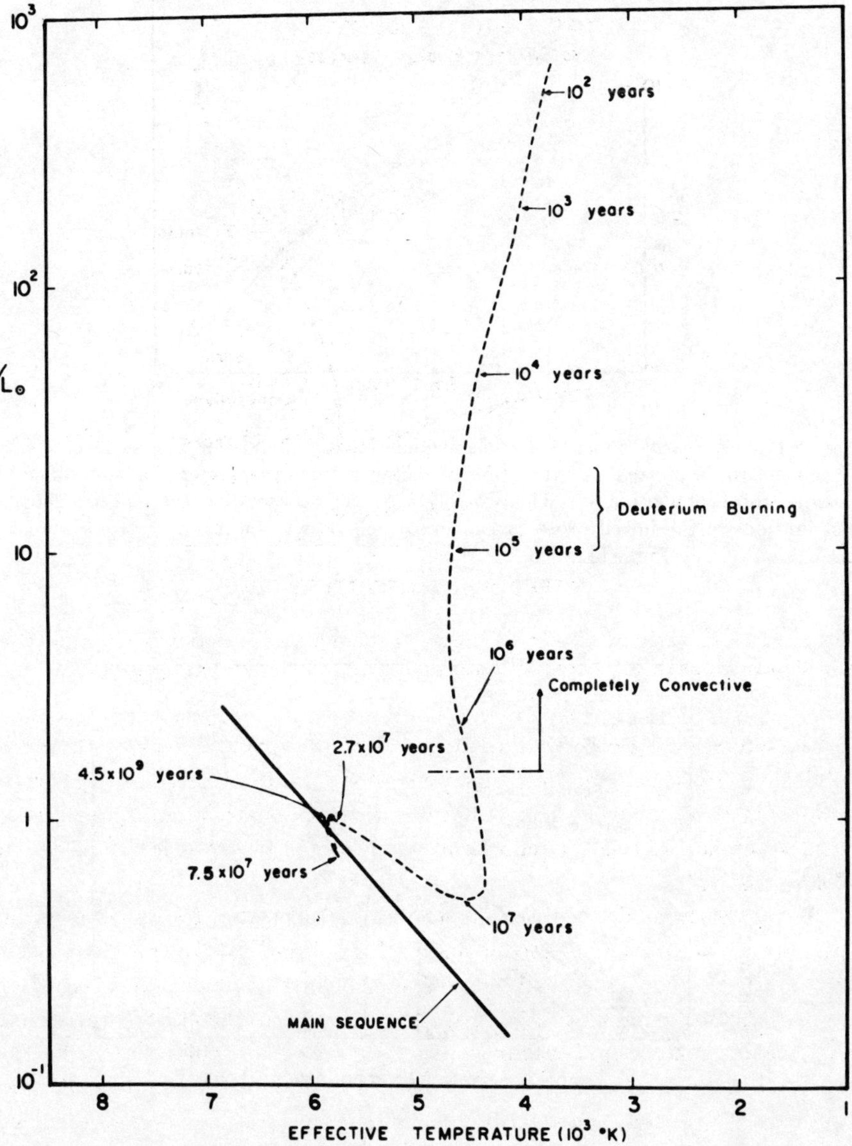

Figure 12.2. An improved calculation of the evolution of the Sun during pre-main-sequence phases (Ezer and Cameron 1966). Reprinted by permission. ©Plenum Press.

one to the region where metallic iron grains exist. Convection, along with other mechanisms, was discussed by Cameron and Pine in connection with angular momentum transport in the nebula and the evolution of the surface density distribution. In several later papers Cameron made further advances in this area by calculating evolving models of the solar nebula.

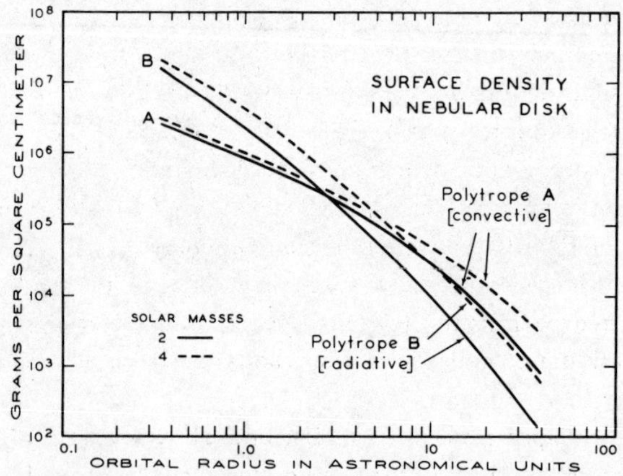

Figure 12.3. An early model of the solar nebula (Cameron 1962*b*). The surface density of the nebula is plotted as a function of distance from the center, for two different assumed total nebular masses. The two polytropes refer to two different assumed density distributions in the initial protostar, before it collapses to form a disk. Reprinted by permission. ©Sky Publishing Co.

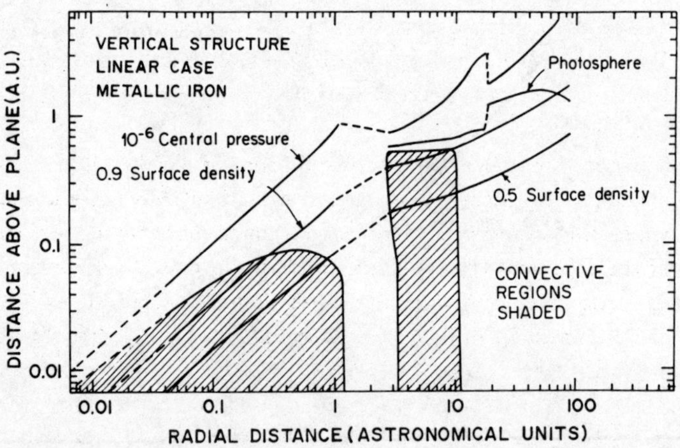

Figure 12.4. A two-dimensional map of a solar nebula model (Cameron and Pine 1973). "Linear case" refers to the density distribution in the protostar before it collapses to form the nebula. "Metallic iron" refers to the type of grains that were used to compute the opacity. Reprinted by permission. ©Academic Press.

12.2. Giant Planets: General Considerations

The question of the origin of the planets themselves is one to which Cameron has devoted considerable thought and effort. The present discussion is concerned mainly with the giant planets, although Cameron's work itself ranged over all aspects of solar system formation. He investigated two principal hypotheses of the formation of these planets. The first of these is based on the idea of a giant gaseous protoplanet. In this picture the following course of events is envisaged.

a) The solar nebula becomes unstable to ringlike perturbations, which then condense to form gravitationally bound objects.

b) The protoplanets, which may be considerably more massive than the present planets, then contract slowly in hydrostatic equilibrium for a time of 10^3–10^7 yr, depending on the mass.

c) When the central temperature reaches approximately 2000 K, the molecular hydrogen near the center begins to dissociate, thereby inducing a hydrodynamic collapse of the whole object on a time scale of less than a year.

d) After collapse proceeds to high enough densities and temperatures so that most of the hydrogen is dissociated and nonideal effects become important in the equation of state, the object regains hydrostatic equilibrium and begins its final phase of very slow contraction and internal cooling, on a time scale of 5×10^9 yr.

In the second hypothesis, a somewhat different sequence of evolutionary phases occurs.

a) Small dust particles in the solar nebula gradually accumulate and build up in mass, on a time scale of 10^6–10^8 yr, to form a small protoplanet composed mainly of rock and ice.

b) When the object has grown to about an Earth mass, it is able to attract nebula gas around it to form a low mass envelope in hydrostatic equilibrium.

c) The core continues to accrete solid matter slowly, and as it does so the envelope also increases in mass. A critical core mass is reached, after which the accretion of gas becomes much more rapid than the accretion of the core, and the envelope builds up in mass to its present value.

d) Accretion is terminated, and after a short period of readjustment, the object contracts and cools on a long time scale (5×10^9 yr) to its present size.

12.3. Early Work on the Core Accretion Model

The key piece of observational evidence that these hypotheses must explain is the fact that the measured mean densities and gravitational moments of the present giant planets imply the existence of dense cores. The core accretion hypothesis explains this fact in a natural way: in the hypothesis of gaseous protoplanets the presence of the core must be explained by the precipitation of grains or accreted planetesimals toward the center during phase (*b*), during which internal temperatures are presumably not hot enough to melt the solid material. Cameron's earlier work on this subject in fact was directed toward the core accretion model. In a paper on particle accumulation (Cameron 1973) based on physical conditions in the Cameron and Pine solar nebula model, he suggested "Jupiter might form when an earthlike planetary nucleus has accumulated something like 10 earth masses." The idea was that the surrounding nebula gas would then collapse hydrodynamically onto the core and that the rate of mass increase thereafter, which could be calculated with the Bondi accretion formula, would be very rapid. At about the same time, Podolak and Cameron (1974) calculated static models of the present giant planets and found that rock or ice cores of 20 and 40 Earth masses were required to explain the mean densities of Jupiter and Saturn, respectively. "This is quite in accordance with the picture of planetary formation ... grains of condensate were able to accumulate into cores of rock ... When these ... cores grew large enough they were able to capture an atmosphere from the surrounding nebula." There has been extensive work on models of the giant planets by several authors (see, for example, Hubbard 1980), and the conclusion still remains that the cores must be present.

Perri and Cameron (1974) attacked another aspect of this problem by calculating static models of protoplanets having a core-envelope structure. The envelope was assumed to follow an adiabat, and its outer boundary, assumed to lie at the tidal radius of the protoplanet, was given physical conditions obtained from a solar nebula model. They reached the significant conclusion that as the assumed core mass was increased, there was a maximum value, beyond which an equilibrium envelope structure did not exist. Beyond the critical point the envelope was unstable to dynamical collapse because of a mean $\Gamma_1 < 4/3$, a consequence of the dissociation of molecular hydrogen. Figure 12.5 illustrates the phenomenon of the critical core mass on the basis of one particular set of parameters assumed by Perri and Cameron. Here the value of this mass, 120 Earth masses, is found to be quite insensitive to the parameter A, which essentially refers to the position of the protoplanet in the solar nebula. With the assumption of a lower assumed adiabat for the solar nebula model the critical core mass came down to 60 Earth masses, still somewhat high when compared with the values deduced from observations. Later work by Harris (1978) and Mizuno (1980), based on the same general approach but with improved physics in the protoplanetary model, brought the value down to the range 10–20 Earth masses.

12.4. Giant Gaseous Protoplanets

Further work by Cameron on the problem of the evolution of the solar nebula led to a change in his approach toward the question of the origin of the planets. The work of Lynden-Bell and Pringle (1974) established the basic principles of the evolution of thin disks, driven by viscous dissipation. Cameron (1978*a*) made an extensive series of numerical calculations in which he applied these principles to the case of the solar nebula. He tested these models for global instability and found that in most cases they were unstable to the formation of massive rings. The rings he presumed would break up into subcondensations in the nebula, or, in other words, giant gaseous protoplanets. He decided to investigate the early evolution of such objects in the mass range 1–30 Jupiter masses, to determine

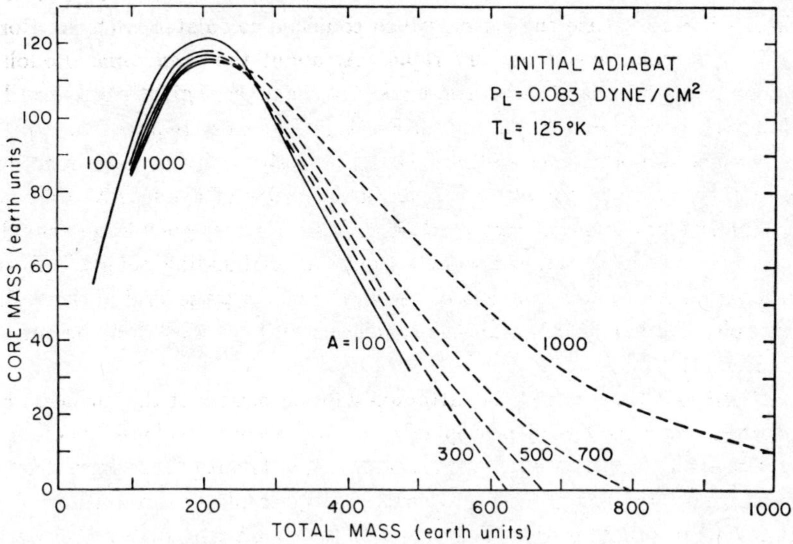

Figure 12.5. The core mass of an equilibrium, adiabatic model of a protoplanet plotted as a function of its total mass (Perri and Cameron 1974). Solid lines refer to hydrodynamically stable envelopes; dashed lines refer to unstable ones. P_L and T_L refer to background values of pressure and temperature, respectively, and A is a parameter depending upon the position of the model in the solar nebula. For this set of parameters, no equilibrium solution exists for a core mass greater than 120 Earth masses. Reprinted by permission. ©Academic Press.

whether all planets could form in this manner. The terrestrial or giant planets could be explained by a three-step process:

1. Gravitational condensation of 10 or more Jupiter masses,
2. Core formation during the early contraction phase by precipitation and accretion, and
3. Removal of the outer gaseous envelope at a later time by thermal effects in the hot inner region of the solar nebula or by the increasing tidal effects of the Sun, which meanwhile would have condensed in the center of the nebula.

To test these ideas DeCampli and Cameron (1979) made detailed spherically symmetric calculations of the evolution of giant gaseous protoplanets, under the assumption that they were isolated from any effects of the forming Sun or from the solar nebula itself. They solved the full set of stellar structure equations and reached several interesting conclusions.

1. Protoplanets have internal convection zones during the early contraction phase.
2. Iron grains can liquify and precipitate to the center in protoplanets of

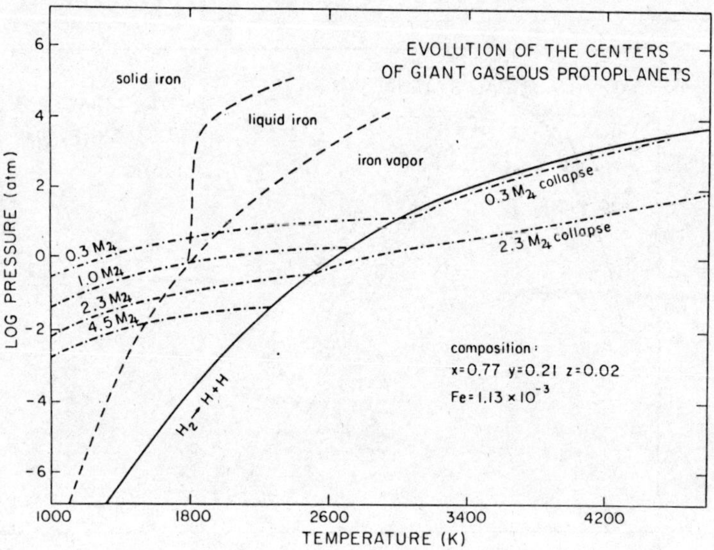

Figure 12.6. The central pressure plotted as a function of the central temperature during the evolution of giant gaseous protoplanets in their early phases (DeCampli and Cameron 1979). Reprinted by permission. ©Academic Press.

less than 1 Jovian mass. The growth time for the liquid droplets as well as their infall time and the time for convection to circulate grains into the inner regions were all found to be short compared with the evolutionary time over which the conditions for liquification existed. The evolutionary tracks for the central regions of protoplanets of various masses are shown in Figure 12.6. The only problem with this finding is that the rather massive protoplanets that were envisaged do not form cores by this process. Therefore accretion of planetesimals and their settling would have to occur rapidly in order to form the core, since the evolution times for the higher masses in the early contraction phase are short—only a few thousand years at 10 Jovian masses.

3. Giant gaseous protoplanets can be tidally stable in the primitive solar nebula, before the formation of the Sun. This point is illustrated in Figure 12.7, which shows the tidal radii as well as the actual sizes of protoplanets as a function of time. During the early stages of the evolution of the massive nebula (Cameron 1978*a*) the matter is not strongly concentrated toward the center and the tidal radius at Jupiter's position is comparable to the size of a protoplanet of 1 Jovian mass where it first becomes gravitationally bound. As the Sun forms in the center, the tidal radius shrinks, but this process occurs on the time scale for viscous evolution of the nebula, which turns out to be comparable to the contraction time of the protoplanet. Thus the actual radius stays close to or inside of the tidal radius. The time scale for evolution of proto-

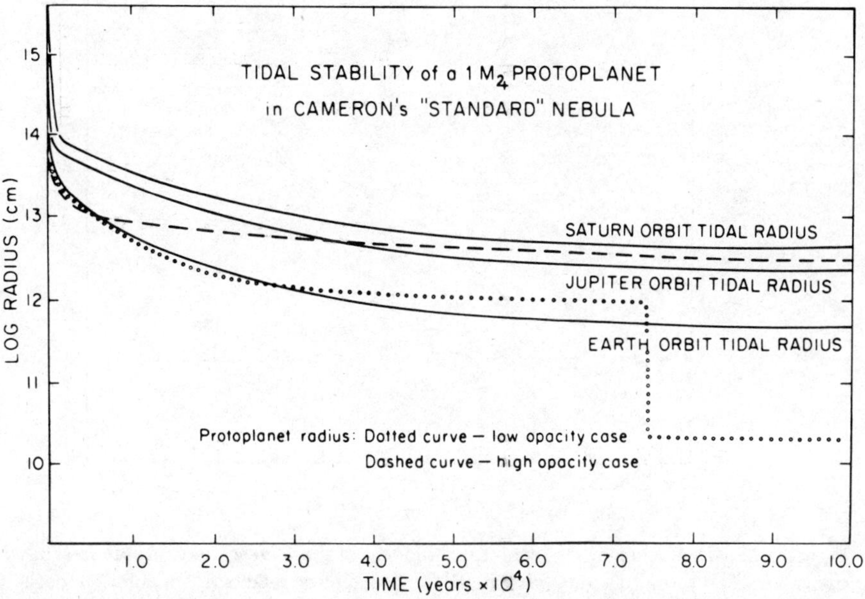

Figure 12.7. The tidal radius plotted as a function of time for a protoplanet of 1 Jovian mass at various positions in the solar nebula (*solid curves*). The nebula evolves according to the model of Cameron (1978*a*). The calculated protoplanetary radius (for 1 Jovian mass) is shown as the dashed and dotted curves (DeCampli and Cameron 1979). Reprinted by permission. ©Academic Press.

planets up to the time when molecular hydrogen begins to dissociate at the center is 4×10^5 yr for Jupiter and 4×10^6 yr for Saturn.

The full evolution of isolated giant gaseous protoplanets of Jovian and Saturnian masses and uniform solar composition is illustrated in Figures 12.8 and 12.9. The early portions of these curves, up to the point of molecular dissociation in the center, are essentially the same as those calculated by DeCampli and Cameron (1979). The evolution was extended by Bodenheimer et al. (1980), through the hydrodynamic collapse, to the point where hydrostatic equilibrium in the entire model was regained (asterisks in Figures 12.8 and 12.9). There the solutions were matched to those for the final cooling phase of the evolution, as calculated by Graboske et al. (1975) for Jupiter and by Pollack et al. (1977) for Saturn. A more detailed review of the properties of these solutions may be found in Bodenheimer (1985).

The next step of the calculations was to combine the evolution of a protoplanet with the evolution of the solar nebula. An example of one of Cameron's evolving nebulae is shown in Figure 12.10. The temperature at the mid-plane is shown as a function of time for various radial positions. Similar plots are also available for the pressure. The initial rapid temperature rise corresponds to the accretion time scale of matter onto the nebula from the surrounding protostellar cloud. The second phase of nebular evolution corresponds to a decrease in mass

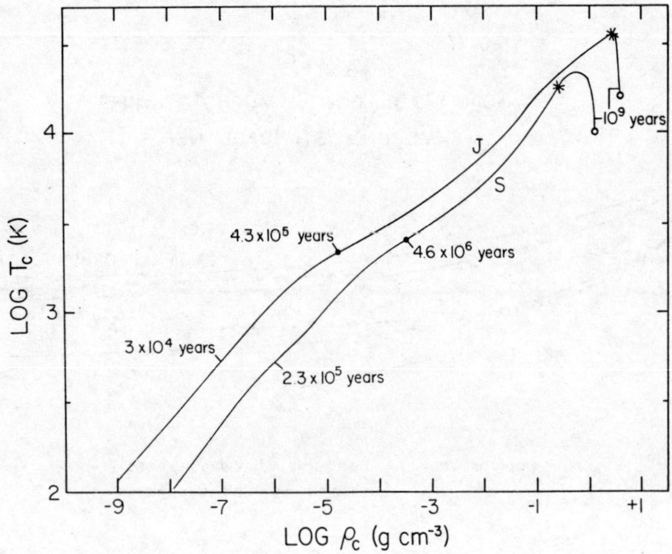

Figure 12.8. Evolution of the centers of models of giant gaseous protoplanets having the mass of Jupiter (J) and Saturn (S). Filled circles refer to the point of onset of hydrodynamic collapse; asterisks, to the beginning of the final phase of contraction and cooling; and open circles, to the positions of the present planets. From Bodenheimer (1985). Reprinted by permission. ©The Arizona Board of Regents.

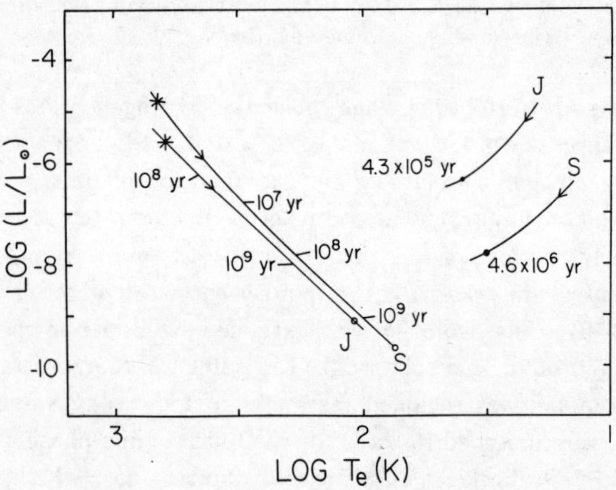

Figure 12.9. Evolution of the surfaces of giant gaseous protoplanets having the mass of Jupiter (J) and Saturn (S). Symbols have the same meaning as in Figure 12.8. From Bodenheimer (1985). Reprinted by permission. ©The Arizona Board of Regents.

due to accretion of nebular material into the Sun and by loss of mass by a coronal "wind" from the surface of the nebula: the time scale is a few times 10^5 yr. Such temperature and pressure variations were used as surface boundary conditions for evolving protoplanets (Cameron, DeCampli, and Bodenheimer 1982). Figure 12.11 shows characteristics of an evolving protoplanet with a boundary condition known as a "thermal bath." The temperature at the surface is assumed to increase at the rate of 10^{-3} K yr^{-1}, similar to that in the evolving solar nebula at Jupiter's position during the earlier mass accretion phase. The pressure is held constant, however, under the assumption that the protoplanet's tidal effect has opened up a gap at its position in the nebula (Lin and Papaloizou 1979). The surface of the protoplanet therefore is affected only by the radiation from the nebula but not by its change in pressure with time. The temperature variation with time in this calculation corresponds to Cameron's "high adiabat" nebular model, in which the temperature at Jupiter's position rises to a maximum of 166 K. The protoplanet first contracts and the central values of T and ρ increase. After 3×10^4 yr expansion results as a consequence of the fact that energy flows from the nebula into the protoplanet and increases its thermal energy. Disruption of the protoplanet occurs on a time scale of 10^5 yr. The effect occurs in general when the time scale for increase of the surface temperature is faster than the protoplanet's contraction time. The effects of using Cameron's "low adiabat" solar nebula were also investigated. In this model the viscous dissipation efficiency is somewhat less than in the case of the high adiabat, and temperatures at a given pressure are lower. The maximum temperature at the Jupiter orbital position is 40 K. Under these conditions a protoplanet can survive the "thermal bath" boundary condition. However, even for the low adiabat, a protoplanet in the asteroid region or near Earth's orbit will not survive: the envelope will expand and be lost. Thus the mechanism has some promise in explaining the formation of the inner and the outer planets. However, it has not yet been shown that the appropriate parameters exist. The instability mechanism in the solar nebula as well as the requirement for a relatively large core of solid material suggests that the initial protoplanet must be of high mass, say 30 Jovian masses. However, the contraction time for such a mass to the point of the onset of dissociation is very short, much shorter than the time scale for increase in the nebular temperature except perhaps for the high adiabat in the inner region of the nebula. Thus envelope evaporation after core formation has not yet been proved to be a likely sequence of events for the inner or outer planets. Note that the envelope loss mechanism does depend on the "thermal bath" outer boundary condition. Calculations of evolving protoplanets with nebular variations in both temperature and pressure taken into account in the surface boundary condition result in survival of the protoplanet at the Jupiter position, at the asteroid position, and at the Earth position. The high surface pressure at the time of maximum temperature is sufficient to prevent rapid expansion.

As an explanation for the origin of all types of planets the model of giant gaseous protoplanets runs into certain difficulties. In addition to the ones mentioned above, there is the problem of the solubility of solid or icy matter in the gaseous envelope. Core formation in giant protoplanets must apparently take place by accretion of planetesimals, followed by their settling toward the center, since in general the mass will be too high to allow the liquid droplet mechanism

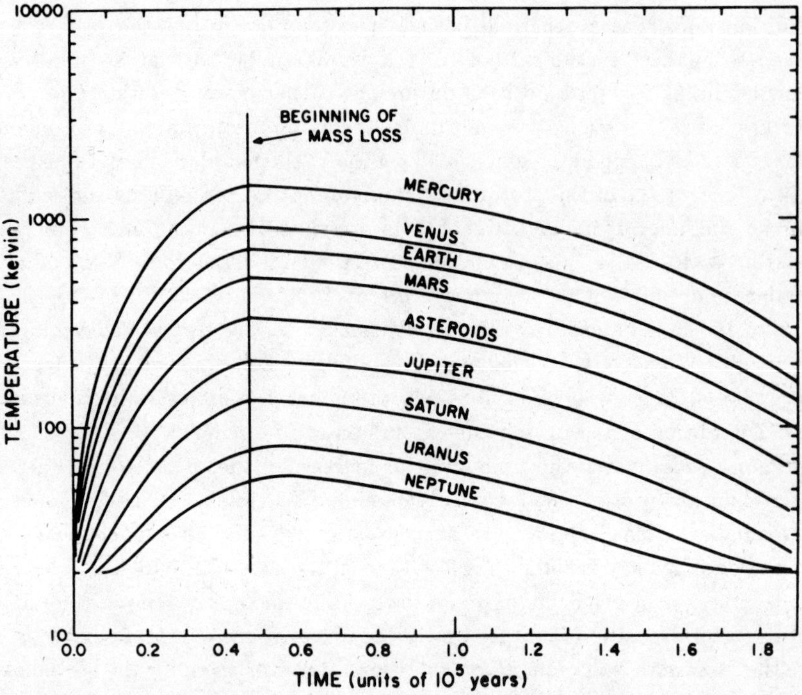

Figure 12.10. The temperature as a function of time in the solar nebula, at the position of the orbits of the present planets, as calculated according to the standard model of Cameron (1978*b*). Reprinted by permission. ©The Arizona Board of Regents.

to work. However, Stevenson (1982*a*) has pointed out that the grain material which is added to the gaseous envelope is, by and large, soluble in the gas. Thus core formation would be very difficult. In view of these problems it seems worthwhile to pursue the second alternative to the formation of giant planets, and much of the work in recent years has been directed toward this model. In fact, Al Cameron in a remark made during the "Protostars and Planets" conference in Tucson, January 1984, apparently favored this view.

12.5. Later Work on the Core Accretion Model

The alternative model requires that the core form first, by gravitational instability in the dust layer of the solar nebula followed by gradual accretion of matter to sizes on the order of the terrestrial planets. The Perri and Cameron study of the critical core mass required for rapid accretion of gas had the distinct advantage that it predicted this critical mass to be almost independent of position in the nebula, in agreement with observational data which suggest that the masses of the cores of all four giant planets fall in the range 15–25 Earth masses. However, the values they obtained were a factor 3–5 too large. Subsequent work by Harris (1978) and Mizuno (1980), based on improved physical assumptions in the calculations, brought the critical core masses down. The Mizuno model has had a significant influence in this field; thus a few details

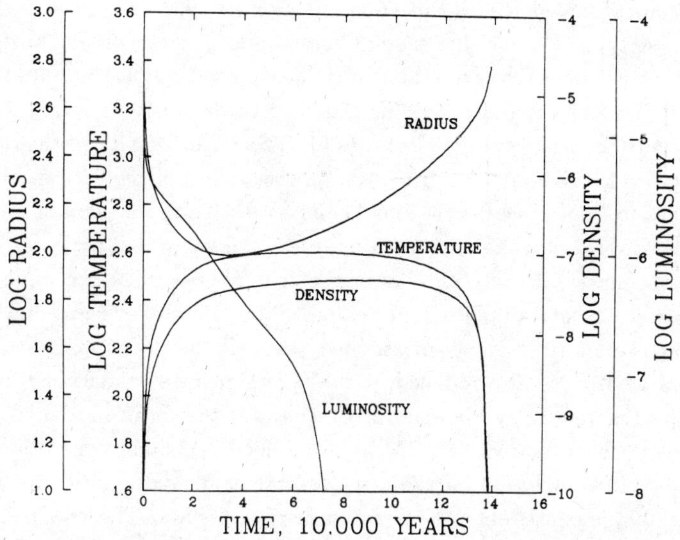

Figure 12.11. The evolution of the radius and luminosity (in solar units), and the central temperature and density (in K and g cm^{-3}, respectively), of a giant gaseous protoplanet of 1 Jovian mass with a surface boundary temperature that increases at a constant rate of 10^{-3} degrees per year ("thermal bath" condition). From Cameron, DeCampli, and Bodenheimer (1982). Reprinted by permission. ©Academic Press.

regarding its assumptions are given here. The core is assumed to be a solid object of density 3 g cm^{-3}; its total mass is a parameter, and its structure is not calculated in detail. Its mass grows in time at a constant rate, typically 10^{-6} Earth masses per year, by accretion of small solid particles from the nebula. The structure of the gaseous envelope, which is assumed to have solar composition, is calculated in detail. It is assumed to be in strict hydrostatic equilibrium, with the gas pressure gradient balancing gravity, and in strict thermal equilibrium, with the energy radiated at the surface balanced by the rate of dissipation of kinetic energy of the planetesimals as they accrete onto the core. The spherically symmetric stellar structure equations are solved, taking into account both radiative and convective energy transport. The opacity comes from atomic and molecular processes at temperatures above about 1800 K, and from grains below that value; in the grain region the opacity is taken to be independent of temperature and density. The grain-to-gas concentration is assumed to be a certain fraction of the standard interstellar value, where the fraction is a parameter, assumed to vary from 0 to 1. The equation of state is assumed to be that of an ideal gas. The outer boundary condition is applied at the tidal radius, given by $R_t = D(m/3\ M_\odot)^{1/3}$, where D is the distance to the Sun and m is the instantaneous protoplanetary mass. At that point the model is matched to nebular values of density and temperature, for example 1.5×10^{-10} g cm^{-3} and 97 K, respectively, at Jupiter's position. These values are assumed to be time-independent.

Mizuno obtained a series of static models in which the core mass was increased. A typical model has a relatively condensed central region ($\rho \approx 0.1$ g cm^{-3}) that is convectively unstable, a radiative zone starting at about $\rho = 10^{-3}$ g cm^{-3}, $T = 2300$ K and extending outward to the surface, and an outer zone near the surface that is nearly isothermal. The mean density in the outer envelope is quite low, about 10^{-8} g cm^{-3}. At low values of the core mass (1 Earth mass) the envelope mass is very small compared with the core mass, while as the core mass approaches the critical value, the envelope mass becomes comparable to the core mass. Mizuno found that the critical core mass, $M_{\rm crit}$, is about 12 Earth masses for a grain abundance equal to the interstellar value and for a core accretion rate of 10^{-6} Earth masses per year. If the grain opacity is reduced, the value of $M_{\rm crit}$ is also reduced; if the accretion rate is increased by a factor 10, $M_{\rm crit}$ is increased by about a factor 2; and if the mean molecular weight in the envelope is increased, then $M_{\rm crit}$ is reduced. A simple analytic model by Stevenson (1982*a*), based on radiative transfer and hydrostatic equilibrium in the envelope, is consistent with these numerical results. The new results show, as Perri and Cameron had already found, that the critical core mass is practically independent of the position of the model in the solar nebula, as reflected by the assumed values of R_t and the boundary values of ρ and T.

The physical nature of the instability when the critical core mass is reached, as well as the long-term evolution of giant planets under the core-accretion gas capture hypothesis, has been recently examined by Bodenheimer and Pollack (1984, 1985). The goal of these calculations, which are ongoing, is to calculate the evolution of the giant planets over the full 4.6×10^9 yr from the time of origin to the present. The full time-dependent equations of stellar structure are employed, in the form given by Bodenheimer (1985). The energy sources include gravitational contraction and energy deposited by accreting planetesimals. Detailed tables of the opacity, including grains, as well as of the equation of state in the region where nonideal effects become important, are included (Bodenheimer et al. 1980). Hydrostatic equilibrium is assumed, and the rocky core, which provides the inner boundary condition, is treated in the same manner as it was by Mizuno. Both radiation and convection are included in the energy transport. The outer boundary condition is applied at the accretion radius ($R_a = Gm/c^2$, where c is the sound speed in the surrounding nebula) or at R_t, whichever is smaller. The assumption is made, therefore, that a particle of gas is not gravitationally bound to the protoplanet if it falls outside R_a. At the appropriate outer boundary, density and temperature are held constant in time and, as in Mizuno's calculation, they are given a range of nebular values. As mass is added to the core, both R_a and R_t increase; therefore, at every time step in the calculation, mass is added to the envelope so that the model will actually fill the required volume. R_a turns out to be less than R_t during the earlier part of the evolution up to the time when the critical core mass is reached; thereafter R_t is generally smaller.

Table 12.1
Evolutionary Calculations

Case	T_b	ρ_b	$\dot{M}$	M_{crit}
M	100	1.2(−10)	10^{-6}	16.6
J	150	5.8(−11)	10^{-5}	29.0
J6	150	5.8(−11)	10^{-6}	16.8
JF	50	5.8(−11)	10^{-6}	16.5
JG	150	5.8(−10)	10^{-6}	16.6
U	50	1.2(−12)	10^{-6}	—
J7	150	5.8(−11)	10^{-7}	11.2
OP	150	5.8(−11)	10^{-6}	12.3

The calculations are generally started at the point where the core mass has about 0.5 Earth masses. The envelope then is optically thick, so the stellar structure equations can be applied, but it has negligible mass. The parameters that are varied during the various runs include the boundary values ρ_b and T_b, the accretion rate of particles onto the core ($\dot{M}$), the position in the solar nebula, and the assumed grain opacity. Table 12.1 gives a summary of the calculated cases. Here $\dot{M}$ is given in Earth masses per year and $M_{\rm crit}$ is the core mass, in Earth masses, beyond which gas accretion becomes very rapid. All calculations in the table were carried out at Jupiter's position in the solar nebula except case U, in which Uranus' position was taken. All calculations assumed a standard interstellar grain abundance except for case OP, in which the grain opacity was reduced by a factor 50. The first row in the table, case M, refers to a calculation with parameters very close to those of Mizuno's model. The properties of this model at 1.66×10^7 yr are in good agreement with those of Mizuno's near the critical core mass, except that the actual value of $M_{\rm crit}$ is 25% higher. It is clear that a change in ρ_b or T_b at a given $\dot{M}$ has little effect on the value of the critical core mass (cases M, J6, JF, and JG). However, a change in $\dot{M}$, other parameters being the same, results in substantial effects on $M_{\rm crit}$ (cases J, J6, and J7). The reduction in grain opacity with other parameters held constant resulted in a reduction in $M_{\rm crit}$ by 25%. The deduced core masses for Jupiter and Saturn fall in the range 20–25 Earth masses, suggesting a mean core accretion rate of 3×10^{-6} Earth masses per year as well as full interstellar grain abundances. For Uranus and Neptune, on the other hand, the core masses are about 13 Earth masses, corresponding to a lower accretion rate of close to 3×10^{-7} Earth masses per year.

Up to the point when the critical core mass is reached, the accretion luminosity dominates the energy production. Near that point the accretion luminosity becomes insufficient to keep the more and more massive envelope in thermal equilibrium; thus it must contract and liberate gravitational energy. The contraction further enhances the rate at which mass must be added to the envelope to keep the outer radius equal to R_t. As the envelope gains mass, the tidal radius increases, further enhancing the rate at which mass must be added. Quickly the rate of accretion of gas far exceeds the rate of accretion of planetesimals onto the core, and, on the time scale of further gas accretion, the core mass hardly changes. After the critical core mass has been reached, however, the envelope does not collapse dynamically; although its contraction becomes very fast, it

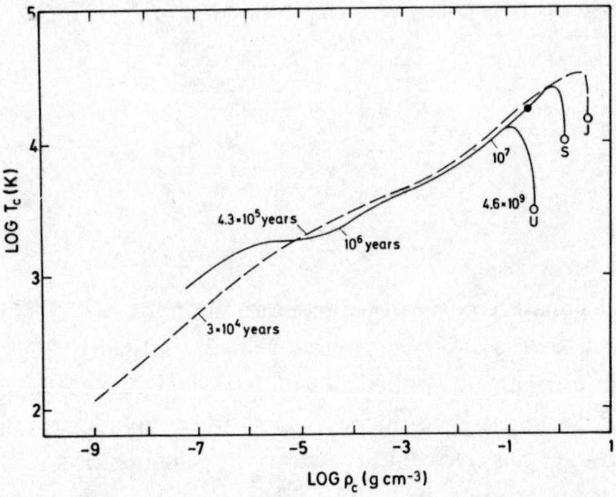

Figure 12.12. The evolution of the central conditions of protoplanets in the core-accretion gas capture model is given by the solid curves. For comparison, the evolution of a giant gaseous protoplanet of uniform composition and 1 Jovian mass is given by the dashed curve (Bodenheimer et al. 1980). Open circles denote the positions, after 4.7×10^9 yr, of models with the masses of Uranus (U), Saturn (S), and Jupiter (J).

remains in hydrostatic equilibrium.

Two of the cases were calculated well beyond the time when envelope accretion begins to dominate. Case J was evolved until the total mass reached 1 Saturn mass. Then the accretion of both envelope and core was stopped and the protoplanet allowed to evolve further at constant mass. Similarly, case U, which is assumed to lie at Uranus' position in the solar nebula, was evolved to the point where the core mass reached 13.2 Earth masses; this mass turned out to be slightly smaller than $M_{\rm crit}$ for the accretion rate chosen. The envelope mass at that time was 1.3 Earth masses; both values agree well with the suggested values for the planet Uranus itself. Again the model was evolved at constant mass beyond that point. In both cases the boundary condition was changed: the outer radius was no longer required to reach out to R_t or R_a, and application of the nebular boundary condition was discontinued. The object was assumed to be an isolated protoplanet that radiates as a blackbody.

The results of these calculations are shown in Figures 12.12 and 12.13. In Figure 12.12 the central conditions in the Saturn and Uranus runs are compared, as a function of time, with the evolution of a giant gaseous protoplanet of 1 Jovian mass, evolved at constant total mass and with uniform chemical composition, neglecting the effects of the core. In the models with cores the quantities plotted refer to the outer edge of the core. The Saturn model evolves to a central temperature (T_c) maximum of 27,000 K; then, shortly after accretion is shut off, it begins to cool, on an ever increasing time scale. At about 4.3 $\times 10^9$ yr the value of T_c has been reduced to 11,000 K, consistent with current models of Saturn at its present state (Stevenson 1982*b*). Similarly, the Uranus model attains a maximum value for T_c of 13,000 K, and at 4.5 $\times 10^9$ yr it has

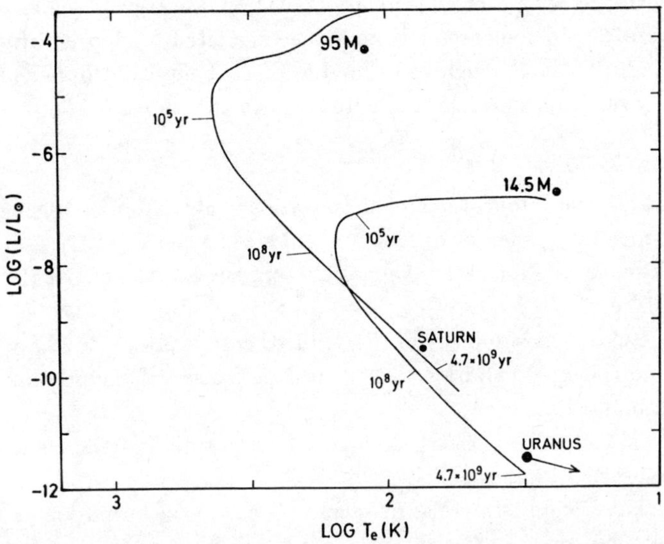

Figure 12.13. The evolution in the theoretical Hertzsprung-Russell diagram of protoplanets that have evolved according to the core-accretion gas capture model. The tracks start at the time when accretion is terminated, so that the objects are evolving with constant masses equal to those of Uranus (14.5 Earth masses) and Saturn (95 Earth masses). Filled circles give the present observed positions of the two corresponding planets; in the case of Uranus this is an upper limit.

been reduced to 3200 K at the core edge, again consistent with current models of the present planet. Figure 12.13 shows the Hertzsprung-Russell diagram for the two objects after accretion has been shut off (before that time surface temperatures are constant and the luminosity increases with time). Times marked on the curves also refer to the postaccretion phase. During the early part of the evolution the contraction is very rapid as the radius decreases from R_t (2×10^{12} cm) to values characteristic of the present giant planets (10^{10} cm). This phase lasts only 10^5 yr, and during it the luminosities are relatively high, $10^{-7}\ L_\odot$ for Uranus and $10^{-4}\ L_\odot$ for Saturn. These values are of interest with regard to possible detection of protoplanets associated with young stars.

The evolutionary time scale then lengthens as the objects approach their limiting radius. The only remaining energy source is then cooling of the hot interior; thus the tracks in the H-R diagram fall practically on lines of constant radius. Although these approximate models do not have accurate surface boundary conditions as compared with those used for models of the present planets, the fit with present observed parameters of Saturn and Uranus is reasonably good. After 4.7×10^9 yr the calculated luminosity for Saturn is somewhat too low but the radius is correct to within a few percent. The discrepancy is similar to that obtained in previous calculations of the late phases of evolution for Saturn (Grossman et al. 1980). One possible explanation is that the limited solubility of He in metallic hydrogen results in precipitation of the helium to the center with the release of additional gravitational energy (Smoluchowski 1967; Salpeter 1973). In the case of Uranus the calculated luminosity after 4.7×10^9 yr is about 10^{-12} $L_\odot$, slightly below the present observed upper limit for the luminosity of Uranus. Again, the agreement in the radius is good.

12.6. Discussion

What are the implications of these various calculations with regard to the origin of the giant planets? Among the issues to be considered are the following:

1. What is the formation time scale compared with that of the terrestrial planets?
2. How can the similarity of the deduced core masses (15–25 Earth masses) and the wide variation of the envelope masses of the present planets be explained?
3. Can the present observed chemical composition of the giant planets be explained by the models?
4. What role does the solar nebula itself play in the formation process? Can any information be deduced about the properties of the nebula from the present characteristics of the giant planets?

A good discussion of the two basic hypotheses for planet formation in the light of these questions appears in Pollack (1985); what follows is a brief summary.

With regard to the formation time scale, the giant gaseous protoplanet (GGP) hypothesis has the distinct advantage, since the formation process occurs on a dynamic time scale during the very early history of the solar nebula. The early existence of Jupiter, with substantially its present mass, is regarded as a necessary condition for the explanation of the lack of a full-sized planet in the asteroid belt and for the relatively low mass of Mars. In the core-accretion gas capture (CAGC) model, the time for accretion of a core of 15 Earth masses at Jupiter's position is almost certainly much longer than that for accreting 1 Earth mass at the Earth's position, or even at Mars' position, so the above properties of the inner solar system cannot be explained. Present versions of accretion theories have difficulty in forming the Jupiter or Saturn cores in a few million years, a condition that is necessary if substantial nebular gas is still to remain once the critical core mass has been reached. The situation is even more difficult for Uranus and Neptune, where core accretion times exceed the age of the solar system. A number of suggestions have been put forward with regard to the question of these long accretion times (see Pollack 1985), but the problem remains unsolved.

The similarity of the core masses of the giant planets can be explained nicely by CAGC. The critical core mass, which is determined by the point where the gravitational influence of the accreting core first becomes strong enough to attract a comparable amount of gas from the nebula, is the crucial point in this regard. Also, the computed value of the critical mass (10–40 Earth masses) falls exactly in the observed range. In the GGP model, there is no obvious reason why the core masses should be similar. The model in fact has difficulties in forming cores at all, particularly because rock and ice are now regarded to be soluble in the hydrogen envelope (Stevenson 1985). Note, however, that this problem is not entirely absent in the CAGC model. Even before the critical core mass is reached, the accreted solid material has to pass through a gaseous envelope of considerable depth and with temperatures ranging up to 10^4 K. Unless the accreted particles are very massive, there is danger that they will be evaporated and dissolved before they even reach the core. The wide variation in the envelope masses among the giant planets can be explained in a general way by both models, although no specific calculations have been made which predict actual planet masses. In the GGP model the radial variation of density in the solar nebula must have influenced the masses of the fragments that form. However, it is still difficult to explain the fact that the gaseous envelope of Uranus and Neptune have relatively small mass. In the CAGC model the variation in envelope mass must have resulted from a radial increase in core accretion times, so that because of the dissipation of the solar nebula there was little gas left to accrete by the time the outer planets approached their critical core masses.

With regard to question (3) above, the observed overabundances of C/H in the atmosphere of all four giant planets, in comparison with the solar value, could be explained under either theory. Similarly, the observed depletion of nitrogen in the atmosphere of Uranus and Neptune does not discriminate between the two models (Pollack 1985). The final question, whether we can infer any properties of the solar nebula from the known facts about the giant planets, is a very challenging one and will not be resolved before more detailed protoplanetary models are constructed and their evolution calculated in conjunction with that of the surrounding nebula. One approach, of course, is based on the equilibrium condensation model for solid particles (Lewis 1974), from which approximate nebular temperatures as a function of radius at the time of planet formation can be deduced. Another interesting approach, in the context of the CAGC model, is based on the concept of tidal truncation (Lin and Papaloizou 1979). The accretion of gas onto the protoplanet will be terminated once it reaches a critical mass, proportional to the inverse of the local Reynolds number in the nebula. After that time the tidal influence of the protoplanet opens up a gap in the nebula, since the time for tidal transfer of angular momentum becomes shorter than the viscous time scale in the nebula. Thus accretion is self-limiting. If Jupiter's accretion was terminated in this way, then from the deduced Reynolds number one can infer the viscous properties, and therefore the evolution time of the nebula. The analysis leads one to the conclusion that this time is about 10^6 yr, which leads to the difficulty mentioned above, that core accretion times for Jupiter are estimated to be considerably longer. Further development of the theory of the origin of the solar system, as well as future observations of young stellar objects where planets are in the process of formation, and, furthermore, detailed observations of the chemical composition of all of the present outer planets should provide some further information regarding the question of the nature of the solar nebula.

Although the preponderance of the evidence now favors the CAGC model over GGP, the latter is still under investigation, although in a rather different context than that of the preceding discussion. Cameron (1985) has considered the question of the origin of Mercury. He suggests that it would be difficult to explain the chemical composition of that planet by an origin from accretion of small particles. Rather, he puts forward the view that Mercury formed as a giant gaseous protoplanet, substantially more massive than the present planet. After precipitation of heavy elements to the center, the heating from the solar nebula resulted in a planetary wind that blew off the outer layers of the planet, including the rocky mantle, leaving behind just the iron core. In that paper Cameron states: "Although the theory is specific to the formation of just one planet, it holds implications for the formation of planets generally, and may provide a means for choosing between such theories." Clearly this field of research is open for numerous surprises in the future.

Acknowledgments

Work by the author reported in this article was supported by National Science Foundation grant AST 83-01229. The assistance of the Max-Planck-Institut für Physik und Astrophysik (Garching, FRG) in the preparation of this manuscript is gratefully acknowledged.

References

Bodenheimer, P. 1985, in *Protostars and Planets,* vol. **2,** ed. D. Black and M. S. Matthews, p. 873 (Tucson: University of Arizona Press).

Bodenheimer, P.; Grossman, A. S.; DeCampli, W. M.; Marcy, G.; and Pollack, J. B. 1980, *Icarus,* **41,** 293.

Bodenheimer, P., and Pollack, J. B. 1984, *Bull. AAS,* **16,** 702.

——. 1985, in preparation.

Cameron, A. G. W. 1962*a, Icarus,* **1,** 13.

——. 1962*b, Sky and Telescope,* **23,** 244.

——. 1963, *Icarus,* **1,** 339.

——. 1973, *Icarus,* **18,** 407.

——. 1978*a, Moon and Planets,* **18,** 5.

——. 1978*b,* in *Protostars and Planets,* ed. T. Gehrels, p. 453 (Tucson: University of Arizona Press).

——. 1985, preprint.

Cameron, A. G. W., DeCampli, W. M., and Bodenheimer, P. 1982, *Icarus,* **49,** 298.

Cameron, A. G. W., and Pine, M. R. 1973, *Icarus,* **18,** 377.

DeCampli, W. M., and Cameron, A. G. W. 1979, *Icarus,* **38,** 367.

Ezer, D., and Cameron, A. G. W. 1963, *Icarus,* **1,** 422.

——. 1966, in *Stellar Evolution,* ed. R. F. Stein and A. G. W. Cameron, p. 203 (New York: Plenum Press).

Graboske, H. C.; Pollack, J. B.; Grossman, A. S.; and Olness, R. J. 1975, *Ap. J.,* **199,** 265.

Grossman, A. S.; Pollack, J. B.; Reynolds, R. T.; Summers, A. S.; and Graboske, H. C. 1980, *Icarus,* **42**, 358.

Harris, A. W. 1978, *Abstract 9th Lunar Planet. Sci. Conf.*, 459.

Hayashi, C. 1961, *Pub. Astr. Soc. Japan,* **13,** 450.

Henyey, L. G., Forbes, J. E., and Gould, N. L. 1964, *Ap. J.,* **139,** 306.

Hubbard, W. B. 1980, *Rev. Geophys. Space Phys.,* **18,** 1.

Lewis, J. S. 1974, *Science,* **186,** 440.

Lin, D. N. C., and Papaloizou, J. 1979, *M.N.R.A.S.,* **186,** 799.

Lynden-Bell, D., and Pringle, J. E. 1974, *M.N.R.A.S.,* **168,** 603.

Mizuno, H. 1980, *Progr. Theoret. Phys.,* **64,** 544.

Perri, F., and Cameron, A. G. W. 1974, *Icarus,* **22,** 416.

Podolak, M., and Cameron, A. G. W. 1974, *Icarus,* **22,** 123.

Pollack, J. B. 1985, in *Protostars and Planets,* vol. **2,** ed. D. Black and M. S. Matthews, p. 791 (Tucson: University of Arizona Press).

Pollack, J. B.; Grossman, A. S.; Moore, R.; and Graboske, H. C. 1977, *Icarus,* **30,** 111.

Salpeter, E. E. 1973, *Ap. J. (Letters),* **181,** L83.

Schwarzschild, M. 1958, *Structure and Evolution of the Stars* (Princeton: Princeton University Press).

Smoluchowski, R. 1967, *Nature,* **215,** 691.

Stevenson, D. J. 1982*a, Planet. Space Sci.,* **30,** 755.

——. 1982*b, Ann. Rev. Earth Planet. Sci.,* **10,** 257.

——. 1985, *Icarus,* **62,** 4.

13

Planet Formation

William M. Kaula

For the mind, by nature stagey, welds its frame
Tomb-like around each little world of a day;
We jump from picture to picture and cannot follow
The living curve that is breathlessly the same.

Louis Macneice

Abstract

A great variety of dynamical and compositional phenomena constrain the origin of the planets, indicating that it was a rather complex sorting process. Most important is the formation of Jupiter: essentially, what are the properties of a protosolar cloud which lead to a Jupiter-like second largest body? A satisfying answer to this question appears to require three-dimensional hydrodynamic collapse modeling starting rather early in the process. Once Jupiter is formed, there are several mechanisms, involving resonances and mass loss from the nebula, to shake the planetary system down to the present ultrastable configuration of a few planets.

The final stages of terrestrial planet formation probably involved collisions between rather large protoplanetary bodies. The final stages of major planet formation were more dominated by gradual loss of smaller bodies from the solar system, leading to the present widely spaced, nearly circular orbits.

13.1. Introduction

Formation of the planets is a poor scientific problem. We have only one well-observed example, and it is evidently the outcome of a very stochastic process. Hence, developing a theory of solar system origin is somewhat analogous to developing a theory of atmospheric dynamics knowing the global weather pattern at only one instant in time, but worse in that the present solar system is very stable and unchanging in many of its properties, and the problem is to infer the route by which it slipped into this well of stability. But from the point of view of the planetary scientist, there is a veritable zoo of data which can be explained only by appealing to circumstances of origin. Planets, meteorites, and most other objects in the solar system have many properties that are quite unaffected by external influences in the solar system as it is now, and hence must have been for most of its 4.57 billion year history. It is unsatisfying merely to accept the planets as having been born different. So to explain differences among the planets, we must study the relatively short period associated with formation of the entire system, before the planets came to be such distinct entities. To help constrain the nature of this brief period, there are astronomical observations and theory which indicate strongly that the planets are by-products of the collapse which formed the Sun and that, because of the conservation of angular momentum, the forming solar system must have had some dynamical properties in common with several other phenomena, ranging from the Galaxy to Saturn's rings. The papers by Lin and Papaloizou and by Bodenheimer in this volume apply these dynamical insights to solar system formation. This paper takes a more backward-looking approach, attempting to connect the observed structure and composition of the solar system to the astronomically based forward-looking models.

The plan of this paper is tripartite: first, to review the evidence in the solar system relevant to its origin; second, to discuss the formation of Jupiter, the greatest difficulty; and third, to explore how the other planets would have developed in a circumsolar environment already containing a sizable proto-Jupiter. The term "proto-Jupiter" is used rather than "Jupiter" to emphasize that a dominant effect on the dynamics of the circumsolar material from which the planets formed must have been a mass concentration second only to the Sun. That this mass concentration evolved into Jupiter is strongly indicated by it being one of two planets not drastically stripped of volatiles, while the other, Saturn, is only 30% as massive and farther from the Sun.

13.2. Data

Since readers of this paper have often seen lists of the planets in order of their distance from the Sun, in Table 13.1 we adopt a different ordering that is based upon mean density, but ineluctably related to the composition of the planets. This ineluctability depends on the limited palette of materials from which to make planets—in turn consequent on the limited variety of nucleosynthetic processes—and on pressure and temperature being the only factors changing densities significantly from those observed in the laboratory. Some decades ago there were zealous attempts to explain some of the density variations in Table 13.1 as arising from remarkable phase transitions, but these have faded with the advance of condensed matter experiment and theory. Hence, if we define planetary categories by dominant constituents in mass, we have a solar system whose 17 largest bodies fall into four categories. This impression of heterogeneity is reinforced if we attempt a categorization by mass. The 17 bodies in Table 13.1 span more than eight orders of magnitude; if we exclude the Sun as different, the remaining 16 bodies span more than five orders of magnitude. More detailed listings of properties are given for the major planets by Stevenson (1982*a*) and for the terrestrial planets by Kaula (1986). Stevenson (1985) objects to the lumping together of several planets as "ice" because it obscures differences significant to their internal structures. However, for the primarily dynamical problem of formation, this classification is still a useful approximation.

Table 13.1
The 17 Most Massive Bodies of the Solar System

Type and Name†	Mass (Earth Masses*)	Radius (Earth Radii*)	Mean Density ($g\,cm^{-3}$)	Solar Distance (AU*)	Average Incli- nation	Orbit Eccen- tricity
Plasma						
Sun	333,000.	109.2	1.39	—	—	—
Fluid						
Jupiter (P)	317.8	10.97	1.31	5.21	0.4°	0.045
Saturn (P)	95.25	9.07	0.69	9.58	0.9°	0.048
Ice						
Neptune (P)	17.21	3.80	1.66	30.20	0.7°	0.009
Uranus (P)	14.60	3.96	1.19	19.14	1.0°	0.032
Ganymede (S)	0.0249	0.414	1.93	5.21	—	—
Titan (S)	0.0227	0.402	1.93	9.58	—	—
Callisto (S)	0.0178	0.378	1.81	5.21	—	—
Triton (S)	0.0095	0.30	1.98	30.20	—	—
Pluto (P)	0.0016	0.20	1.12	39.44	15.19°	0.242
Rock						
Earth (P)	1.000	1.000	5.52	1.000	1.5°	0.030
Venus (P)	0.815	0.950	5.25	0.815	1.6°	0.036
Mars (P)	0.107	0.532	3.94	1.524	1.7°	0.08
Io (S)	0.0149	0.285	3.55	5.21	—	—
Moon (S)	0.0123	0.273	3.34	1.000	—	—
Europa (S)	0.0081	0.245	3.04	5.21	—	—
Iron						
Mercury (P)	0.0553	0.383	5.42	0.387	7.3°	0.18

*One Earth mass is 6×10^{24} kg. One Earth radius is 6371 km. One AU is 1.5×10^{8} km.
†P=planet; S=satellite.

Table 13.2 also is an arrangement differing from the familiar. It lists the 15 most abundant elements primarily on the basis of the volatility of their commonest chemical forms and secondly on their degree of chemical interactiveness. The depletion factors in Table 13.2 suggest a simpler pattern: the compositions of the planets are essentially determined by volatility, which suggests separation of planetary material by condensation from a vapor of solar composition. Table 13.2 gives the depletion factors only for the Earth. They are estimates in that it is assumed that most nickel and sulfur went into the largely fluid core which, from its density, must be predominantly iron. This is a trivial assumption for the nickel (which is more siderophile than iron), but not for sulfur, which vaporizes at about 700 K if not combined with iron.

Tables 13.1 and 13.2 suggest that for 11 of the 16 bodies (all but Jupiter, Saturn, the Moon, Io, and Europa) composition is determined mainly by condensation from a vapor (popularly known as the solar nebula) whose temperature decreases with distance R from the Sun approximately proportionately to $1/R$, a theme most vigorously pursued by Lewis (1974). This trend is confirmed by data from meteorites (of precisely known chemistry but uncertain dynamical history) and planets (of precisely known dynamics but very limited chemical accessibility). Taking a pressure of 10^{-3} atmospheres and assuming volatiles are removed when the temperature drops below about 1440 K at Mercury, this model accounts for the densities of the 11 bodies rather neatly. But, aside from not explaining the five other bodies, the model must be very neat indeed, since the condensation temperatures of iron and the most abundant silicates differ by only 25 degrees at 10^{-3} atmospheres, and become equal at 7×10^{-5} atmospheres. Hence, while the correlation of volatile abundance with distance from the Sun

Table 13.2
The 15 Most Abundant Elements (and the 26 most abundant nuclei)

Type, Number, Name	Main Isotopes	Solar Abundances: Relative to Silicon	Solar Abundances: Mass %	Estimated Depletion in Earth	Most Common Forms
Hyperactive Gas					
1 Hydrogen (H)	1*,2	27,000.	75.7	5×10^{-7}	H_2O, CH_4, NH_3
Active Gas					
6 Carbon (C)	12*,13	11.6	0.36	5×10^{-4}	CO_2, CH_4
7 Nitrogen (N)	14	2.4	0.095	3×10^{-5}	NH_3, NO, N_2
8 Oxygen (O)	16	19.0	0.87	0.2	Many: see above & below
Inert Gas					
2 Helium (He)	3,4*	2000.	22.5	10^{-11}	He
10 Neon (Ne)	20*,22	2.65	0.15	10^{-11}	Ne
18 Argon (Ar)	36*,38	0.10	0.010	10^{-10}	Ar
Lithophile					
11 Sodium (Na)	23	0.06	0.004	0.2	Na_2O
12 Magnesium (Mg)	24*,25,26	1.07	0.073	1.0	MgO
13 Aluminum (Al)	27	0.085	0.0063	1.2	Al_2O_3
14 Silicon (Si)	28*,29,30	1.000	0.079	1.0	SiO_2
20 Calcium (Ca)	40	0.062	0.007	1.2	CaO
Siderophile					
26 Iron (Fe)	54,56*	0.900	0.141	1.2	Fe, FeO, Fe_2O_3
28 Nickel (Ni)	58	0.048	0.008	1.2	Ni, NiO
Chalcophile					
16 Sulfur (S)	32*,34	0.5	0.045	0.3	SO_2, H_2S, FeS

*Most abundant isotope.

is a very real constraint, several other factors must be taken into account for a satisfying theory of planetary formation.

Given the unavoidable complexity of the chemistry of solar system formation, abundances insensitive to chemical interactions are stronger constraints. Two are most notable: the inert gas abundances and the isotope ratios of oxygen (the only abundant gas to have three isotopes). The inert gas data are shown in Figure 13.1, and the oxygen isotope data in Figure 13.2.

The classical interpretation of the inert gas data is that they were removed from the nebula when the Earth's material was in bodies as small as the meteorite parent bodies (Wasson 1969), and hence that the Earth accumulated in a gas-free environment. Subsequent to the formulation of this hypothesis, it was found that Mars is less abundant than the Earth in primordial argon (isotopes 36 and 38) by a factor of 1/250, while Venus is more abundant by a factor of 80. The Mars argon difference can be explained as evidencing very little outgassing from that planet, but the Venus argon difference must reflect origin circumstances. If the argon abundances reflect the nebula and condensation of small planetesimals, then there must have been both extraordinary gradients of conditions in this nebula and extraordinary suppression of subsequent dynamical mixing of protoplanetary bodies among zones.

The oxygen isotope data in Figure 13.2 require preservation of compositional heterogeneities from nucleosynthesis through planet formation (although photochemical effects have been shown to produce fractionations disproportionate to mass [Thiemens, Heidenreich, and Lundgerg 1983]). While these isotope ratios are strong qualitative constraints, they are difficult to quantify because of the uncertainties of the dynamical histories of meteorites. It is impressive that there is a range of isotope ratios within the Pueblo Allende meteorite requiring

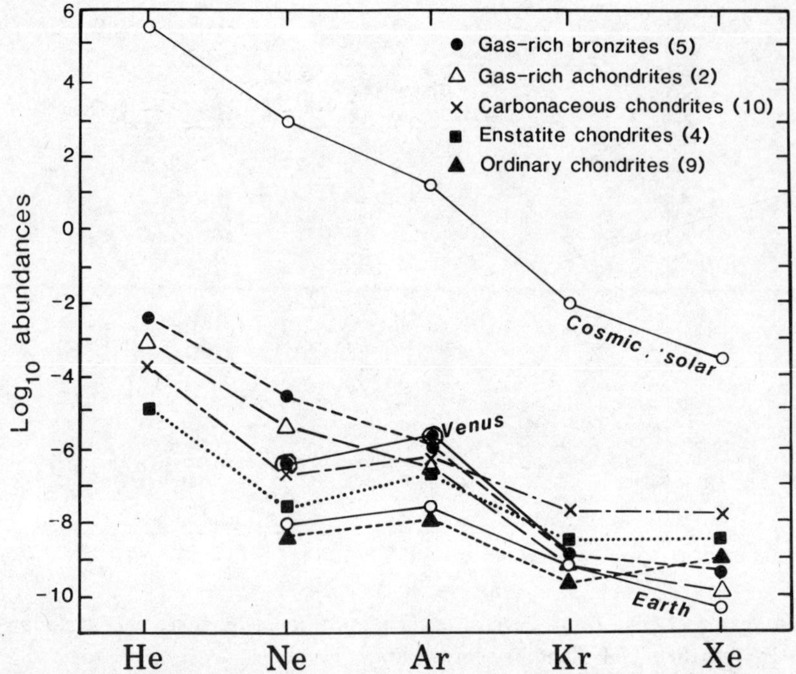

Figure 13.1. Abundances of inert gases. The abundances are in cubic centimeters per gram, which roughly equal atomic abundances normalized to 100 for silicon. The Venerean abundances of Kr and Xe are the same as those of the gas-rich bronzites within the uncertainties. The figure is based on Kaula (1968) and Donahue and Pollack (1983).

at least two sources, while all those measured on Earth rocks can be explained as arising from mass-dependent transport effects from a single nucleosynthetic source. Probably most significant cosmogonically is the overlap on the isotope plot of lunar rocks and ultrabasic rocks—closest to mantle composition—of the Earth. Even the one category of meteorites which falls on the terrestrial line, the eucrites, is displaced along the line from these rocks most nearly representative of the Earth's bulk composition.

There are some other differences in volatile content among the four bodies sampled—Earth, the Moon, Mars, and Venus—but they appear to be consequences of evolution, such as the loss of water through photochemical dissociation on Venus and the loss of ^{14}N relative to ^{15}N on Mars. It may eventuate, however, that the differing volatile evolutions of the Earth and Venus depend on differing degrees of loss of primordial atmospheres because of great impacts during the formation phase.

In addition to the mean densities, the moments-of-inertia of the planets are important constraints on their origins and early evolutions. Venus and Mercury rotate too slowly for their moments to be measurable, but the moments for the Earth and Mars indicate that almost all metallic iron separates into a core in

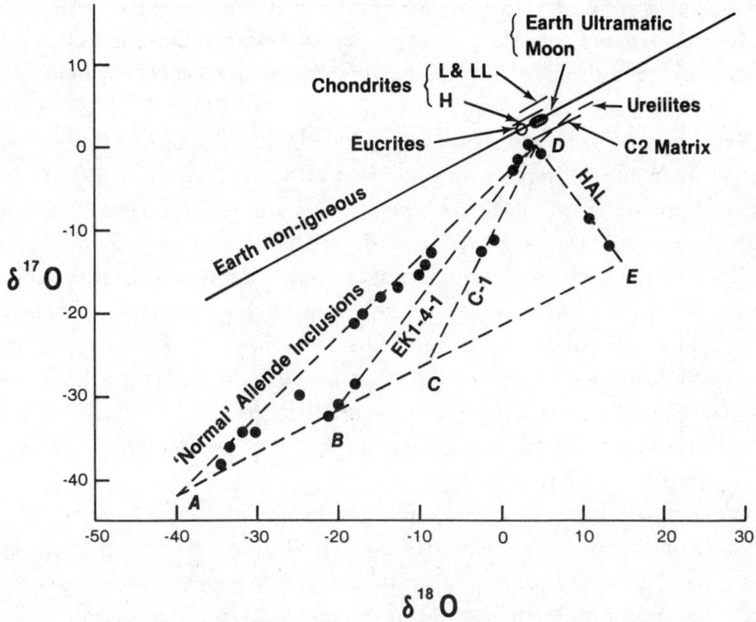

Figure 13.2. Oxygen isotope ratios. The ratios are expressed as differences from the ratio SMOW (standard mean ocean water), $\delta^I\text{O} = 1000([^I\text{O}/^{16}\text{O}] - [^I\text{O}/^{16}\text{O}]_S)/[^I\text{O}/^{16}\text{O}]_S$, where I denotes 17 or 18 and S denotes SMOW. The ratios in Allende inclusions are correlated with mineral phases: low ratios are in spinels; intermediate ratios, in pyroxenes; high ratios, in melilites. From Clayton and Mayeda (1977) and Lee, Mayeda, and Clayton (1980).

terrestrial planets. The moments of Jupiter and Saturn indicate that they both must have icy or rocky cores of about 15 $M_\oplus$ mass, comparable to the masses of Uranus and Neptune. This coincidence is sometimes interpreted as suggesting a similar mode of formation (Stevenson 1982*b*). A more certain inference is that Jupiter and Saturn differ significantly from the Sun in the ratio of their hydrogen and helium to other elements. From the mass abundances in Table 13.2, the material of Jupiter's core must have lost about 66% of its hydrogen plus inert gas complement, while that of Saturn must have lost about 91%.

Discussions of the elemental abundances for the terrestrial planets (e.g., Wänke, Dreibus, and Jagowitz 1984) usually conclude by assuming a mixture of a refractory component and a volatile component. The latter is commonly assumed to have a composition like carbonaceous chondrites. However, as indicated in Figure 13.1, carbonaceous chondrites have high abundances of inert gases: in particular, argon. Hence, either the amount of such volatile material added to the Earth after its material suffered loss of argon must be very limited, or impacts have effects on inert gases Ne, Ar, etc., greatly different from those on compounds of the chemically active gases C, N, and O.

The semimajor axes in Table 13.1 are traditionally expressed as "Bode's law": $a_{n+1}/a_n \approx 1.75 \pm 0.20$. However, dynamical considerations, as discussed in §13.4, suggest that the ratio of mean motions is more significant: $\Omega_n/\Omega_{n+1} \approx 2.3 \pm 0.4$. Corroboratory of this suggestion of resonant interactions, together with dissipative processes, are the extremely small eccentricity of Neptune's orbit—0.009—and the survival of the test particle Pluto in a 2:3 resonance with Neptune. As with any resonance, the Pluto:Neptune lock indicates that either (1) many states were sampled and only Pluto survived, or (2) there were "nudging" processes in the formation of the outer solar system which could push Pluto into the resonance. Associated with such processes would be outward transfer of angular momentum and loss of energy, either by dissipation or ejection of matter from the system. (Forbiddingly improbable, of course, is that Pluto was once a satellite of Neptune, requiring the double chance of something to break it away from Neptune and a quick means of shifting it into the resonance before it collided with Neptune.) However, the tiny eccentricity of Neptune's orbit indicates that angular momentum transfer was predominantly outward to Neptune, thus favoring the "nudging."

Further indicative of orbit:orbit resonances and shifting loci thereof due to mass loss from the solar system are (1) the extraordinary emptiness of the spaces between the planets and (2) the appreciable eccentricities and inclinations in the asteroid belt: much more than could be generated by self-interaction.

The closely associated phenomena of planetary rotations and satellite systems are also diagnostic of solar system origin circumstances; for reviews see Harris and Ward (1982) and Stevenson, Harris, and Lunine (1985). Most striking, of course, is the contrast between Earth and Venus, two planets similar in size and density but one rotating 224 times as fast as the other. Furthermore, the Earth's Moon has by far the largest mass ratio of a satellite to its planet (excepting tiny Chaeron to Pluto, plus possible asteroid couplings). These circumstances are strongly suggestive that the final stages of terrestrial planet formation entailed very large impacts.

13.3. Formation of Proto-Jupiter

The usual model of formation of the solar system is a collapse leading, because of conservation of angular momentum, to an axisymmetric accretion disk around a central mass, followed by gravitational instabilities in this disk. Common to these models is some sort of balance between mass and angular momentum infall into the disk and the transfer of mass inward, angular momentum outward by viscous processes within the disk (Lynden-Bell and Pringle 1974). Debate is about (1) the mass of the disk relative to that of the central mass; (2) the phasing of the gravitational instabilities relative to the formation of the central mass; (3) whether the instability is of the gas or of solids within the gas; and (4) the role of thermal convection in mixing and turbulence within the disk. These differing properties are coupled. Thus massiveness of the disk is associated with early instability of the gas (Cameron 1978), while a lower density disk is associated with relatively late instability of solids (Lin and Papaloizou 1985). Convection is regarded as an inevitable consequence of the opacity resulting from the abundance of smoke (Morfill 1983), but quantitative inferences therefrom are difficult. Other properties which seem correlated with the massiveness of the disk are short time scales of collapse and preservation of heterogeneities, both consequences of a high ratio of the rate of angular momentum infall to the rate of angular momentum transfer within the disk (Cassen and Summers 1983; Cassen, Shu, and Terebey 1985).

But the model property of axisymmetry maintained until a central condensation is well under way and the disk radius is less than 300 AU may be deceptively unrealistic for inferences as to planet formation. The Sun's mass is much too small to form by a symmetric collapse from a gas and dust cloud of typical density and temperature, which suggests that irregularities played a necessary role in solar collapse from the beginning. Collapse processes may greatly enhance irregularities existing in a typical cloud—say, 10^{-19} g cm^{-3} in density and 10,000 AU in radius—in the process of its shrinking to the size of a protosolar system: 100 AU. A calculation pursuing this hypothesis is by Boss (1985), who applied a three-dimensional hydrodynamic code to a slowly rotating cloud of 2 $M_\odot$ with $\pm 5\%$ irregularities in density about a mean of 10^{-15} g cm^{-3}. "Slowly rotating" was defined by the ratio β of rotational to gravitational energy being less than 10^{-3}:

$$\beta = \Omega^2 R^3 / 2\, GM_\odot < 10^{-3},$$

where Ω is the rotation rate, R is the radius, G is the gravitational constant, and M is the mass. These cloud models collapsed to form single central concentrations surrounded by barlike mass distributions. Strong gravitational torques developed leading to outward transfer of angular momentum, permitting collapse of a central core. The calculation was not carried beyond one free-fall time, at which the core had about 0.03 $M_\odot$, because of the coarseness of the computational grid. However, there were already great intensifications of density variations in the outer part of the cloud. For example, at a radius of 50 AU the ratio of maximum to minimum density was 40. Such intensifications would manifestly alter the conditions for gravitational instability leading to planet formation.

The confinement of Boss's calculation to low values of the parameter β was based upon earlier modeling that indicated more rapidly rotating clouds would break into several fragments of subsolar mass (Boss 1983). However, these models were started at a much higher density, 2.5×10^{-14} g cm^{-3}, leaving open the question of whether the angular momentum transfer necessary to form a solar-sized star occurs at an earlier stage. Also still to be taken into account are magnetic field effects, which tend to suppress fragmentation and transfer angular momentum.

The systematic mapping of the coupled problems of fragmentation and angular momentum transfer in a collapsing cloud remains to be done, awaiting the application of larger computer facilities. It is still quite unsure how exceptional is a system of low angular momentum density and high first to second body mass ratio like the Sun and Jupiter. What seems certain, however, is that any collapse of a stellar mass cloud enhances asymmetries, leading to transfer of angular momentum outward and fragmentation into a mass ratio varying inversely with the angular momentum density.

Models of instabilities in axisymmetric accretion disks may improve general insight, but understanding of the formation of Jupiter is more likely to be advanced by three-dimensional models of cloud collapse. Collapse processes must have influenced the development of density irregularities leading to protoplanetary clumpings within the outer parts.

Heterogeneities of collapse as suggested by the Boss (1985) models would lead to gaseous protoplanets, which may be far from identical with the contemporary planets Jupiter and Saturn: in particular, there is the problem of forming their rock and ice cores, and of removing most of the hydrogen and inert gases. Indeed, Stevenson (1982*b*) has suggested that "rainout" of these cores in a gaseous protoplanet may be impossible at pressures exceeding about 10 megabars, because the suppression of bonding differences may make the materials entirely miscible. The argument is plausible in principle, but detailed confirmation is yet to come, so the pressure limit may vary appreciably. Ten megabars is the central pressure of an icy body of about 40 $M_{\oplus}$, more than enough to acquire a Jupiter-like hydrogen-helium envelope. See also Hubbard (1984, p. 260) on this topic.

What has been described here is similar to stage 1 of Cameron (1985*a*). But he hypothesizes that the gravitational instabilities of the gas which would occur would be so massive that Stevenson's (1982*b*) stricture would apply, so that ice and rock cores as massive as 15 $M_{\oplus}$ could not precipitate; the limit would be nearer 1 $M_{\oplus}$. Hence subsequent heating by the forming Sun is invoked to strip off the gas so that these solid bodies could merge to form the 15 $M_{\oplus}$ cores inferred from observation. This model requires the cores to be formed in the inner solar system and then to move outward to acquire their gaseous envelopes from cooler material.

The outward motion of the major planets by tidal interaction with the gaseous disk within their orbits is quite plausible, as argued in §13.4. But the Cameron (1985*b*) scenario does not address the marked difference in character of Neptune and Uranus from Jupiter and Saturn; one would expect the outermost planets to move outward first and hence have the greatest chance of acquiring gaseous envelopes. The evidence in §13.2 suggests that a proto-Jupiter had a great influence on the nature of the planetary system beyond its orbit as well as that within. Hence it seems more plausible that the instabilities which occurred in the collapse phase were confined to a limited zone, which corresponded to the eventual Jupiter and Saturn. The size and nature (gas or dust or a mixture thereof) of instabilities forming in a barlike nebula is not at all clear, but there remain a variety of possibilities of how the present core-and-envelope structures of Jupiter and Saturn could be formed.

The essential point of this section is that the inevitable strong asymmetry of the envelope around the forming Sun would have led to the formation of largely gaseous protoplanets which could evolve into planets like Jupiter and Saturn, thus explaining the observational evidence of early formation of these bodies, prior to the loss of appreciable volatiles from the nebula.

13.4. Formation of Other Planets

Cameron (1978) long hypothesized that if one or two planets could be formed by gaseous instabilities as described in §13.3, then so would eight or nine. Subsequently, there was necessarily a rainout of any ice and rock cores in these protoplanets, followed by dissipation of their gaseous envelopes, most obviously by tidal effects from the growing Sun. These hypotheses were unpopular mainly because of the massive nebula which must be lost, but a more fundamental difficulty is that the removal of the volatiles must be extraordinarily efficient indeed to achieve the drastic losses of inert gases in Figure 13.1.

More popular have been planetesimal hypotheses, in which the planets grew out of solid bodies much smaller than themselves. These hypotheses were originated by Chamberlin (1904), and their analytic development was most extensively carried out by Safronov (1969). In these models the phases of orbital perihelia and the nodes of planets and planetesimals are taken as random with respect to each other, so planetary growth is an elaboration of particle-in-box models. The principal difficulties are (1) excessively long accumulation times for the outer planets and (2) the need for mechanisms to suppress growth of all bodies in a zone but the protoplanet itself. For difficulty (1), Safronov (1969) suggested a combination of (*a*) a greatly enhanced density of bodies in the Uranus-Neptune zones due to perturbations by Jupiter and (*b*) extremely low relative velocities (less than 10% of Neptune's escape velocity), because any higher velocity objects would have been ejected from the solar system. For difficulty (2), Safronov (1969) suggested an outstripping of the largest body in a zone compared to the second because of its enhanced capture cross section due to gravitational attraction, and Öpik (1972) suggested tidal disruption of the larger planetesimals by the protoplanet.

Safronov's solution to difficulty (1) depends on the earlier formation of a massive Jupiter, discussed in §13.3. But his solution to difficulty (2) has now been tested by computer experiments (Greenberg et al. 1978; Wetherill 1980), with negative conclusions: many sizable planetesimals form, with zones of influence too small for them to interact sufficiently to reach the end of only four terrestrial planets. Meanwhile, the initial computer experiments on tidal disruption (Mizuno and Boss 1985; Kaula and Beachey 1986) indicate that if collisions are dissipative, tidal disruption will not occur.

The most recent computer experiments on dynamical interactions in a swarm of planetesimals by Wetherill (1985, 1986) indicate, however, that growth can reach as few as four terrestrial planets provided that (1) there is no tidal disruption and (2) the swarm is initially confined to a narrow zone, about 0.4

AU in width, only 25% of the area eventually covered by the terrestrial planets (allowing for secular perturbations of the eccentricities of Mercury and Mars [Brouwer and van Woerkom 1950]). What is therefore needed is a mechanism of confinement. The obvious source of such a mechanism is a preexisting massive Jupiter.

As discussed by Lin and Papaloizou (1979) and Goldreich and Tremaine (1980), a single body in orbit will push a continuous disk of matter away from itself, provided the viscosity of the disk is low enough. This condition is expressed by Lin and Papaloizou as $qRe > 1$, where q is the mass ratio of the body to the primary and Re is the Reynolds number of the disk. Furthermore, this effect will occur, though more weakly, at resonances with the body's orbit: at a 2:1 resonance, the condition is $q^2Re > 1$. Associated with these effects are wave structures, and hence the piling up of matter, as discussed by Lin and Papaloizou in this volume.

The detailed analysis by Goldreich and Tremaine (1980) indicates that if the resonances are "unsaturated"—i.e., if the enhancement of density is not truncated by overlap of matter—then the eccentricity of the outermost body in a system of planets and disks will decrease. In other words, in such systems there is characteristically an outward transfer of angular momentum disproportionate to the outward transfer of energy. The angular momentum, or second Delaunay action variable, depends on $[a(1-e^2)]^{1/2}$, whereas the energy relates directly to the first variable, $a^{1/2}$. Such a property obviously arises from dissipation, since energy is lost to heat, but angular momentum is conserved in the system. Rounding of planet orbits also arises from their ejecting bodies by gravitational perturbations, because the ejected bodies in acquiring positive energy and large eccentricities for escape must carry away angular momentum disproportionately small compared to their energy. The most extreme example of disproportionate transfer is a system dominated by secular resonances, dependent on the commensurabilities of perihelion precession rates (Ward 1981). In a secular resonance there is zero transfer of energy but the eccentricities of the inner bodies are increased. Hence there must be an outward transfer of angular momentum, leading to a decrease in the eccentricities of the outer bodies.

Thus Neptune's orbit probably became so round—differing only 4×10^{-5} (since $f \approx e^2/2$) from a perfect circle, finer than most machine shop specifications—through gravitational interactions with smaller bodies and viscous matter interior to its orbit. Angular momentum transfer by such interactions would have been enhanced as time went on, because the influence radius of Neptune increased relative to its capture radius, whereas the trend from a gaseous nebula to a swarm of planetesimals would have increased the Reynolds number Re as viscosity decreased.

It is difficult to imagine objections to a scenario in which evolution of the nebula subsequent to the formation of Jupiter and Saturn is dominated by a gradual transition from gas to planetesimals to planets, and most of the gases and some of the planetesimals are lost from the nebula. The absence of a planet from the asteroid belt, the reduction of the balance of the system to only eight planets, the smallness of Mars, the seeming overstability of the system, the emptiness of the zones between the planets, and the ratios of successive mean motion of 2.0 or greater all support the hypothesis.

Sweeping by resonances depends on the gradual loss of nebula matter and would tend to break down if this loss ceased, leading to a Safronov-like situation, where there is an increase of eccentricities with the agglomeration of planetesimals into planets (however, Neptune indicates planetary formation must have been virtually complete before loss from the planetary system ended). The models of Wetherill (1985, 1986) suggest such a trend; he imposes a sharp confinement, mimicking resonance effects, as a starting condition and then lets the system go as purely interactive among many planetesimals. In actuality, the terrestrial zone was a transition from Jupiter-dominated in the outer part to self-interacting in the Earth-Venus part.

As Weidenschilling (1975) suggested, there may be a significant transition between Mars and Earth orbits in the Jupiter influence because bodies perturbed within this limit by Jupiter will have enough energy to be ejected from the solar system on subsequent encounters with Jupiter.

Certainly indicative of hierarchical growth of the terrestrial bodies are the marked differences between the Earth and Venus: the argon abundances, rotation rates, and bigness of the Moon suggest a penultimate state of a few large bodies.

A particular problem is the small size and iron composition of Mercury. Probably all researchers concur that formation of a planet close to the Sun was delayed by turbulence, but there remain a variety of viable hypotheses as to how the planet eventually got together. Weidenschilling (1978) proposed that the differing strengths of iron and silicate planetesimals were crucial to determining which material was drawn into the Sun by gas drag and which remained to form the planet. Cameron (1984) hypothesizes that Mercury is a cinder remanent from a nebula zone hot enough to vaporize silicates. Wetherill (1986) suggests Mercury might be a fragment from a great collision involving a planet which had already differentiated an iron core.

The Wetherill (1985, 1986) models lead to appreciable heating upon the formation of terrestrial planets, because large impacters bury their energy (Barrell 1918; Safronov 1969; Kaula 1979). The largest such event may have occurred late in the growth of the Earth, leading to the loss of primordial inert gases (Cameron 1984) and the formation of a severely devolatilized, iron-poor Moon (Cameron 1985*b*; Stevenson 1986). The need for such cataclysmic events has not been invoked for the major planets, and the modest eccentricities of planetary orbits in the outer solar system suggest that the major planets formed from much smaller bodies. However, the much greater ratio of influence radii to collision radii makes it likely that the small eccentricities arose from the mechanisms described above: angular momentum gains disproportionately large compared to energy gains by interaction with viscous matter interior to their orbits, and angular momentum losses disproportionately small by the ejection of bodies from the system.

13.5. Discussion and Conclusions

This review has slighted several problem areas which have been investigated appreciably in recent years: hydromagnetic effects; gas-dust interaction and other small scale physics affecting initial agglomeration; the consequent instabilities and thermal convection; the effects of a gaseous nebula on the growth of a protoplanet; the contraction of a gaseous protoplanet; solar effects on the nebula; compositional variations within the classes defined in Table 13.1; and the formation of satellite systems and planetary atmospheres. This slighting is mainly because these aspects do not appear critical to explaining the main facts set forth in Table 13.1. Thus accumulation of terrestrial bodies in the presence of a dense gaseous nebula, emphasized by Hayashi, Nakazawa, and Nakagawa (1985), appears to make little difference after planetesimals are tens of kilometers in size (Wetherill 1985, 1986) and fails to explain the inert gas depletions persuasively. Gas-dust interactions in the nebula are important in explaining various meteorite properties (Morfill 1983; Weidenschilling 1984) but give little indication that planetesimal growth could be limited to the required eight planet cores. Hence collisions of massive protoplanets were important to planet formation. Differing hypotheses of loss of the gaseous nebula have significant implications for the nature of the early Sun but appear to make little difference in the main nebular effect on planet formation: a gradually decreasing source of gravitational attraction.

The most serious neglect is discussion of hydromagnetic effects, since in the early stages of collapse of the protosolar cloud they may have greatly affected fragmentation and hence the development of a 1000:1 ratio of largest to second largest mass. In this area the work of Mouschovias on ambipolar diffusion (Mouschovias et al. 1985 and citations therein) seems most important.

But the main message is that any meaningful solution to the formation of the planets must be hierarchical, formation of the smaller bodies being necessarily dependent on the three most massive components of the system: the proto-Sun, proto-Jupiter, and the gaseous nebula. Furthermore, by the stage that there was sufficient mass concentration to initiate collapse of a proto-Jupiter, hydromagnetic effects were minor so that the physics of planet formation is primarily mechanical and secondarily thermal. The main need is three-dimensional computer modeling of hydrodynamic collapse, but dynamical insight will continue to be necessary because of the probable importance of resonances and nebular loss to evolution after proto-Jupiter is more than a few Earth masses in size.

Acknowledgments

This work is partly supported by NASA grant NGL 05-007-002.

References

Barrell, J. 1918, *The Evolution of the Earth and Its Inhabitants* (New Haven: Yale University Press).

Boss, A. P. 1983, *Icarus,* **55**, 181.

——. 1985, *Icarus,* **61**, 3.

Brouwer, D., and van Woerkom, A. J. A. 1950, *Astr. Papers Am. Ephem. Naut. Alm.,* **13(2)**, 79.

Cameron, A. G. W. 1978, *Moon and Planets,* **18**, 5.

——. 1983, *Icarus,* **56**, 145.

——. 1984, *The Partial Volatilization of Mercury,* Harvard Center for Astrophysics preprint.

——. 1985*a,* in *Protostars and Planets II,* ed. D. C. Black and M. S. Matthews, p. 1073 (Tucson: University of Arizona Press).

——. 1985*b, Icarus,* **62**, 319.

Cassen, P., Shu, F. H., and Terebey, S. 1985, in *Protostars and Planets II,* ed. D. C. Black and M. S. Matthews, p. 448 (Tucson: University of Arizona Press).

Cassen, P., and Summers, A. 1983, *Icarus,* **53**, 26.

Chamberlin, T. C. 1904, *Carnegie Inst. Washington Yearbook,* **3**, 195.

Clayton, R. N., and Mayeda, T. K. 1977, *Geophys. Res. Lett.,* **4**, 295.

Donahue, T. M., and Pollack, J. B. 1983, in *Venus,* ed. D. M. Hunten et al., p. 1003 (Tucson: University of Arizona Press).

Goldreich, P., and Tremaine, S. 1980, *Ap. J.,* **241**, 425.

Greenberg, R.; Wacker, J. F.; Hartmann, W. K.; and Chapman, C. R. 1978, *Icarus,* **35**, 1.

Harris, A. W., and Ward, W. R. 1982, *Ann. Rev. Earth Planet. Sci.,* **10**, 61.

Hayashi, C., Nakazawa, K., and Nakagawa, Y. 1985, in *Protostars and Planets II,* ed. D. C. Black and M. S. Matthews, p. 1100 (Tucson: University of Arizona Press).

Hubbard, W. R. 1984, *Planetary Interiors* (Princeton: Van Nostrand Reinhold).

Kaula, W. M. 1968, *An Introduction to Planetary Physics: The Terrestrial Planets* (New York: Wiley).

——. 1979, *J. Geophys. Res.,* **84**, 999.

——. 1986, in *The Solar System,* ed. M. G. Kivelson (Englewood Cliffs, NJ: Prentice-Hall), in press.

Kaula, W. M., and Beachey, A. E. 1986, in *Origin of the Moon,* ed. W. K. Hartmann, R. J. Phillips, and G. J. Taylor (Houston: Lunar and Planetary Institute), in press.

Lee, T., Mayeda, T. K., and Clayton, R. N. 1980, *Geophys. Res. Lett.,* **7**, 493.

Lewis, J. S. 1974, *Science,* **186**, 440.

Lin, D. N. C., and Papaloizou, J. 1979, *M.N.R.A.S.,* **186**, 799.

——. 1985, in *Protostars and Planets II,* ed. D. C. Black and M. S. Matthews, p. 981 (Tucson: University of Arizona Press).

Lynden-Bell, D., and Pringle, J. E. 1974, *M.N.R.A.S.,* **168**, 603.

Mizuno, H., and Boss, A. P. 1985, *Icarus,* **63**, 109.

Morfill, G. E. 1983, *Icarus,* **53**, 41.

Mouschovias, T. C., et al. 1985, *Ap. J.,* **291**, 772.

Öpik, E. 1972, *Irish Astr. J.,* **10**, 190.

Savronov, V. S. 1969, *Evolution of the Protoplanetary Cloud and Formation of the Terrestrial Planets* (Moscow: Nauka; Trans. NASA TT-F-677).

Stevenson, D. J. 1982*a*, *Ann. Rev. Earth Planet. Sci.,* **10**, 257.

——. 1982*b*, *Planet. Space Sci.,* **30**, 755.

——. 1985, *Icarus,* **62**, 16.

——. 1986, in *Origin of the Moon,* ed. W. K. Hartmann, R. J. Phillips, and G. J. Taylor (Houston: Lunar and Planetary Institute), in press.

Stevenson, D. J., Harris, A. W., and Lunine, J. I. 1985, in *Satellites,* ed. J. Burns and M. S. Matthews (Tucson: University of Arizona Press), in press.

Thiemens, M. H., Heidenreich, J. E., III, and Lundgerg, L. 1983, *Lunar Planet. Sci.,* **14**, 785.

Wänke, H., Dreibus, G., and Jagowitz, E. 1984, in *Archean Geochemistry,* ed. A. Kroner et al., p. 1 (Berlin: Springer-Verlag).

Ward, W. R. 1981, *Icarus,* **47**, 234.

Wasson, J. T. 1969, *Nature,* **223**, 163.

Weidenschilling, S. J. 1975, *Astr. J.,* **80**, 145.

——. 1978, *Icarus,* **35**, 99.

——. 1984, *Icarus,* **60**, 553.

Wetherill, G. W. 1980, *Ann. Rev. Astr. Ap.,* **18**, 77.

——. 1985, *Science,* **228**, 877.

——. 1986, in *Origin of the Moon,* ed. W. K. Hartmann, R. J. Phillips, and G. J. Taylor (Houston: Lunar and Planetary Institute), in press.

14

Comet Formation

Jack G. Hills

Abstract

The zone of comet formation and the mechanism which brought the comets into the Oort Cloud are still not clear. The classical Oort Comet Cloud which extends outward of 20,000 AU from the Sun is just the halo of the comet zone where perturbations by passing stars are frequent enough to allow a steady stream of comets to enter the planetary system. Most long period comets are assumed to lie inside the inner radius of the Oort Halo. The zone of comet formation is likely to have been within the Jeans radius of the proto-Sun, which was about 5000 AU. Observations of highly volatile molecules in comets indicate that the zone of comet formation was at least as far from the Sun as the present orbits of Uranus and Neptune and probably very much farther. While some comets probably formed in the Solar Nebula beyond the orbit of Pluto, it is very difficult for these comets to have been thrown into the Oort Halo by the perturbations of the known planets. A particularly promising zone for comet formation is the outer parts of the collapsing proto-Sun where radiation pressure from the Sun and surrounding coforming protostars may have driven together the dust grains in the protocomet dust clumps.

14.1. Introduction

Comets are the least physically and chemically altered fossils of the formation of the solar system. Comets may have formed well outside the planetary system and perhaps even thousands of astronomical units away in the outermost collapsing layers of the proto-Sun.

A study of the chemical constituents of comets will provide valuable insights into stellar nuclear synthesis, which was the subject of Professor Cameron's first major contribution to astrophysics. Comets may contain nearly unaltered dust grains from a variety of sources including Mira variables, novae, supernovae, planetary nebulae, Wolf-Rayet stars, and other mass-losing stars. These primitive nuclear grains have mantles of more volatile molecules that froze out in the cool environment of the Sun's parent molecular cloud. When comet grains become available for the type of detailed laboratory analysis discussed by Professor Wasserburg at this symposium, they will provide valuable clues about the sites of stellar nuclear synthesis and mass loss. Comet samples may become available relatively soon as, at least some, Earth-crossing asteroids are comets which have lost most of their volatiles. Before this time occurs, it would be desirable to know how and where the grains assembled into comets in order to assess the possibility that some chemical constituents of comets may be overabundant relative to a standard astrophysical mixture due to the nature of the comet formation process. There is also a possibility that comets have locked up a large fraction of the "metals" in the proto-Sun so that the "metals" in the present Sun are underabundant relative to an astrophysical mixture.

14.2. Where Comets Come from Today

The long period comets which enter the inner planetary system today originate in a spherical shell around the Sun having an inner semimajor axis of about 2×10^4 AU and an outer semimajor axis of 1×10^5 AU (Oort 1950). This Oort Cloud is well outside the Jeans radius of the proto-Sun. The low mass density within this spherical shell makes it unlikely that comets originated here. They likely formed somewhere within the Jeans radius of the proto-Sun and subsequently were ejected into these distant orbits.

Larson and Starrfield (1971) found that within a molecular cloud the Jeans radius of a protostar of solar mass is given by

$$R_{\max} = 0.41\, GM_{\odot}/\mathcal{R}T, \tag{1}$$

where $\mathcal{R} = 3.36 \times 10^7$ ergs g^{-1} K is the gas constant. This gives $R_{\max} = 5000$ AU for a temperature $T = 20$ K. The proto-Sun's Jeans radius is well inside the inner radius of the classical Oort Comet Cloud.

We expect any mechanism which perturbed the comets into the Oort Cloud to have smoothly filled the entire range of orbits from the point of origin of the comets within the solar Jeans radius to the outer parts of the Oort Cloud. What causes the outlandish hole in the semimajor axis distribution inside $a = 2 \times 10^4$ AU? This hole results from the infrequency of the close stellar encounters needed to send new comets with semimajor axes less than 2×10^4 AU into the planetary system (Hills 1981). We shall elaborate on this point.

To become conspicuous, a comet has to pass close enough to the Sun to suffer massive sublimation. This occurs if the comet passes within about 2 AU of the Sun, which requires it to cross the orbits of Saturn and Jupiter. On crossing the orbit of Jupiter, even the most tightly bound comets in the Oort Halo, which have semimajor axes of about 2×10^4 AU, suffer an average perturbation in orbital binding energy equal to 15 times their initial orbital binding energy (Hills 1983*b*). About half these comets are immediately ejected into hyperbolic orbits and half are thrown into orbits with much smaller semimajor axes and orbital periods. The comets thrown into the short period orbits are at risk of being ejected into hyperbolic orbits at each perihelion passage. Within a time which is short compared to the original orbital period in the Oort Cloud, most of these short period comets are ejected into hyperbolic orbits.

Without the perturbations due to passing stars, we would know little about comets. (In the total absence of stellar perturbations, the tidal field of the

Galaxy would perturb comets in very distant bound orbits enough that some would be deflected into the planetary system.) Comets in orbits that pass through the planetary system would quickly be ejected into hyperbolic orbits while those comets not in planet-crossing orbits would never enter the planetary system. The greater the semimajor axis of a comet, the bigger the target its orbit presents to passing stars. Also, the larger the semimajor axis of an orbit, the longer its orbital period and the more time available to accumulate perturbations during each orbital period. Beyond a certain critical semimajor axis, the average accumulated orbital perturbation suffered in one revolution is sufficient to refill the loss cone constituting those comets in orbits which pass through the planetary system. In this case, the fraction of the comets of a given semimajor axis in planet-crossing orbits is independent of the rate at which these comets are ejected into hyperbolic orbits by the major planets. This situation occurs for comets with semimajor axes greater than about 2×10^4 AU. The distant comets in this steady-state trickle constitute the Oort or Steady-State Comet Cloud (Hills 1981).

Occasionally, stars may pass sufficiently close to the Sun to perturb even those comets with semimajor axes less than 2×10^4 AU enough that some fraction are forced into orbits that take them into the planetary system. This produces an intense shower of comets which lasts about one orbital period before the major planets eject them into hyperbolic orbits (Hills 1981). After the comet shower, no comets are seen from this inner zone until another star makes a close passage to the Sun. In most comet-formation scenarios, more comets have entered the planetary system from the inner comet cloud where the comets have semimajor axes less than 2×10^4 AU than from the Steady-State Oort Comet Halo.

The tidal field of the Galaxy dissociates comets with semimajor axes greater than about 1×10^5 AU. Torbett and Smoluchowski (1984) have recently studied the effect of the tidal field of the Galaxy on distant stellar companions as a function of the inclination of their orbits with respect to the galactic plane. They find that the maximum allowed semimajor axis is greatest in the galactic plane and is at a minimum for orbits perpendicular to the galactic plane.

A constant fraction, determined by statistical equilibrium, of the comets in the Steady-State Oort Cloud are in orbits that penetrate the planetary system. Combining this known fraction with the observed rate at which comets enter the planetary system allows us to determine the number of comets in the Oort Steady-State Cloud. This is about 1×10^{11} (Oort 1950; Hills 1981, 1986). Because of observational difficulties with low luminosity comets, this applies to comets with radii greater than about 1 km. There may be many more comets with smaller radii (Shoemaker and Wolfe 1982).

A typical comet with a radius of 1 km and a density of 1 g cm^{-3} has a mass of 4×10^{15} g. If we assume this to be the average mass of comets, we find the integrated mass of all comets in the classical Oort Cloud to be about 4×10^{26}

g, or nearly an Earth mass, $M_\oplus$. However, Shoemaker (1982) has noted that the mass spectrum of comets is similar to that of the asteroids. In this situation most of the mass would be in the first few largest objects. Such a mass spectrum would greatly increase the inferred mass of the comet cloud. Shoemaker (1982) estimates the total mass of comets in the classical Oort Cloud to be at least 100 $M_\oplus$, or more than two orders of magnitude greater than simpleminded estimates based on equal-mass comets.

Most proposed mechanisms for forming comets within the proto-Sun result in the accumulation of one to two orders of magnitude more comets in the inner comet cloud than in the Steady-State Oort Cloud (cf. Hills 1981). This corresponds to 10^{12-13} comets within the inner cloud. The mass of the inner cloud would then be about 10–100 $M_\oplus$ in the conservative (simpleminded) model and 10^{3-4} $M_\oplus$ using the Shoemaker mass spectrum.

Most proposed mechanisms for forming comets within the Jeans radius of the proto-Sun result in at least an order of magnitude more comets being ejected into hyperbolic orbits than make their way to the Oort Halo. This suggests that more than 10^{12} comets have been ejected from the solar system into interstellar space. Is this consistent with the fact that no interstellar comets have ever been observed? If the space density of stars in the solar neighborhood is $n_* = 0.1$ pc^{-3} and each star ejects at least 10^{12} comets into interstellar space, then the space density of interstellar comets is $n_c = 10^{11}$ pc^{-3}. The average pre-encounter velocity of an interstellar comet with respect to the Sun should be similar to that of stars in the solar neighborhood, or about $V = 30$ km s^{-1}. The average number of interstellar comets which pass within impact parameter p of the Sun per unit time is given by

$$\frac{dN_c}{dt} = \langle n_c V \rangle \pi p^2. \tag{2}$$

From this equation we find that the average time between encounters with interstellar comets having impact parameters of $p = 2$ AU or less is about 1100 yr. This time interval is consistent with no interstellar comets having been observed to date.

It is obvious that the current dynamical structure of the Oort Comet Cloud was only weakly determined by the comet-formation mechanism. A variety of formation and ejection mechanisms would produce the Oort Steady-State Comet Cloud. We are left with the problem of determining not only how the comets formed but also where they formed. The only major constraints which can be placed on proposed models is the fact that comets contain some very highly volatile molecules and that an ejection mechanism must exist which is capable of transporting the comets from their points of origin to the Oort Cloud.

14.3. Formation Mechanisms and Zones

In this section we shall consider in turn each of the suggested zones of comet formation in order of distance from the Sun.

14.3.1 Jupiter-Saturn Zone in Solar Nebula

The suggestion that comets may have formed in the vicinity of the major planets goes back at least to Kuiper (1951). There are two major problems associated with forming the observed comets in this part of the Solar Nebula: too few comets make it to the Oort Halo, and the observed comets suffer noticeable sublimation in this zone.

Because of the strong gravitational perturbations of Jupiter, only a small fraction of any comets which form in this zone can make it to the Oort Comet Halo. Fernandez (1982) estimates that fewer than 10^{-3} of the comets formed in the Jupiter-Saturn zone reach the Oort Cloud. Most such comets are ejected into hyperbolic orbits. To get 1 $M_\oplus$ in comets into the Oort Halo requires that at least 10^3 $M_\oplus$ in comets be hurled into hyperbolic orbits. This is several times the combined mass of Jupiter and Saturn. Their ejection of such a large mass into hyperbolic orbits would cause large-scale shrinking of the orbits of these planets.

Many new comets such as Kohoutek develop distinct coma even before they cross the orbit of Jupiter, which indicates considerable sublimation. Comet Schwassmann-Wachmann shows flaring episodes even though it is located well beyond Jupiter. The observed sublimation at these large distances is not consistent with comet formation in this zone. The volatile molecules would have left the dust grains long before they accumulated into comets if the temperature in the Solar Nebula was at least as high as the present-day temperature in this zone, which was certainly the case.

The flaring episodes observed in Comet Schwassmann-Wachmann are expected if it is a short period comet which was captured sufficiently recently that its interior has not yet come to the equilibrium temperature expected at its present distance from the Sun. Schwassmann-Wachmann is unusually massive, so it is not surprising that its interior temperature is still below its final equilibrium value. The interior contains a mixture of ices with different sublimation temperatures. Heat diffuses more easily in a solid than do molecules. The progressive heating of the comet causes an increase in the vapor pressure in the interior as the more volatile molecules undergo sublimation. Unless the comet is sufficiently fractured to allow the easy escape of these gases, the vapor pressure builds up until it is able to overcome the tensile strength of the bulk material in the comet. Eventually a sector of the comet will burst like popcorn releasing the trapped gases which will also carry off small chunks of solid debris. These gas-release events may be responsible for the flaring episodes.

We may conclude from the activity of observed comets when they pass through this zone and the inefficiency at getting them to the Oort Cloud from this zone that it is very unlikely that any appreciable fraction of the observed comets formed in the Jupiter-Saturn zone.

14.3.2 Uranus-Pluto Zone in Solar Nebula

Comet formation in this zone is a serious possibility. A high fraction of any comets which formed in this zone reached the Oort Halo, and there is less problem with volatiles than in the Jupiter-Saturn zone.

Fernandez (1982) finds that about 10% of the comets that form in this zone are ejected into long period orbits while the remainder are ejected into hyperbolic orbits. From his data I estimate that about 20–40% of the long period comets reach the Oort Steady-State Comet Halo (semimajor axes $a > 2 \times 10^4$ AU), while the remainder become part of the inner comet cloud. If the total mass of the comets in the Oort Halo is 1 $M_\oplus$, then the total mass of the comets ejected from the Uranus-Pluto zone is in the range 25–50 $M_\oplus$. This mass is of the same order as the combined mass of Uranus and Neptune. The orbits of Uranus and Neptune would have shrunk greatly if the comets formed in this zone. But, as Professor Kaula noted in his lecture at this symposium, Neptune has an unusually small orbital eccentricity. The orbital dissipation produced by the ejection of the comets from this zone by Uranus and Neptune would have tended to circularize as well as shrink the orbits of these planets. The massive comet cloud suggested by Shoemaker is not consistent with comet formation in this zone. If the mass of the Oort Halo were 100 $M_\oplus$, the mass of the comets formed in the Uranus-Neptune zone would have to be about 2500–5000 $M_\oplus$, or about two orders of magnitude greater than the combined mass of Uranus and Neptune.

Volatiles are less of a problem in this zone than in the Jupiter-Saturn zone. Current blackbody temperatures in this zone are about T = 40–60 K. The peak temperature in the Solar Nebula at Neptune, as shown in the plot of Professor Shu at this symposium, was only about 60 K. However, some recent UV observations of comets by Feldman (1983) indicate that the CO abundance varies considerably among different comets while most other molecules tend to have about the same relative abundances in various comets. Yamamoto (1985) interprets this as being the result of the sublimation of CO becoming important at a radius which lies within the comet-formation zone. The sublimation of CO occurs at about 25 K in the Solar Nebula, which would place the comet-formation region well beyond Pluto if this interpretation of the CO abundances is correct.

It may be difficult to form comets in a reasonable time in this zone because of the low dust densities. Simpleminded models of dust accretion in the Solar Nebula suggest formation times of 10^7–10^8 yr (Hills 1973). These results are independent of the velocity dispersion of the dust in the nebula. However, if the dust layer becomes dense enough for gravitational instabilities to occur (such as in the models of Goldreich and Ward 1974), then the formation time scales may not be a problem.

14.3.3 Formation in the Solar Nebula beyond Pluto

There are fewer problems with volatiles in this zone. Even CO and NH_3

froze out within this region of the Solar Nebula. Cameron (1978) and Biermann and Michel (1978) have suggested comet formation in this zone.

Support for comet formation in this region comes from the *IRAS* discovery of infrared emission from extended dust shells around a number of main-sequence stars. Allowing for selection effects, Smith and Terrile (1984) estimate that 10% of all main-sequence stars have remnant dust disks. In the case of β Pictoris they find a flattened dust layer extending out to at least 400 AU. The thickness of this dust layer is about 20% of its radius. They estimate its mass as 0.01–200 $M_\oplus$ depending on the assumed mass spectrum of the dust clumps. The example of β Pictoris shows that extended accretion disks may form around main-sequence stars. Such dust disks provide an opportunity for comet formation in this zone. The large thickness of the disk around β Pictoris makes it unlikely that any gas remains in the disk because the gas would quickly dissipate the kinetic energy of the dust and cause it to settle much closer to the central plane of the nebula. The current dust population may be leftover dust which failed to accumulate into comets, or it may be debris from comet collisions.

The low dust density in this zone makes comet formation difficult. Accretion is too slow, and gravitational instabilities may require an unrealistically massive accretion disk. However, radiation pressure may have allowed the formation of comets in this zone (Hills 1982). Absorption of radiation in a protocomet dust clump produces a net inward radiation pressure which causes the dust to drift together.

The main difficulty with forming the observed comets in this zone is finding a mechanism for ejecting them into the Oort Cloud. Any comets that formed in the Solar Nebula would be in nearly circular orbits, so they would not enter the known planetary system. There are no planets to eject them into the Oort Cloud unless a hypothetical Planet X exists beyond Pluto. Such an outer planet could directly eject comets into the Oort Cloud or pass them on to the inner planets, which would in turn eject them into the Oort Cloud.

Professor Cameron (1978) has proposed another mechanism for the ejection of the comets into the Oort Cloud. This invokes sudden mass loss in the proto-Sun and Solar Nebula. If a comet is initially in a circular orbit, its orbit will become hyperbolic if half or more of the mass interior to the orbit is lost in less than one orbital period of the comet (cf. Hills 1980, 1983*a*). However, this mechanism is not consistent with the existence of the extended dust disks found by *IRAS* around many main-sequence stars. (Of course, we could assume that the 90% of all main-sequence stars not having observed dust disks did have them but they were dispersed when half or more of the mass was lost within a dynamical time scale.)

Another difficulty with the sudden-mass-ejection mechanism is that it requires comets to form in an environment in which the disk dissipates within a dynamical time. It is difficult to form a coherent accretion disk in less than one period of revolution even if energy dissipation in the disk occurs on a time scale which is short compared to the orbital period. The dust cannot settle differentially with respect to the gas to form a thin disk in less than the period of revolution of the disk. It would be very difficult to form the comets by accretion or gravitational instabilities in the disk if it dissipates in less than a dynamical time scale. However, they may still form by radiation pressure (Hills 1982).

14.3.4 Formation in the Outer Collapsing Layers of the Proto-Sun

As noted in the paper by Shu at this symposium, the proto-Sun develops a luminous core on which a steady stream of material rains down for several dynamical (free-fall) time scales. Several mechanisms have been proposed to form comets in this infalling material. The gas and dust are quite cool in the infalling material, so there is no problem with volatiles.

One way of forming comets in this zone has recently been proposed by Bailey (1985). He notes that protostars eventually develop vigorous winds. This outward gas flow opposes the gas and dust raining down on the protosolar core. The dust concentrates in the interface between the two gas streams. He assumes that the comets form by accretion in this concentrated dust layer. This is not a particularly robust mechanism. However, once the dust has been concentrated by the wind, radiation pressure may act to concentrate the dust into comets.

The proposed formation of comets by the forcing together of dust clumps by radiation pressure does have the virtue of being probably the best studied comet-formation model (Hills 1982; Hills and Sandford 1983*a*, *b*). At the time when the Sun was in the accretion phase, the luminosity of its core was about 30 times its present luminosity (Winkler and Newman 1980). In addition, the Sun may have been surrounded by a very luminous cluster of coforming stars. Embryonic clusters such as the Trapezium cluster and the σ Persei Cluster have densities 10^{3-4} times greater than the stellar density in the solar neighborhood (Blaauw 1964; Hills 1980), and the average luminosity per star in such a young cluster with its massive stars is considerably greater than that for stars in the solar neighborhood. Even higher stellar densities may have been attained earlier in the formation of a stellar association (Hills 1980). Such an environment is very conducive to the formation of comets by the driving together of protocomet grain clumps by the net inward radiation pressure induced by radiation from the Sun and surrounding protostars. A grain near the outer surface of a protocomet dust clump feels a net inward radiation pressure because the absorption of radiation by the dust in the clump will cause a greater radiation flux into the dust clump than out of it.

Hills (1982) finds that the time needed to form a comet in the collapsing proto-Sun at 1000 AU from its center is about an order of magnitude less than the free-fall time. The ratio of the formation time to collapse time decreases further if the comets formed closer to the Sun.

Hills and Sandford (1983*b*) studied the formation of comets in an anisotropic radiation field using a two-phase (dust and gas), two-dimensional, radiation hydrodynamic code. They find that comets can form in a highly anisotropic field.

A protocomet dust clump in such an anisotropic field shows the greatest initial compression in the direction of the greatest radiation flux. However, this compression causes an increase in the optical depth in the direction perpendicular to the initial compression, which eventually results in the radiation pressure in this direction becoming comparable to that in the direction of the largest radiation flux. Eventually, all parts of the dust clump are driven to its center by the radiation pressure.

There are two potential problems with this model. The degradation of the radiation from the Sun and surrounding protostars into the far-infrared by dust absorption and re-emission makes it difficult for small grains (smaller than the wavelength of light) to be pushed about by the radiation field. This is less of a problem near the Sun where the radiation field is very strong or in the outermost parts of the collapsing proto-Sun where the dust density is low, the absorption is small, and the radiation from surrounding protostars may be important.

Turbulence may also prevent comet formation if the turbulent velocity exceeds the draft velocity of the dust. However, for fully developed turbulence the turbulent velocity only increases as the third root of the radius of the protocomet while the radiation-induced draft velocity is directly proportional to its radius, so turbulence is never a problem for large comets. Turbulence may explain the lack of very small comets, although a cluster of small comets may form in a large protocomet dust clump which has large density inhomogeneities within it. Radiation pressure would cause this large dust clump to shrink, which would increase its average density. The overall shrinking of the cloud would cause even greater density enhancements within the subclumps. A highly inhomogeneous protocomet dust cloud of sufficient size to suffer significant shrinking in the radiation field could fragment into a number of small comets. The small cluster of comets produced by this mechanism would easily be broken up by the tidal field of the Sun before the clump enters the planetary system. Such comet clumps could not exist among the comets in the Steady-State Oort Halo since these comets presumably have already passed through the planetary system in order to have been ejected by planetary perturbations into these distant orbits. However, comet clusters may exist in the inner comet cloud.

There are several attractive features of models in which comets form in the outer proto-Sun. There is no problem with volatiles. The temperature in this zone is the coolest of any part of the proto-Sun. The large initial semimajor axes of the comets in this zone result in their being relatively close to the Oort Halo in terms of orbital binding energy, so relatively mild orbital perturbations are enough to bring them into the Oort Halo. The high eccentricities of the orbits of the comets which form in the collapsing proto-Sun allow a high percentage of these comets to pass close enough to the Sun for planetary perturbations to eject them into the Oort Halo.

The high orbital eccentricities of the comets which form in the outer, collapsing layers of the proto-Sun may also allow a high percentage to make their way to the Oort Halo if the proto-Sun suffers any significant mass loss while the comets are collapsing toward the proto-Sun. This mechanism is basically a modification of the ejection model that Professor Cameron (1978) proposed for circular orbits. Let a_0 be the semimajor axis of a comet prior to the proto-Sun mass loss. If the mass loss occurs instantaneously while the comet is at a distance r from the Sun, its orbit will become parabolic if the fractional mass loss suffered by the proto-Sun exceeds

$$\Delta M/M_\odot > \frac{1}{2}(r/a_0) \tag{3}$$

(Hills 1980, 1983*a*). If a comet has a circular orbit so that $r = a_0$, then the requirement for dissociation is $\Delta M/M_\odot \geq 0.5$; however, if the orbit is highly eccentric and $r \ll a_0$, then much less mass is needed for dissociation (or causing a significant increase in the semimajor axis of the orbit). If the mass loss starts abruptly and persists for less than an orbital period, then some comets will find themselves near perihelion when the mass loss commences and will suffer severe inflation of their orbits even if $\Delta M/M_\odot$ is relatively small.

If the frequency distribution of the orbital eccentricities of the comets which formed in the outer proto-Sun followed statistical equilibrium and the comets which reached the Oort Halo were ejected there by planetary perturbations, then the total mass of the comets which formed in the collapsing proto-Sun with semimajor axes greater than 500 AU was about 10^2 times greater than the total comet mass in the Oort Halo (Hills 1981). This corresponds to about 10^{13} comets with a minimum integrated mass of 10^2 $M_\oplus$. However, if the initial comet orbits were more nearly radial than the distribution given by statistical equilibrium, then a higher fraction of the comets in the inner cloud would have entered the planetary system and so be ejected into the Oort Halo. In this case, fewer comets would be required in the inner cloud to populate the observed outer, steady-state, comet (Oort) halo. This is also true if comets diffused into the Oort Halo from the inner cloud as a result of impulsive proto-Sun mass loss when the comets were near perihelion.

If there are 10^{13} comets in the inner cloud with semimajor axes of 3×10^3 AU and less, then a collision will occur among these comets about every 100 yr. The debris produced would include dust, which through subsequent collisions would settle into progressively smaller (lower energy) orbits until it is constrained by angular momentum. The very large dust disk around β Pictoris may have been produced by such a mechanism rather than being the remnant of the original accretion disk around β Pictoris.

14.3.5 Formation in a Satellite Accretion Disk

Cameron (1973) suggested that comets may form in a companion accretion disk. He suggested that it had a mass of 4×10^{32} g $= 0.2$ $M_\odot$, which is orders of magnitude greater than the total mass of the comets in the Oort Cloud. The assumption is made that the companion nebula broke apart without producing a single massive companion. Today we might propose that the companion disk became Nemesis, the hypothetical solar companion. This comet-formation mechanism would allow most comets to form directly within the Oort Cloud Halo. In this case, there would be few comets with semimajor axes less than the initial semimajor axis of the companion nebula. There may be no massive inner comet cloud in this model.

14.3.6 Formation in Interstellar Molecular Clouds

It has been suggested that comets may form in molecular clouds and then be captured by the Sun (cf. McCrea 1975). I find this mechanism to be highly unlikely. The low densities in molecular clouds make it much more difficult to form comets here than in the proto-Sun. It would be very difficult for the Sun to have captured any significant number of interstellar comets. Capture would be virtually impossible unless the velocity of the Sun with respect to the interstellar comets is less than the orbital velocity of comets in the Oort Cloud, or about 0.1 km s^{-1}. But the velocity of the Sun with respect to a typical molecular cloud is about 20 km s^{-1}. The chance that the relative velocity could be this small is virtually nil. This is especially true in the light of the fact that molecular clouds occur in spiral arms and stars have a streaming velocity with respect to the gas in the arms (cf. Rohlfs 1977) which effectively rules out a relative velocity as small as 0.1 km s^{-1}.

14.3.7 Summary

In my opinion the most likely comet-formation zones are the Uranus-Pluto region, the Solar Nebula beyond Pluto, and the collapsing outer layers of the proto-Sun.

The basic question about the Uranus-Neptune zone is whether it was cool enough to have allowed the condensation of the more volatile constituents of comets. Also, even if the mass of the Oort Halo is only 1 $M_{\oplus}$, the mass of the comets in this formation zone had to be one or two orders of magnitude greater than the combined mass of Uranus and Neptune. This would cause considerable shrinking of their orbits. A massive Oort Cloud such as that proposed by Shoemaker is not consistent with this model. In conclusion, this model could be eliminated if sufficiently volatile constituents were found in comets or if the Oort Halo were shown to be sufficiently massive.

I feel that cometlike objects almost certainly formed in the Solar Nebula beyond Pluto. The fossil dust ring around β Pictoris shows that massive, extended stellar nebulae do form. However, I question whether the observed long period solar comets formed in such a region. If they formed in a dynamically relaxed Solar Nebula, the comets in this zone would have nearly circular orbits which would not take them into the planetary zone. Unless Planet X, or Nemesis, exists to perturb their orbits, it would be very difficult for the comets in this zone to make their way to the Oort Halo. However, comets may still exist in this zone. Tracking data from the *Pioneer* and *Voyager* space probes may eventually allow the presence of comets in this zone to be sensed from their integrated gravitational attraction on the space vehicles.

The collapsing outer layers of the proto-Sun are an attractive place to form comets. There is no problem with volatiles, and the nearly radial orbits of these comets would allow a high percentage of them to pass through the planetary system, which would allow some to be ejected into the Oort Halo. Ejection into the Oort Halo would also occur if solar mass loss occurred while these comets were near perihelion.

Comets now in the inner comet cloud would be directly observable at Pluto (Hills 1986). While the comets which cross the orbit of the Earth have been strongly perturbed by Jupiter so that we do not see such comets unless they have recently been perturbed into Jupiter-crossing orbits by passing stars, the orbits of comets with perihelia on the order of 40 AU or greater have been little perturbed by the known planets. Except for perturbations due to the occasional close encounters of passing stars, the orbits of these comets have not been altered since the formation of the solar system. A comet detector at Pluto could directly measure the comet flux in the inner comet cloud.

14.4. Total Mass of All Solar Comets

The total mass of the solar comet cloud could be an appreciable fraction of the mass of the Sun especially if the comets formed by the coagulation of dust grains by radiation pressure, which would allow the formation of comets within the low density, collapsing layers of the proto-Sun. These comets could have absorbed an appreciable fraction of the metals in the proto-Sun.

We noted earlier that Shoemaker has postulated that the mass of the classical Oort Comet Halo could be up to 100 $M_\oplus$ (it must be at least 1 $M_\oplus$). We also noted that the mass of the inner cloud beyond 500 AU may be up to 10^2 times as massive as the Oort Halo, which suggests an inner cloud as massive as 10^4 $M_\oplus$. The mass of the comets in the zone from Pluto to 500 AU is difficult to estimate. Comets in this zone are sufficiently outside the planetary system that they do not perturb the planetary orbit significantly, and they are sufficiently close to the Sun that no passing star is likely to have perturbed them (Hills 1984, 1986). Any comets formed in this zone are likely to have remained in their original orbits. However, the total mass of these comets may eventually be found from their gravitational attraction on space probes.

It is possible to place some limits on the integrated mass of comets in these various zones from the lack of clear-cut hyperbolic comets. If there were a sufficiently massive cloud of comets, comets that passed through it from the outer parts of the Oort Halo would appear to be in hyperbolic orbits since their observed velocity would be too high for them to be gravitationally bound by the combined mass of the Sun and planets.

Some comet orbits do appear to be slightly hyperbolic. It is usually assumed that this results from a rocket effect produced by nonisotropic emission of gases from the comets as they pass near the Sun. It is customary for observers to parameterize the excess kinetic energy of such comets by determining the semi-major axis a_H of a comet having a binding energy equal to the excess kinetic energy of the comet at infinity. The most hyperbolic comets found after allowing for the rocket effect and planetary perturbations have $a_H = -10^4$ AU (Marsden and Sekanina 1973). We can estimate the maximum mass, M_{shell}, of a shell of comets of radius R_c by assuming that its gravitational attraction on incoming comets made these comets appear hyperbolic. This gives

$$\frac{M_{\text{shell}}}{M_\odot} < \frac{1}{2}\left|\frac{R_c}{a_H}\right|. \tag{4}$$

For $|a_H| = 10^4$ AU, we find that $M_{\text{shell}} < 1 \times 10^4\ M_\oplus \simeq 0.025\ M_\odot$ for $R_c = 500$ AU and $M_{\text{shell}} < 1.5 \times 10^3\ M_\oplus = 0.005\ M_\odot$ for $R_c = 100$ AU. It is clearly possible for a sizable fraction of the protosolar metals to be locked up in the comet cloud. The possibility that this may occur was suggested by Tinsley and Cameron (1974) and Tinsley (1974). [I recently found that Whipple (1975) went through a similar estimate of the maximum mass of the comet cloud.]

References

Bailey, M. E. 1985, preprint.

Biermann, L., and Michel, K. W. 1978, *Moon and Planets,* **18**, 447.

Blaauw, A. 1964, *Ann. Rev. Astr. Ap.,* **2**, 229.

Cameron, A. G. W. 1973, *Icarus,* **78**, 407.

——. 1978, *Moon and Planets,* **18**, 5.

Feldman, P. D. 1983, *Science,* **219**, 347.

Fernandez, J. A. 1982, *Astr. J.,* **86**, 1318.

Goldreich, P., and Ward, W. R. 1974, *Ap. J.,* **183**, 1051.

Hills, J. G. 1973, *Icarus,* **78**, 505.

——. 1980, *Ap. J.,* **235**, 986.

——. 1981, *Astr. J.,* **86**, 1730.

——. 1982, *Astr. J.,* **87**, 906.

——. 1983*a*, *Ap. J.,* **267**, 322.

——. 1983*b*, *Astr. J.,* **88**, 1269.

——. 1984, *Astr. J.,* **89**, 1559.

——. 1986, in *The Galaxy and the Solar System,* ed. M. S. Mathews (Tucson: University of Arizona Press), in press.

Hills, J. G., and Sandford, M. T. 1983*a*, *Astr. J.,* **88**, 1519.

——. 1983*b*, *Astr. J.,* **88**, 1522.

Kuiper, G. P. 1951, in *Astrophysics,* ed. J. A. Hynek, p. 357.

Larson, R. B., and Starrfield, S. 1971, *Astr. Ap.,* **13**, 190.

Marsden, B. G., and Sekanina, Z. 1973, *Astr. J.,* **78**, 1118.

McCrea, W. H. 1975, *Observatory,* **95**, 239.

Oort, J. H. 1950, *Bull. Astr. Inst. Netherlands,* **11**, 91.

Rohlfs, K. 1977, *Lectures on Density Wave Theory* (New York: Springer-Verlag).

Shoemaker, E. M. 1982, private communication.

Shoemaker, E. M., and Wolfe, R. E. 1982, in *Satellites of Jupiter,* ed. D. Morrison, p. 277 (Tucson: University of Arizona Press).

Smith, B. A., and Terrile, R. J. 1984, *Science,* **226,** 1421.
Tinsley, B. M. 1974, *Ap. J.,* **192,** 629.
Tinsley, B. M., and Cameron, A. G. W. 1974, *Ap. Space Sci.,* **31,** 31.
Torbett, M. V., and Smoluchowski, R. 1984, *Nature,* **311,** 641.
Whipple, F. 1975, *Astr. J.,* **80,** 525.
Winkler, K.-H., and Newman, M. J. 1980, *Ap. J.,* **236,** 201.
Yamamoto, T. 1985, *Astr. Ap.,* **142,** 31.

Contributors and Editors

W. David Arnett
Astronomy and Astrophysics Center
University of Chicago
5640 South Ellis Avenue
Chicago, IL 60637

P. Bodenheimer
Lick Observatory
Board of Studies in Astronomy and Astrophysics
University of California
Santa Cruz, CA 95064

A. G. W. Cameron
Harvard College Observatory
60 Garden Street
Cambridge, MA 02138

Donald D. Clayton
Department of Space Physics and Astronomy
Rice University
Houston, TX 77001

Stirling A. Colgate
Los Alamos National Laboratory
T-6, MS B275
P.O. Box 1663
Los Alamos, NM 87545

William A. Fowler
W. K. Kellogg Radiation Laboratory 106-38
California Institute of Technology
Pasadena, CA 91125

Carl J. Hansen
Department of Astrogeophysics
Joint Institute for Laboratory Astrophysics
University of Colorado
Boulder, CO 80309

Wolfgang Hillebrandt
Max-Planck-Institut für Physik und Astrophysik
Institut für Astrophysik
Karl-Schwarzschild-Strasse 1
D-8046 Garching bei München
Federal Republic of Germany

Jack G. Hills
Theoretical Astrophysics Group
Los Alamos National Laboratory
T-6, MS B288
P.O. Box 1663
Los Alamos, NM 87545

Icko Iben, Jr.
Department of Astronomy
University of Illinois
341 Astronomy Building
Urbana, IL 61801

William M. Kaula
Department of Earth and Space Sciences
University of California
Los Angeles, CA 90024

D. N. C. Lin
Board of Studies in Astronomy and Astrophysics
Lick Observatory
University of California
Santa Cruz, CA 95064

Chris C. Meisl
W. K. Kellogg Radiation Laboratory 106-38
California Institute of Technology
Pasadena, CA 91125

B. E. J. Pagel
Royal Greenwich Observatory
Herstmonceux Castle
Hailsham, East Sussex BN27 1RP
England

J. Papaloizou
School of Mathematical Sciences
Queen Mary College
London
England

William H. Press
Harvard-Smithsonian Center for Astrophysics
60 Garden Street
Cambridge, MA 02138

Martin J. Rees
Institute of Astronomy
Madingley Road
Cambridge CB3 0HA
England

M. Ruderman
Stanford Linear Accelerator Center
Stanford University
Stanford, CA 94305

Frank H. Shu
Astronomy Department
University of California
Campbell Hall
Berkeley, CA 94720

James W. Truran
Department of Astronomy
University of Illinois
341 Astronomy Building
Urbana, IL 61801

Sachiko Tsuruta
Department of Physics
Montana State University
Bozeman, MT 59717